Innovative Structural Applications of High Performance Concrete Materials in Sustainable Construction

Innovative Structural Applications of High Performance Concrete Materials in Sustainable Construction

Editors

Fausto Minelli
Enzo Martinelli
Luca Facconi

MDPI • Basel • Beijing • Wuhan • Barcelona • Belgrade • Manchester • Tokyo • Cluj • Tianjin

Editors
Fausto Minelli
University of Brescia
Italy

Enzo Martinelli
University of Salerno
Italy

Luca Facconi
University of Brescia
Italy

Editorial Office
MDPI
St. Alban-Anlage 66
4052 Basel, Switzerland

This is a reprint of articles from the Special Issue published online in the open access journal *Sustainability* (ISSN 2071-1050) (available at: https://www.mdpi.com/journal/sustainability/special_issues/structural_application).

For citation purposes, cite each article independently as indicated on the article page online and as indicated below:

LastName, A.A.; LastName, B.B.; LastName, C.C. Article Title. *Journal Name* **Year**, *Volume Number*, Page Range.

ISBN 978-3-0365-4461-8 (Hbk)
ISBN 978-3-0365-4462-5 (PDF)

Contents

About the Editors

Fausto Minelli

Fausto Minelli is a Full Professor in Structural Engineering at the University of Brescia, Department of Civil Engineering, Architecture, Land, Environment and of Mathematics (DICATAM) since Nov 2021. His research interests include experimental, analytical and numerical investigations on fibre reinforced concrete and high-performance concrete structures, reinforced concrete structures with very little cement content, non-linear modelling of r.c. structures, assessments and rehabilitation of existing structures and infrastructires.

The research activity is mainly focused on the structural level, with a number of studies at the material level. Four main research topics can be outlined:

- Assessment and structural retrofitting of existing structures and infrastructures, with special emphasis on r.c. bridges and masonry structures.

- Mechanical characterization of FRC, HPC materials and concretes with little or no cement content.

- Shear behaviour of members without transverse reinforcement.

- Structural Applications of Fibre-Reinforced-Concrete.

Fausto Minelli is author of more than 150 scientific papers, 80 of which are published in international journals or in international conferences.

Enzo Martinelli

Enzo Martinelli is an Associate Professor of Structural Analysis and Design at the Department of Civil Engineering of the University of Salerno, Italy. He became Full Professor in France in 2013 and in Italy in 2017. In the last twenty years, he has been working on various subjects, including the experimental characterisation and theoretical modelling of concrete structures, the seismic response of structures, and the mechanics of fiber-reinforced composite materials. More recently, he has been the Principal Coordinator Contact in the EU-funded projects "Environmentally-friendly solutions for Concrete with Recycled and natural components" (EnCoRe, FP7-PEOPLE-2011 IRSES, n. 295283) and "Sustainability-driven international/intersectoral Partnership for Education and Research on modelling next generation CONCRETE" (SUPERCONCRETE, H2020-MSCA-RISE-2014, n. 645704). He is co-author of more than 300 papers published both in international journals and conferences, he has supervised and co-supervised a total of 12 PhD theses. Dr. Enzo Martinelli is currently Associate Editor of the European Journal of Civil and Environmental Engineering and member of the Editorial Board of several journals, including the *Journal of Structural Engineering* (ASCE) and *Engineering Structures* (Elsevier).

Luca Facconi

Luca Facconi is an Assistant Professor working at the Department of Civil, Environmental, Architectural Engineering and Mathematics of the University of Brescia (Italy). He obtained his Ph.D. in "Rehabilitation of historical and modern buildings" from the University of Brescia in 2012. His main fields of activity include the experimental and numerical study of reinforced concrete, fiber-reinforced concrete (FRC) and masonry structures. He has been involved in many studies aiming at investigating the behavior of masonry and infilled reinforced concrete structures under seismic loading in order to develop innovative retrofitting techniques adopting fiber reinforced mortar coating. He has proposed analytical models for predicting the response of masonry structures under static and cyclic loading as well as models able to assess safety requirements for FRC structures. He is also co-responsible for research projects focusing on structural assessment and health monitoring of existing concrete and masonry bridges. Prof. Facconi has authored or co-authored more than 25 scientific papers in the field of structural engineering, with emphasis on the behavior of masonry and FRC structures under static and seismic loading.

Editorial

Innovative Structural Applications of High Performance Concrete Materials in Sustainable Construction

Fausto Minelli [1,*], Enzo Martinelli [2] and Luca Facconi [1]

1 DICATAM—Department of Civil, Environmental, Architectural Engineering and Mathematics, University of Brescia, 25123 Brescia, Italy; luca.facconi@unibs.it
2 Department of Civil Engineering, University of Salerno, 84084 Fisciano, Italy; e.martinelli@unisa.it
* Correspondence: fausto.minelli@unibs.it

Citation: Minelli, F.; Martinelli, E.; Facconi, L. Innovative Structural Applications of High Performance Concrete Materials in Sustainable Construction. *Sustainability* **2021**, *13*, 12491. https://doi.org/10.3390/su132212491

Received: 20 October 2021
Accepted: 21 October 2021
Published: 12 November 2021

It is well-known that concrete is the most widely utilised construction material in the world. Therefore, any action intending to enhance sustainability of the construction industry cannot help but consider the supply chain, production, distribution demolition and eventual disposal, landfilling or recycling of this intrinsically composite material. The use of High-Performance Concrete, although it may sound counterintuitive at first, can be one of the most effective, though technologically challenging, options to make the construction sector more sustainable. Indeed, high-performance should not only be intended in terms of mechanical properties, but also in terms of durability and capacity of the materials to cope with harsh environmental exposure conditions.

In this light, even the use of Recycled Concrete Aggregates (RCAs) in the production of Recycled Aggregate Concrete (RAC), which is generally accepted as the most realistic solution to reduce the environmental impacts of concrete productions, should guarantee a sufficiently high-performance, regardless of the origin of RCAs [1]. To this end, specific processing procedures and mix-design formulations are requested with the aim to obtain limited loss in performance when RAC is exposed to severe environmental conditions, like water immersion [2] or freeze-thaw cycles [3].

Similarly, the partial replacement of Ordinary Portland Cement (OPC), which is notoriously the most environmentally harmful among concrete constituents, with other Supplementary Cementitious Materials (SCMs) or Alternative Binders (ABs) should undergo careful research work intended at assessing the actual performance of the resulting cementitious composites. In this respect, promising results have been obtained by considering several SCMs or ABs, such as Municipal Solid Waste Incinerator Bottom Ash [4], Belitic Calcium Sulfoaluminate Cement [5] or Steel Slag [6], the latter consisting of either ground granulated blast furnace slag (GGBFS) or unprocessed ladle furnace slag (LFS).

Moreover, the use of High-Performance Cementitious Composites requires further investigations into their mechanical behaviour, especially in the cases where they are supposed to be almost perfectly bonded to other materials, such as normal-strength concrete of existing RC members [7,8] or masonry walls [9], also considering their use with the twofold objective of structural (and, specifically, seismic) and energy retrofitting of buildings. In this respect, retrofitting existing buildings can oftentimes be more cost-effective than constructing a new facility. In fact, besides the reduced energy consumption due to the adoption of energy conservation retrofits, existing buildings can take advantage from innovative seismic retrofitting interventions due to their efficient design, minimal maintenance and disruption for installation. Therefore, when sustainability initiatives, such as those described above, are taken into consideration in designing renovations and retrofits for existing buildings, operation costs and environmental impacts are reduced, thus leading to increased building adaptability, durability, and resiliency.

The challenge of enhancing sustainability by raising durability of concrete structures is particularly relevant in those applications where maintenance is particularly expensive and impactful, in terms of both direct intervention costs and indirect costs deriving from downtime, like in the case of Geothermal Power Plants [10].

The present Special Issue, entitled "Innovative Structural Applications of High-Performance Concrete Materials in Sustainable Construction", aims at providing readers with the most recent research results on the aforementioned subjects and further fostering a collaboration between the scientific community and the industrial sector on a common commitment towards sustainable concrete constructions.

Author Contributions: Data curation, writing—original draft preparation, writing—review and editing, supervision, F.M., E.M. and L.F. All authors have read and agreed to the published version of the manuscript.

Funding: This research received no external funding.

Institutional Review Board Statement: Not applicable.

Informed Consent Statement: Not applicable.

Data Availability Statement: Not applicable.

Conflicts of Interest: The authors declare no conflict of interest.

References

1. Pani, L.; Francesconi, L.; Rombi, J.; Mistretta, F.; Sassu, M.; Stochino, F. Effect of Parent Concrete on the Performance of Recycled Aggregate Concrete. *Sustainability* **2020**, *12*, 9399. [CrossRef]
2. Sharaky, I.; Issa, U.; Alwetaishi, M.; Abdelhafiz, A.; Shamseldin, A.; Al-Surf, M.; Al-Harthi, M.; Balabel, A. Strength and Water Absorption of Sustainable Concrete Produced with Recycled Basaltic Concrete Aggregates and Powder. *Sustainability* **2021**, *13*, 6277. [CrossRef]
3. Santana Rangel, C.; Amario, M.; Pepe, M.; Martinelli, E.; Toledo Filho, R.D. Durability of Structural Recycled Aggregate Concrete Subjected to Freeze-Thaw Cycles. *Sustainability* **2020**, *12*, 6475. [CrossRef]
4. Faleschini, F.; Toska, K.; Zanini, M.A.; Andreose, F.; Settimi, A.G.; Brunelli, K.; Pellegrino, C. Assessment of a Municipal Solid Waste Incinerator Bottom Ash as a Candidate Pozzolanic Material: Comparison of Test Methods. *Sustainability* **2021**, *13*, 8998. [CrossRef]
5. Markosian, N.; Tawadrous, R.; Mastali, M.; Thomas, R.J.; Maguire, M. Performance Evaluation of a Prestressed Belitic Calcium Sulfoaluminate Cement (BCSA) Concrete Bridge Girder. *Sustainability* **2021**, *13*, 7875. [CrossRef]
6. Rubio-Cintas, M.D.; Parron-Rubio, M.E.; Perez-Garcia, F.; Bettencourt Ribeiro, A.; Oliveira, M.J. Influence of Steel Slag Type on Concrete Shrinkage. *Sustainability* **2021**, *13*, 214. [CrossRef]
7. Javidmehr, S.; Empelmann, M. Shear Bond between Ultra-High Performance Fibre Reinforced Concrete Overlays and Normal Strength Concrete Substrates. *Sustainability* **2021**, *13*, 8229. [CrossRef]
8. Nishiwaki, T.; Mancinelli, O.; Fantilli, A.P.; Adachi, Y. Mechanical and Environmental Proprieties of UHP-FRCC Panels Bonded to Existing Concrete Beams. *Sustainability* **2021**, *13*, 3085. [CrossRef]
9. Facconi, L.; Lucchini, S.S.; Minelli, F.; Grassi, B.; Pilotelli, M.; Plizzari, G.A. Innovative Method for Seismic and Energy Retrofitting of Masonry Buildings. *Sustainability* **2021**, *13*, 6350. [CrossRef]
10. Al-Obaidi, S.; Bamonte, P.; Animato, F.; Lo Monte, F.; Mazzantini, I.; Luchini, M.; Scalari, S.; Ferrara, L. Innovative Design Concept of Cooling Water Tanks/Basins in Geothermal Power Plants Using Ultra-High-Performance Fiber-Reinforced Concrete with Enhanced Durability. *Sustainability* **2021**, *13*, 9826. [CrossRef]

Article

Durability of Structural Recycled Aggregate Concrete Subjected to Freeze-Thaw Cycles

Caroline Santana Rangel [1], Mayara Amario [1], Marco Pepe [2,3], Enzo Martinelli [2,3,*] and Romildo Dias Toledo Filho [1]

1 Department of Civil Engineering, COPPE, Federal University of Rio de Janeiro, Rio de Janeiro 21941-972, Brazil; carolrangel@poli.ufrj.br (C.S.R.); mayara_amario@poli.ufrj.br (M.A.); toledo@coc.ufrj.br (R.D.T.F.)
2 Department of Civil Engineering, University of Salerno, 84084 Fisciano (SA), Italy; mapepe@unisa.it
3 TESIS srl, 84084 Fisciano (SA), Italy
* Correspondence: e.martinelli@unisa.it; Tel.: +39-081-96-4098

Received: 7 July 2020; Accepted: 8 August 2020; Published: 11 August 2020

Abstract: The increasing global demand for natural resources and the extensive production of construction and demolition waste (CDW) raise concerns for both the economic and environmental consequences that they can induce. Several efforts are being made with the aim to promote sustainable practices in the construction industry. In this context, one of the most relevant options refers to reusing CDW in new construction: specifically, the use of recycled concrete aggregate (RCA) is attracting a growing interest. Unfortunately, although the behavior of recycled aggregate concrete (RAC) has been widely investigated in the last few years, there are still knowledge gaps to fill on various aspects of the RAC performance, such as its durability in extreme conditions. The present study deals with the freeze-thaw performance of normal- (C35) and high-strength (C60) RAC produced with RCAs derived from different sources. Specifically, ten concrete mixtures were subjected to a different number of freeze-thaw cycles (namely, 0, 150 and 300), with the aim of analyzing the degradation of key physical and mechanical properties, such density, compressive strength, elastic modulus and tensile strength. Based on the obtained experimental results, a novel degradation law for freeze-thaw cycles is proposed: it unveils a relationship between open porosity of concrete, which is directly correlated to the peculiar properties of RCAs, and the corresponding damage level determined on RAC specimens.

Keywords: recycled concrete aggregate; recycled aggregate concrete; durability; freeze-thaw cycles; mechanical properties

1. Introduction

Millions of tons of waste are produced annually as a result of construction and demolition, which become a significant concern, due to their significant impact on the preservation of the environment [1]. Moreover, concrete is the most used construction material for both structural and non-structural applications and, as a consequence, also the depletion of natural resources needed for its production is becoming a relevant issue [2].

Therefore, in the last few decades, the scientific community have been analyzing various technical solutions, with the aim of mitigating the environmental impact of the concrete industry. In this context, replacing the ordinary aggregates with recycled concrete aggregate (RCA) and, hence, producing recycled aggregate concrete (RAC), is one of the most promising solutions [3]. As a matter of the principle, this option can not only reduce the demand for natural resources, but it can also lead to easing the pressure on landfills of construction and demolition waste (CDW) [4].

Several studies demonstrate the feasibility of using RAC also for structural purposes [5–9], but a limited number of researches investigate its long-term performance when exposed to extreme environmental conditions [10,11], which is an essential aspect for any construction material to be adopted in real-world applications. It is worth highlighting that the durability of concrete structures can be undermined by several degradation processes possibly occurring in concrete, which often propagate from the exposed surfaces throughout the cortical layers of concrete through its porous micro-structure [12].

1.1. The Effect of Freeze-Thaw Cycles on Concrete

Concrete mixtures, even presenting adequate mechanical properties, need to be additionally tested for their durability performances. In fact, the exposure to freeze-thaw cycles is one of the most severe and detrimental environmental actions on concrete [13]. In cold regions, the damage of concrete structures induced by the sequence of freezing and thawing is a serious problem, and it can cause high levels of concrete deterioration.

The freeze-thaw resistance of concrete is controlled by several physical properties, among which matrix porosity and aggregate properties [14]. Shang et al. [15] explain into details the processes developing inside concrete subjected to freeze-thaw cycles: initially, the water expands throughout pores after freezing and, if the required volume is larger than the available space, the excess water is eliminated by expansion pressure; when the pressure exceeds the material resistance local cracks appear. However, repeated cycles in a humid environment result in water entering cracks during the defrost stage and freeze again later, with progressive degradation as freeze-thaw cycles proceed. Cheng et al. [16] also remark that the internal structure of the mixtures is damaged during the action of freeze-thaw cycles due to volume expansion and temperature stress.

In their review of the literature on durability of concrete, Li et al. [17] describe that damage is due to the expansion of water molecules beyond the volume restrictions of concrete in the freezing phase of the degradation. As a result, the typical deterioration of concrete by freezing and thawing includes random cracking and surface scaling, which have an adverse effect on the mechanical properties and permeability of the material.

1.2. The Freeze-Thaw Impact on the Physical Properties of RAC

In recent years, several researchers investigated the freeze-thaw resistance of RAC and concentrated their analysis on the degradation of physical and mechanical performances. In fact, being composed of attached mortar (AM) and old natural aggregates, RCAs are generally characterized by a higher porosity than the companion natural aggregates [18] and, consequently, their inclusion within the cement-based matrix can have a relevant effect on the resulting concrete durability.

The freeze-thaw resistance of concrete with 0%, 20%, 40% and 60% of RCA was investigated by Tuyan et al. [19] after 300 cycles (where each cycle consisted of freezing the samples from 5 to −18 °C within 3 h and then thawing in water at 5 °C within 1 h). The authors reported that the freeze-thaw resistance was lower in the tested RAC mixtures than in the reference ones: more cracks were detected in RAC specimens, resulting in greater mass loss. As expected, mass loss increased with the number of freeze-thaw cycles and this effect was more pronounced in mixtures with higher water-to-cement (w/c) ratio and, at a lower extent, higher RCA content. An increase in w/c ratio increases the number and the volume of capillary pores, as well as the water present inside the cement paste, which is the main cause of the expansive internal pressure during freezing [20]. In addition, Tuyan et al. [19] observed a stable relationship between freeze-thaw damage and water absorption of concrete mixtures.

Diagne et al. [21] carried out an experimental investigation of the use of RCA in RAC specimens subjected to 0, 5, 10 and 20 freeze-thaw cycles (where each cycle consisted of 24 h at −23 °C and then 23 h at 21 °C). The authors confirmed that the material durability is directly correlated to the number of the freeze-thaw cycles. They found that water drainage is faster in specimens with higher percentage of RCAs, due to the higher porosity of the overall composite.

Wawrzenczyk et al. [22] explain that in the initial stages of degradation the mass of the concrete first increases and, hence, it decreases; in addition, they observe that concretes with less resistance to freeze-thaw processes present greater mass losses and linear expansion at the end.

The experimental results by Zhu et al. [23] show that the freeze-thaw resistance of RAC is lower than that of ordinary concrete and, moreover, the weights of both natural and recycled concrete initially decreased (up to 200 cycles), and then increased with increasing freeze-thaw cycles. The mass variation of RAC after freeze-thaw cycles were caused by deterioration of the sample surface and the water absorption. This fact is consistent with the point of view of Wu et al. [24].

1.3. The Freeze-Thaw Impact on the Mechanical Properties of RAC

Richardson et al. [10] evaluate the degradation of recycled concrete specimens subjected to up to 56 freeze-thaw cycles (with freezing performed at −18 °C in air and thawing in water at 20 °C until the core temperature reached 6 °C) and they concluded that the strength of the concrete at the beginning of the freeze-thaw test determines its ability to resist the hydrostatic pressures created by the action of the degradation process. The authors explain the type of RCA, their quality and the possible processing procedure have a role on the resulting performance of RAC. For instance, RAC produced with treated RCA (e.g., washed before mixing) is more durable than natural concrete mixtures, which, in the authors' interpretation, can be explained by a higher initial strength (and compactness), which has a beneficial effect in mitigating damage induced by the freeze-thaw cycles.

Li et al. [14] explain that, when water freezes in the saturated state, its volume in the concrete pores expands and the adjacent particles are pressed against each other. Consequently, the concrete particles break down, resulting in a reduction in the internal compaction of the sample. Since RCA mixtures tend to have lower specific density and higher water absorption, as well as a higher specific porous structure, they are more prone to damage caused by freeze-thaw cycles in concretes produced with this raw material. Based on the experimental results, Shang et al. [15] also affirm that the compressive strength and tensile strength of concrete decreased as freeze-thaw cycles were repeated.

Wang et al. [25], after investigating the mechanical properties of concretes with 30 MPa subjected to various freeze-thaw cycles (0, 5, 15, 30, 50, 75 and 100 cycles), highlight that existing micro-cracks tend to expand as concrete is subjected to repeated free-thaw cycles. Furthermore, the residual compressive strength is negatively affected by both the number of free-thaw cycles and the minimum freezing temperature.

Zhu et al. [23] observe that in the initial freeze-thaw period (100 cycles), the internal structures of RAC remains relatively compact, with only a few micropores and cracks. Conversely, after 400 cycles, the bond between aggregates and mortar became weaker, the ITZ became less distinct and a higher number of both microcracks and pores appeared. When samples failed in the late freeze-thaw period (800 cycles), it was no longer possible to clearly distinguish the ITZs within the mixture, showing that the concrete had become very porous. However, all the tested RAC mixtures meet the minimum strength requirement for at least 500 cycles and, although the freeze-thaw resistance of RAC was lower than that of natural concrete, the study suggests that recycled concrete could be used for at least 50 years in severe cold areas.

Yang et al. [26] explain that the peculiar material properties of RCA, such as higher water absorption, old paste attachment and initial cracks, make freeze-thaw cycles a serious durability problem for RAC. In addition, they demonstrate that the RAC specimens exhibited good resistance to freezing, but the peak strain of the recycled concretes was higher than that of natural concretes.

The study by Ren et al. [27] presents results for compressive strength of concrete with a 60% recycled coarse aggregate replacement ratio after 0, 25, 50 and 75 rapid freeze-thaw cycles in water (with sample temperature variation from 8 °C to −17 °C and then again to 8 °C in 2.5–3 h), which indicates that the compressive strength of the samples decreases up to failure as the number of FT cycles increases.

1.4. Significance of the Research

The concise state-of-the-art review presented in the previous subsections demonstrates that the research topic proposed in the present manuscript is worthy of investigation as only a limited number of recent studies are available in the literature and, in addition, the possible influence of RCAs on the resulting performance of RAC subjected to freeze-thaw cycles has not been completely understood.

Moreover, one of the main issues on the use of RCAs in concrete is related their significant heterogeneity depending on both the origin concrete from which they are obtained and the treatments they actually undergo.

Therefore, the RAC mixtures analyzed in this study are produced with RCAs derived from two different sources: concrete debris produced in the laboratory (with controlled and know properties of original concrete) and concrete waste derived from field demolition operation (with uncontrolled and unknow properties of the original concrete). Specifically, the amount of attached mortar (AM), which is an essential parameter controlling the RAC properties [28], is determined for the RCAs employed in the present research with the aim to investigate possible correlations between AM and the observed level of damage induced by freeze-thaw cycles.

Finally, both reference concrete and RAC mixtures investigated in this research are purposely designed with the aim for them to achieve a given target strength: this result is achieved by applying a mix-design technique developed by the authors for RAC, which takes into account the properties of RCAs [29,30]. Specifically, both normal- (35 MPa) and high-strength (60 MPa) reference and RAC mixtures are considered in the present study, with the aim of investigating whether concrete with significantly different initial (28 days) strengths result in different evolution of damage induced by freeze-thaw cycles. Conversely, the majority of studies available in the literature are based on simplistically replacing part of the ordinary constituents with recycled ones, with no considerations about the resulting properties of RAC.

2. Materials and Methods

2.1. Materials

Natural sand (nominal diameter smaller than 4.75 mm) was used as fine aggregate and granite type stones, with nominal diameter ranging from 9.5 to 4.75 mm and 19 to 9.5 mm were employed as natural "coarse aggregate 0" (Nat_C0) and "coarse aggregate 1" (Nat_C1), respectively.

The recycled concrete aggregates (RCAs) employed herein were obtained from two different sources [28]: the debris of concrete produced in laboratory, labelled "laboratory produced" waste (L-waste), and the debris of concrete elements derived from a recycling plant, labelled "demolition" waste (D-waste). Both residues were broken into smaller pieces by means of a jaw crusher. Then, the recycled aggregates were sieved in the two above-mentioned sizes, "coarse aggregate 0" (RCA_C0) and "coarse aggregate 1" (RCA_C1). In the final step, the RCAs were homogenized by longitudinal blending bed technique [31].

In order to characterize both natural and RCAs particles, the following tests were performed: particle size distribution [32], density and water absorption capacity at 24 h [33–35], attached mortar content by thermal shock (method described by Rangel et al. [28]), "Los Angeles" abrasion [36] and actual packing density [37].

The properties of the natural and recycled aggregates are summarized in Table 1. The data reported in Table 1 highlight the different properties (i.e., in terms of density, water absorption capacity and AM content) for RCAs employed in this study, which were derived from two different sources (L-waste and C-waste). As is also well demonstrated in the literature, these properties depend on both the processing procedure adopted for concerting concrete debris into RCAs as well as on the properties of the original concrete [28,38].

Table 1. Properties of the natural and recycled aggregates used in this study.

Properties		Sand	Nat_C0	RCA_L_C0	RCA_D_C0	Nat_C1	RCA_L_C1	RCA_D_C1
Maximum grain size (mm)		4.8	9.5	9.5	9.5	19.0	19.0	19.0
Density (kg/m^3)		2447	2662	2178	2168	2636	2105	2255
Water absorption (%)		0.5	1.5	7.3	7.6	1.3	8.2	6.1
Attached mortar content (%)		-	-	44.3	46.2	-	64.8	35.1
Abrasion wear (%)		-	39.5	41.2	41.5	36.1	46.7	46.3
Actual packing density	Class 1	0.70	0.68	0.66	0.57	0.60	0.56	0.57
	Class 2	0.52	0.54	0.60	0.55	0.56	0.55	0.57
	Class 3	0.68	0.55	0.60	0.57	0.54	0.60	0.58

The cement used in this study was "high initial strength Portland cement", labelled CPV-ARI, according to the National Brazilian Standard (NBR) 16697 (2018) [39], with a specific gravity of 3181 kg/m^3 and 28-day compressive strength of 40.6 MPa. A polycarboxilate polymer superplasticizer "MC Powerflow 1180" with a solid concentration content of 35% and specific mass of 1.070 g/cm^3 was used in all mixes for workability control. The superplasticizer saturation dosage is 1.5% of solids based on the cement weight.

2.2. Concrete Mixtures Composition and Mixing Procedures

The concrete mixtures composition was performed by the use of the well-known compressive packing model (CPM) that was originally developed for conventional structural concrete, but has been recently extended for concrete mixtures with RCA by Amario et al., 2017 [29] and Rangel et al., 2017 [30]. It is worth highlighting once again that adopting this methodology is one of the specific features of the present study, as each mixture is designed specifically for given properties (e.g., workability at the fresh state and compressive strength at 28 days of curing), and taking into account the inherent properties of constituents.

In order to take into account of the high absorption of RCAs, the recycled particles were added in dry condition during mixing and the expected absorption rate of coarse RCAs was considered in the mix-design calculation. Specifically, the absorption value of 50% of the total absorption obtained experimentally (24 h) was used, as, according to previous studies [29,40], RCAs absorb about 50% of their total absorption value during mixing.

Concrete mixtures were designed for two classes of compressive strength: normal strength with 35 MPa and high strength with 60 MPa. All mixtures were produced with 30% of paste (sum of the free mixing water, cement and superplasticizer) in relation to the total volume of concrete.

The two mixtures with only natural aggregates were named "CX-NAT", where X indicates the strength class (35 or 60). The RCA mixtures were named "CX-Y-Z", where X indicates the strength class (35 or 60), Y indicates the origin of RCA (L for laboratory concrete residue or D for demolition concrete residue) and Z indicates the RCA fraction that was used (C0 for "coarse aggregate 0" and C1 for "coarse aggregate 1").

Table 2 summarizes the compositions of the ten concrete mixtures and highlight (in the last column) the overall amount of total mortar volume ($V_{M,tot}$), which is a relevant parameter for RAC, as it is determined as the sum of the volume of the attached mortar (V_{AM}) present in RCAs (Table 1) and the new mortar volume ($V_{M,new}$) given by the sum of sand, cement, water and superplasticizer:

$$\begin{gathered} V_{M,tot} = V_{AM} + V_{M,new} \\ V_{AM} = \frac{W_{RCA,C0}}{\gamma_{RCA,C0}} \cdot AM_{RCA,C0} + \frac{W_{RCA,C1}}{\gamma_{RCA,C1}} \cdot AM_{RCA,C1} \\ V_{M,new} = \frac{W_{water}}{\gamma_{water}} + \frac{W_{cement}}{\gamma_{cement}} + \frac{W_{SP}}{\gamma_{SP}} + \frac{W_{sand}}{\gamma_{sand}} \end{gathered} \tag{1}$$

where W_i represents the weight used for each fraction (see Table 2) and γ_i is the corresponding fraction's density (see Table 1).

Table 2. Mix proportions of normal and recycled aggregate concretes.

Mixtures	w/c_{eff}	Materials (kg/m³)											$V_{M,tot}$ (%)
		C1			C0			Sand	CEM	SP	w_{eff}	w_{tot}	
		NAT	RCA_L	RCA_D	NAT	RCA_L	RCA_D						
C35-NAT	0.60	452	0	0	457	0	0	868	325	1.86	196	212	65.5
C35-L-C0	0.57	451	0	0	0	373	0	866	338	1.93	194	217	74.9
C35-L-C1	0.57	0	361	0	456	0	0	867	336	1.92	191	216	77.6
C35-D-C0	0.55	451	0	0	0	0	371	866	345	1.97	191	214	75.1
C35-D-C1	0.57	0	0	384	453	0	0	862	341	1.95	194	216	73.1
C60-NAT	0.32	448	0	0	452	0	0	860	448	19.20	145	150	65.5
C60-L-C0	0.31	448	0	0	0	371	0	861	458	19.62	141	152	74.8
C60-L-C1	0.30	0	356	0	450	0	0	856	461	19.76	138	151	77.3
C60-D-C0	0.29	448	0	0	0	0	369	860	464	19.89	134	145	74.6
C60-D-C1	0.30	0	0	382	451	0	0	857	463	19.84	137	147	72.7

As also well demonstrated in the literature [28,40], the AM content in coarse RCAs can be directly correlated to the open porosity (w) of the employed aggregates:

$$w_{RCA} = w_{NAT} \cdot (1 - AM_{RCA}) + w_{AM} \cdot AM_{RCA} \tag{2}$$

where:

- w_{RCA} is the open porosity (i.e., water absorption capacity) of the RCA;
- w_{NAT} (equal to 1%) and w_{AM} (equal to 15%) represent the porosity of the two phases present in the RCAs: natural aggregates and attached mortar, respectively;
- AM_{RCA} is the volume of AM present in each RCA fraction.

Consequently, in the case of RACs, the higher the replacement ratio, the lower the quantity of natural stones and the higher the overall amount of mortar: this increment is directly correlated to the "quality" of the employed RCA (see Equation (2)). In the present study, it is noted the total mortar volume increases from 65.5% to around 78% for concretes of both C35 and C60 classes (see Table 2).

Due to the high water absorption capacity of RCAs, a specific methodology was adopted for the mixing process. Specifically, the total water was divided into two equal parts, and the addition of the parts was performed at different times of the mixing. This methodology was chosen based on the two-stage mixing approach (TSMA) method proposed by Tam et al. [41] and Tam & Tam [42,43] for concrete containing RCAs. The mixing procedure was performed in the following order:

1. Placement of all coarse and fine aggregates and mixing for 1 min for their homogenization;
2. Addition of half of the total amount of water and mixing for 1 min;
3. Addition of cement and mixing for 1 min with aggregates;
4. Addition of remaining water (second half) and all superplasticizer and mixing for 8 min, for full action of the superplasticizer and completion of the concrete mixing process.

The casting of concrete into the molds was executed in two layers followed by mechanical compaction by 30 s. The specimens were demolded after 24 h and conducted in a wet chamber cure (relative humidity of 100% and temperature of 21 °C ± 1 °C).

2.3. Testing Procedures

The fresh concrete properties were investigated through a slump test using the Abram cone, according to NBR NM 67 [44]. Compressive strength and splitting tensile strength tests were carried out on cylindrical specimens with 75 mm of diameter and 150 mm of height after 28 days of curing, according to NBR 5739 [45] and NBR 7222 [46], respectively. The specimens were tested on a 1000 kN Shimadzu testing machine at a rate of axial displacement of 0.1 mm/min for the compressive tests and 0.3 mm/min for the splitting tensile tests.

Freeze-thaw resistance was assessed by evaluating the compressive strength, elastic modulus, splitting tensile strength and mass loss of the produced mixtures after being subjected to 0, 150 and

300 freeze-thaw cycles which were performed in a climatic chamber presenting an automatic control of the temperature variations.

The freeze-thaw cycles (started at the age of 28 days) were carried out based on ASTM C666 [47]. Initially, the specimens were kept immersed in water at temperature of 20 ± 2 °C for 48 h and then the cycles started. Each cycle consisted of reducing the temperature from 4 °C to −18 °C and then reheating to 4 °C for a total cycle time of 5 h (Figure 1).

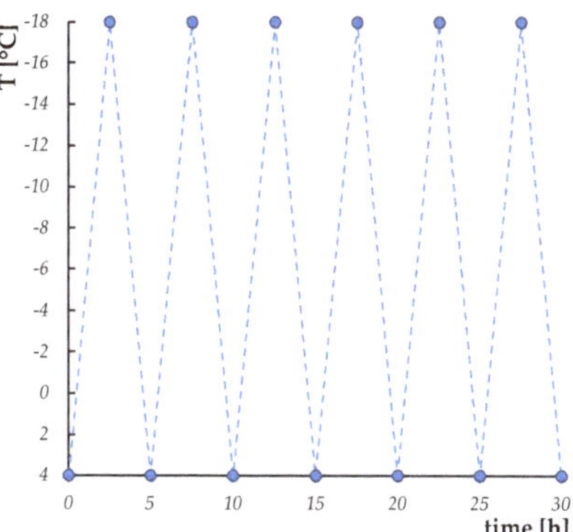

Figure 1. Freeze-thaw degradation cycles.

The weighing of cylindrical specimens was carried out in order to measure the mass variation throughout the application of the freeze-thaw protocol. It is worth highlighting that, for all mixtures and cycle levels, non-degraded samples were kept in a humid chamber (21 °C temperature and 100% humidity) and tested as reference samples at the same age as the degraded samples.

Moreover, the physical properties of the produced concretes were measured at 28 days in accordance with the NBR 9778 [48] for determining their water absorption capacity (herein defined as concrete open porosity, w_{open}), voids index and concrete density.

3. Results and Discussion

3.1. Physical and Mechanical Properties of RAC at 28 Days

Table 3 summarizes the results obtained for workability, physical and mechanical properties of the ten concrete mixtures at 28 days.

Table 3. Workability, physical and mechanical properties of concrete mixtures at 28 days.

Mixture	Slump (mm)	Water Absorption (w_{open}) (%)	Voids Index (%)	Density (kg/m^3)	$f_{c,28}$ (MPa)	$E_{c,28}$ (GPa)	$f_{t,28}$ (MPa)	$f_{t,28}/f_{c,28}$ (%)
C35-NAT	175	3.0	7.0	2303	34.2	21.3	2.7	7.9
C35-L-C0	180	3.6	8.1	2239	35.7	22.1	2.7	7.6
C35-L-C1	165	3.9	8.6	2221	35.3	21.2	2.9	8.2
C35-D-C0	165	3.7	8.4	2238	34.4	21.7	2.9	8.4
C35-D-C1	195	3.5	7.8	2247	33.5	20.9	2.6	7.8
C60-NAT	165	1.1	2.7	2411	60.1	29.1	3.9	6.5
C60-L-C0	180	1.7	4.0	2354	60.5	29.8	4.0	6.6
C60-L-C1	170	1.9	4.3	2339	61.9	30.1	4.4	7.1
C60-D-C0	165	1.6	3.8	2361	62.6	31.0	4.4	7.0
C60-D-C1	160	1.4	3.3	2376	59.7	29.5	4.1	6.9

All concrete mixtures presented a slump value of 180 ± 20 mm, considered satisfactory, since it allows an excellent casting of the concrete: this demonstrate that it is possible to produce RAC which present similar workability in comparison with conventional concrete mixture containing only natural aggregates. This was possible due to the specific mix-design methodology adopted in the present study, in which a testing step was performed to ensure the desired slump value of each mixture.

Regarding the physical properties, in the literature, there is a consensus among authors that RACs present higher total water absorption (i.e., concrete open porosity, w_{open}), higher voids index and lower density in comparison with the corresponding natural concretes [28]: also, the results obtained in the present research, in terms of physical properties, are in accordance with the literature.

As for the total water absorption, as expected, the high strength concrete samples present lower values than the normal strength ones. Moreover, in both cases, the reference mixture produced with only natural aggregate was the one with the lowest water absorption. The RCA leading to mixtures with the highest total absorption value was RCA_L_C1 for both classes, which leads to an increase in absorption of 30% for C35 and 72% for C60, in relation to the absorption of the companion natural concrete.

As well known, the open porosity of concrete also depends on the constituents employed within the mixture, and, in the specific case of RACs, it is influenced by the presence of attached mortar of the RCAs particles. This is confirmed by the fact that the aggregates RCA_L_C1, which caused the highest variation in the absorption of concrete, are characterized by the highest value of attached mortar content (Table 1). Therefore, when producing mixtures of a high- and a normal-strength class, the influence of this aggregate type in the final absorption of concrete is not the same: the results show that the influence of the aggregates is greater in the high-strength class. Therefore, the use of RCA, and the consequent increase in the total mortar volume, caused proportionally greater variation in the absorption of the high strength class concrete.

The voids index, which is directly related to the total water absorption of concrete, followed the same trend of the absorption results: the lowest voids index values correspond to the high strength class and the greater values correspond to the normal-strength class. In addition, the concrete mixtures with 100% natural aggregate have the lowest values for this property in each class, and mixtures with 100% RCA_L_C1 have the highest values, that represent a proportional increase in voids index of 23% for C35 and 59% for C60 compared to the natural concretes.

As expected, the density values of RACs are lower than the values for natural concretes for both strength classes. The largest difference in C35 from *C35-NAT* was obtained by *C35-L-C1* with a decrease of 3.6%. In C60, the highest proportional decrease was of 3.0% of *C60-L-C1*, compared to natural concrete. Thus, the proportional variation between recycled and natural concretes was similar, regardless of strength class.

Figure 2 describes the resulting relationship between total mortar volume and the corresponding concrete open porosity: it shows that these two parameters are strongly connected to each other.

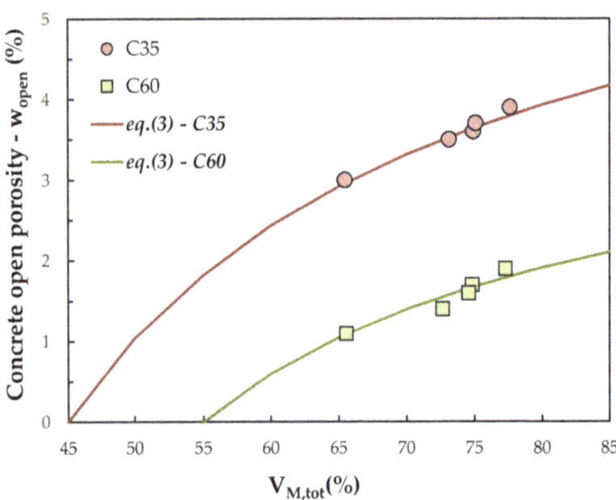

Figure 2. Relation between total mortar volume ($V_{M,tot}$) and concrete open porosity (w_{open}).

The rising curves in Figure 2 highlight that an increase in total mortar volume causes an increase in total absorption and the following equation can be proposed for describing this trend:

$$w_{open} = \frac{k_1 \cdot (V_{M,tot} - V_{M,0})}{k_2 + (V_{M,tot} - V_{M,0})} \quad (3)$$

where k_1, k_2 and $V_{M,0}$ represent constant values which were calibrated for both C35 and C60 mixtures (see Table 4).

Table 4. Calibration of numerical parameters of Equation (3).

Concrete Class	k_1	k_2	$V_{M,0}$
C35	7.3	30.0	45.0
C60	4.2	30.0	55.0

The 28-days compressive strength results ($f_{c,28}$) show that the RACs obtained values similar to natural concretes for the two strength classes analyzed herein, confirming the initial assumption of this study. This is because the premise of this study is to understand the influence of the use of RCAs in concretes of the same resistance subjected to degradation by rapid freeze-thaw cycles. In C35, the mixtures showed a maximum variation of 4% (for *C35-D-C1*) compared to the required value of 35 MPa. In C60, the maximum range was also 4% (for *C60-D-C0*) from the desired value of 60 MPa. This scenario was made possible by properly considering the specific characteristics of the RCAs in the scientific mix-design stage, with proper consideration of RCAs absorption water capacity (50% of the total absorption value), as well as the individual determination of each proportion. These results confirm the excellent capability of CPM to predict the compressive strength of recycled and natural concretes of different strength classes.

The results of elastic modulus at 28 days ($E_{c,28}$) for C35 of RACs showed a decrease of 1.8% and an increase of 3.7% compared to the value of natural concrete for *C35-D-C1* and *C35-L-C0*, respectively. In C60, the recycled concretes ranged from 1.3% and 6.5% increase compared to *C60-NAT*, for *C60-D-C1* and *C60-D-C0*, respectively. This highlights that, in C60 class, all concrete containing RCAs obtained greater elastic modulus than the reference natural mixture.

The tensile strength results ($f_{t,28}$) of C35 showed that *C35-D-C1* had the lowest value, with a decrease of 3%, compared with the natural mixture, and *C35-L-C1* and C35-D-C0 mixes had the highest values, with an increase of 7%, compared with *C35-NAT*. A clear trend was observed: *C35-D-C1* (lowest $f_{t,28}$) was also the mixture with the lowest $f_{c,28}$. For the high strength class, all concrete containing RCAs had higher tensile strength value than the natural concrete. The mixtures *C60-L-C1* and *C60-D-C0* showed better tensile behavior than the others (as for C35), with a proportional increase of 12% compared to the result of natural concrete. Normally, as the compressive strength increases, the tensile strength also increases, but at a decreasing rate. As a consequence, the tensile-to-compressive strength ratio for normal-strength mixtures is higher than the tensile-to-compressive strength ratio for high performance concretes. Therefore, as expected, the experimental values obtained for the ratio $f_{t,28}/f_{c,28}$ were higher for C35 than for C60.

3.2. The Influence of the Freeze-Thaw Cycles on the Physical and Mechanical Performances of RAC

The nomenclature "FT" was adopted to represent the expression "freezing-thawing". The degraded samples were labelled by the number of cycles to which they were subjected: "FT150" and "FT300" for 150 and 300 cycles, respectively. Moreover, reference samples were not degraded and tested at the same age as the degraded samples: the abbreviations "REF(FT150)" and "REF(FT300)", according to the samples they serve as reference, were used.

3.2.1. Visible Surface Damage

Figure 3 shows surface damage caused by repeated freeze-thaw cycles. For C35 class, after 150 freeze-thaw cycles, some fragmentation was observed on the specimen surfaces, both on the upper and lower sides of the samples (Figure 3a). At 300 cycles, the superficial fragmentation pattern remained similar to what observed after 150 cycles. However, after 300 degradation cycles of C35, an increase in the size of some pores on the lateral surface of the samples became more evident (Figure 3b). It is worth highlighting that the small holes present on the surfaces of the produced samples are due to the entrapped air on the contour during the casting process.

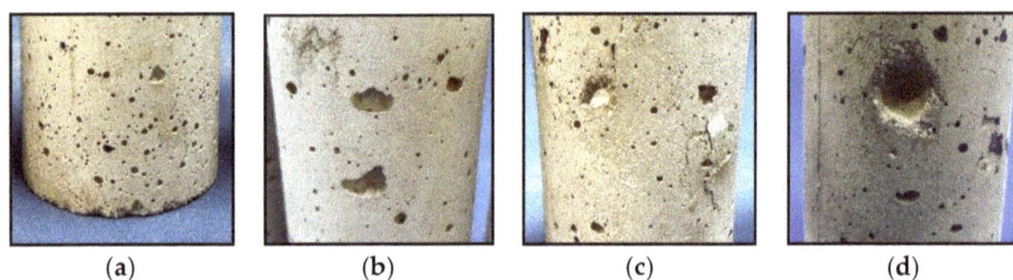

(a) (b) (c) (d)

Figure 3. Typical surface damage of samples after freeze-thaw cycles: (**a**) C35 after 150 cycles; (**b**) C35 after 300 cycles; (**c**) C60 after 150 cycles only for recycled aggregate concretes (RACs); (**d**) C60 after 300 cycles only for RACs.

No significant variations were identified between natural and recycled concretes. Therefore, for the normal strength class, the visible damage suggests that the degradation process mainly affected the (new) concrete mortar and not the coarse aggregates present within the mixture (for both 150 and 300 cycles). This indicates that, in this case, the (new) normal strength mortar is less resistant to degradation than both natural and recycled coarse aggregates. However, surface damage of C60 class does not indicate the same behavior. In fact, surface damage after freeze-thaw cycles for class C60 shows that, unlike the samples of C35, no fragmentation of the surface has appeared. This indicates that the mortar of high strength concrete suffers less freeze-thaw damage than the mortar of normal strength concrete. This fact can be explained by the higher porosity of mortars with higher water-to-cement ratio. On the other hand, aggregate fragmentation was observed on the side of C60 samples after 150 cycles, but only in RAC samples (Figure 3c). After 300 cycles, the surface damage pattern of C60 follows the same pattern identified for 150 cycles, but the recycled mixtures had a higher amount of (complete) fragmentation, with more evident "holes" (Figure 3d). The natural mixture (*C60-NAT*) showed no visible damage after 150 and 300 freeze-thaw cycles. This indicates that the aggregates that suffered this type of degradation were only RCAs. As the difference between the RCA and natural aggregate structures lies in the presence of the aged adhered mortar around the RCA grains, it can be assumed that this fragmentation behavior of RCAs when exposed to the degradation process occurred mainly in the attached mortar. This can be explained by the higher porosity of the existing mortar. During freeze-thaw cycles, as the high strength concrete mortar showed no visible damage, it was likely that the (more fragile) attached mortar could not withstand the internal stresses on the concrete structure generated by the increase in water volume when it turns to ice. Thus, the attached mortar cracked first, releasing the generated pressure. Similarly, it can be concluded that, in the case of C35 class, it was the new concrete mortar that cracked first, not reaching the point at which the fragmentation of the adhered mortar of RCAs occurs.

3.2.2. Mass Loss

Figure 4 shows the mass loss results for natural and recycled concretes after the two different numbers of freeze-thaw cycles.

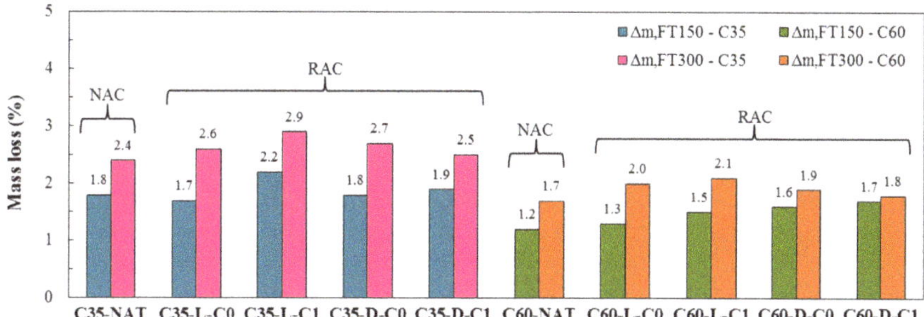

Figure 4. Mass loss after freeze-thaw degradation.

According to Wu et al. [24], the main reason for the mass decrease of concrete (natural and RAC) when subjected to repetitive freezing and thawing processes is due to the fact that, when the mortar fragmentation and the appearance of internal cracks (caused by freezing pressure of internal water) occur, the concrete becomes weaker and more fragmented with the development and extension of these cracks, which consequently makes the mass of the samples decrease.

C35 class had higher mass loss values than C60 class for both 150 and 300 cycles. This is related to the higher porosity of C35, together with the lower resistance of its internal structure to freeze-thaw cycles. In both classes, concrete mass loss rates decrease with increasing number of cycles. That is, the loss in the first stage (0–150) was greater than the loss in the second stage (150–300). At the end of 300 cycles, the concretes with RCA_L_C1 had the highest mass loss and the natural mixtures (*C35-NAT* and *C60-NAT*) had the lowest mass loss results for both classes. This fact can be explained by the greater porosity of RCA_L_C1 and the lower porosity of the natural aggregates, respectively.

3.2.3. Compressive Stress-Strain Behavior

The typical stress-strain responses obtained in the compressive strength tests of C35 and C60 in the freeze-thaw durability study are shown in Figure 5.

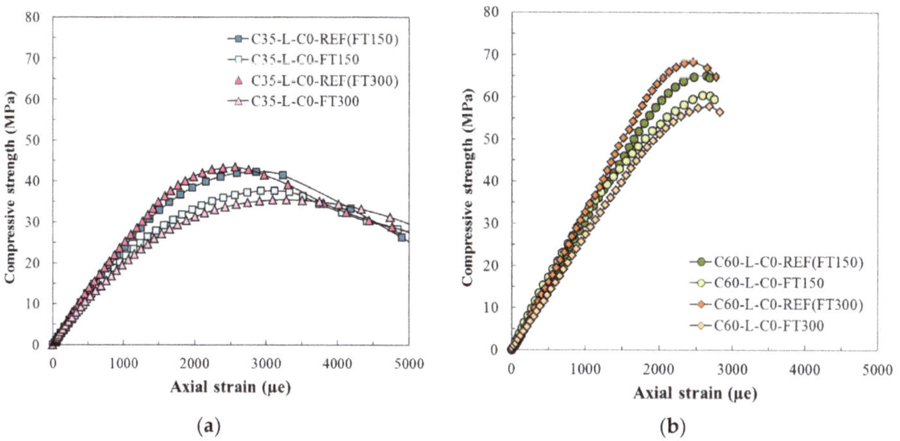

Figure 5. Typical compressive behavior of reference and after freeze-thaw cycles: (**a**) C35; (**b**) C60.

All concrete mixtures of C35 showed a variation in compressive behavior after degradation by 150 freeze-thaw cycles, and this can be verified by comparing the results obtained for REF(FT150) and FT150 (Figure 5a). Individually, for each mixture, the reference curves achieved higher strength values and lower peak strain values than the curves of degraded samples. The same was observed for 300 cycles stress-strain curves but, in this case, REF(FT300) samples show higher resistance than REF(FT150) samples, due to the higher age of REF(FT300) (higher cement hydration). FT300 samples showed lower resistance compared to FT150, and this is explained because FT300 samples suffered 150 freeze-thaw cycles more. For C60, the same comments can be made, since FT150 curves and FT300 curves behaved differently from their respective references, REF(FT150) and REF(FT300) (Figure 5b).

Regarding the type of rupture, the normal strength concretes have well-defined diagonal cracks for the reference samples (REF(FT150) in Figure 6a and REF(FT300) in Figure 6c), similar to what was observed at 28 days. However, the cracks appeared more uncoordinated for the degraded samples (FT150 in Figure 6b and FT300 in Figure 6d), without forming a clear diagonal, but the appearance of cracks in various directions. This fact can be explained by the fragmentation of the mortar and the pore volume variation (due to the degradation process), already mentioned above, which cause modification in the direction of the cracking process during the test. The rupture line, in fact, occurs in the weakest "line" of the concrete, that is, in the weakest path of the internal structure, and can therefore be different from the main diagonal. The rupture line focuses on the defects of the material structure, which, in this case, were produced by the freeze-thaw degradation.

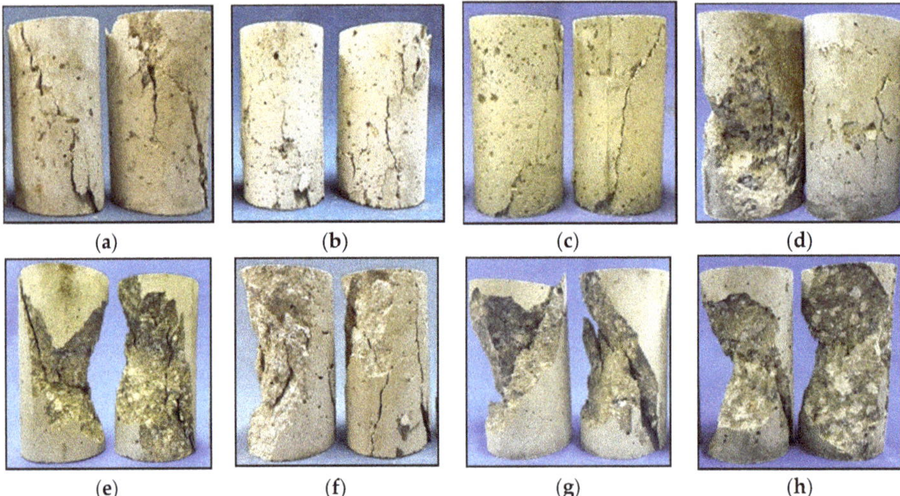

Figure 6. Sample of the typical rupture of the C35 and C60 in the compressive strength test in the freeze-thaw durability study: (**a**) C35-REF(FT150); (**b**) C35-FT150; (**c**) C35-REF(FT300); (**d**) C35-FT300; (**e**) C60-REF(FT150); (**f**) C60-FT150; (**g**) C60-REF(FT300); (**h**) C60-FT300.

Examples of cracking patterns occurred in each class of specimen are shown in Figure 6 for C35 and C60.

High strength concretes, as expected from the 28-day compressive strength test, exhibit less ductile (in comparison with C35) rupture upon reaching the maximum stress for the reference samples (REF(FT150) in Figure 6e and REF(FT300) in Figure 6g), but still being possible to identify a main diagonal of rupture. The degraded samples (FT150 in Figure 6f and FT300 in Figure 6h) appear to have a very similar cracking pattern with fragile rupture. It can only be commented that, in some specimens, this diagonal is less evident, being possible to identify cracks in other directions as well. It is also worth noting that, for all the rupture patterns discussed above, it was possible to identify an association between the rupture mechanism and the concrete damage. The rupture development occurred by "following" the weakest regions of the concrete, as aggregate fragmentation in RACs and pore expansion in all mixtures (as seen in the Figure 3).

Table 5 presents the mechanical properties results for the samples degraded by 150 and 300 freeze-thaw cycles (FTC), as well as their respective references (control samples) for compressive strength, elastic modulus and tensile strength.

Table 5. Mechanical properties in the freeze-thaw durability study.

Class	Mixture	Group ID	Compressive Strength (MPa)		Elastic Modulus (GPa)		Tensile Strength (MPa)	
			150 FTC	300 FTC	150 FTC	300 FTC	150 FTC	300 FTC
C35	*C35-NAT*	Reference	39.1	39.9	23.5	23.9	2.9	3.1
		Degradation	35.8	33.8	22.2	19.6	2.7	2.6
	C35-L-C0	Reference	42.1	43.3	24.1	24.6	3.2	3.3
		Degradation	37.9	35.5	22.3	19.3	2.9	2.6
	C35-L-C1	Reference	39.9	41.1	23.8	24.3	3.0	3.2
		Degradation	37.5	32.4	22.5	18.5	2.8	2.5
	C35-D-C0	Reference	41.1	42.1	24.4	24.8	3.1	3.2
		Degradation	38.0	33.6	22.8	19.4	2.9	2.5
	C35-D-C1	Reference	39.6	41.1	23.6	24.2	2.9	3.1
		Degradation	36.8	33.8	21.9	19.3	2.7	2.5
C60	*C60-NAT*	Reference	65.0	68.2	32.3	34.1	4.1	4.2
		Degradation	61.2	60.1	30.6	29.0	3.9	3.7
	C60-L-C0	Reference	64.9	67.9	33.1	34.4	4.2	4.3
		Degradation	60.8	57.8	31.8	27.8	3.9	3.7
	C60-L-C1	Reference	68.0	70.3	32.5	34.6	4.5	4.6
		Degradation	65.7	59.1	31.9	27.7	4.3	3.9
	C60-D-C0	Reference	65.6	67.1	32.9	33.9	4.5	4.5
		Degradation	61.9	57.6	31.1	27.7	4.3	3.9
	C60-D-C1	Reference	63.9	65.8	32.5	33.6	4.3	4.4
		Degradation	61.1	57.2	31.3	27.7	4.1	3.8

Moreover, the percentage decreases of the degraded samples in relation to their reference are presented in Figure 7. The obtained results confirm that all samples of both classes suffered compressive strength loss when undergoing to freeze-thaw cycles. The C35 class presents a degradation percentage ranging between 7.1% to 10% after 150 cycles and from 15.3% to 21.2% after 300 cycles (Figure 7a). The increase of the degradation percentage is approximately twice the number of cycles increased from 150 to 300, but the mixtures did not follow the same behavior in both steps. For smaller number of cycles (150), *C35-L-C1* had the best behavior (lowest decrease in resistance) among the five mixtures of this class, however, with increasing cycles (from 150 to 300), the mixture showed to be the most impacted by the freeze-thaw cycles. However, the natural mixture *C35-NAT* showed a similar rate of degradation for both phases, moving from 0 to 150 and from 150 to 300 cycles. Therefore, it can be noted that the resistance capacity of the natural concrete to freeze-thaw degradation is greater for larger numbers of cycles than RAC mixtures. Despite the higher absolute degradation in C60 resistance values compared to C35, the percentage resistance decrease is higher for the normal strength class for both 150 and 300 cycles. That is, the degradation cycles have the greatest impact on the compressive strength (percentage) in the normal strength class. The compressive strength values of C60 correspond to percentage decreases from 3.4% to 6.3% and from 11.9% to 15.9% after 150 and 300 cycles, respectively. The same comments about behavior regarding C35 mixtures can be considered for C60: *C35-L-C1* starts with the best behavior up to 150 cycles, but the best behavior after 300 cycles is the one of the natural concrete *C60-NAT*.

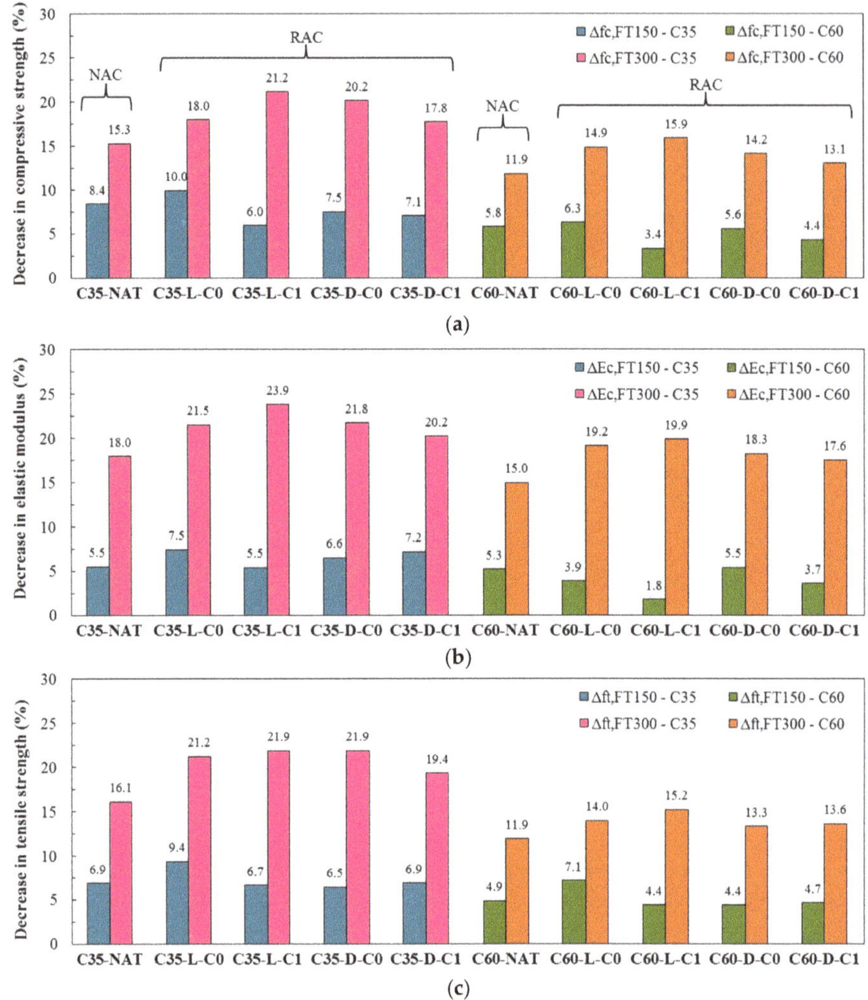

Figure 7. Decrease in (**a**) compressive strength, (**b**) elastic modulus and (**c**) tensile strength after freeze-thaw degradation.

The elastic modulus results for the degraded samples and their respective references are also presented in Table 4 and, Figure 7b summarizes the total decrease values for this property. For the C35 class, the percentage decreases were of 5.5% to 7.5% after 150 cycles and 18% to 23.9% after 300 cycles. In all the cases, the rate of degradation for the second phase of cycles (150–300) was higher than for the first ones (0–150). Mixtures *C35-L-C1* and *C35-NAT* had the best responses up to 150 cycles, but *C35-NAT* was better after 300 cycles, while *C35-L-C1* fell to the worst behavior among the five mixtures. Overall, the high strength class showed lower modulus loss values after degradation cycles. The reductions of C60 were only 1.8% to 5.5% after 150 cycles and 15% to 19.9% after 300 cycles. For this class, natural concrete stands out with the best behavior after 300 FTCs.

Therefore, the results show that the impact of freezing and thawing cycles was higher on elastic modulus than on compressive strength. This can be explained by two factors. The first is that the stiffness of concrete is a more sensitive property than strength. That is, any damage, small defect or small detachment in the ITZs is easily "accused" by the elastic modulus. This is because the stiffness is calculated at lower levels of force, usually at 40% of the ultimate load, which show defects that are not so felt by the compressive strength. So, there are small damages that, despite not being so reflected in the compressive strength, are already able to impact the elastic modulus. This is usually the difference between these properties. The second factor that can be mentioned is that each concrete material has its own elastic modulus. During the thermal variation that the concrete is suffering in the test, and the consequent variation in the volume of water inside its structure, micro displacements and micro cracks are generated between the concrete components (in ITZs). This is because, as each material has a different elastic modulus, a differential deformation between the components of the concrete occurs. This behavior is more clearly evidenced in the stiffness, however, it is less evident in the resistance. In other words, this problem of multi-phase materials with multiple stiffness appears more prominent in the elastic modulus.

3.2.4. Tensile Strength

The results regarding tensile strength in the freeze-thaw durability study are presented in Table 4 and Figure 7c. For C35, the values show a decrease of 6.5% to 9.4% for samples degraded by 150 cycles, whereas after 300 cycles the degradation reach 16.1% to 21.9%. *C35-L-C0* mixture had the highest percentage impact and the others showed similar behavior after 150 cycles, but the natural mixture stands out as the best resistance after 300 cycles. The high strength class presents smaller percentage decreases than the C35 class. For 150 cycles, the percentage loss was between 4.4% and 7.1%, the highest being for *C60-L-C0*, similar to what occurs with this aggregate for C35. After 300 cycles, the results evolved to 11.9% to 15.2% loss. The tensile strength behavior of concretes in the freeze-thaw degradation process was very similar to the compressive strength behavior. Therefore, the comments previously exposed for compressive strength can also be applied to tensile strength.

4. Freeze-Thaw Degradation-Law for RAC

The experimental results presented lead to propose a generalized degradation-law for RAC submitted to freeze-thaw cycles, which unveils the existing relationships between the mechanical properties after freeze-thaw degradation and the initial open porosity of the concrete mixture. The proposed relationship between concrete open porosity (namely, w_{open}) and the three key mechanical properties characterizing the concrete mixtures (compressive strength, elastic modulus and tensile strength) are presented in the following equations:

$$\frac{f_{c,FT}}{f_{c,REF}} = \frac{1}{1 + a \cdot \left(w_{open}\right)^2} \tag{4}$$

$$\frac{E_{FT}}{E_{REF}} = \frac{1}{1 + b \cdot \left(w_{open}\right)^2} \tag{5}$$

$$\frac{f_{t,FT}}{f_{t,REF}} = \frac{1}{1 + c \cdot \left(w_{open}\right)^2} \tag{6}$$

where a, b and c are calibrated parameters related to the properties of compressive strength, elastic modulus and tensile strength, respectively. The calibrated degradation-law curves proposed by the Equations (4)–(6) are presented, together with the experimental results, in Figure 8.

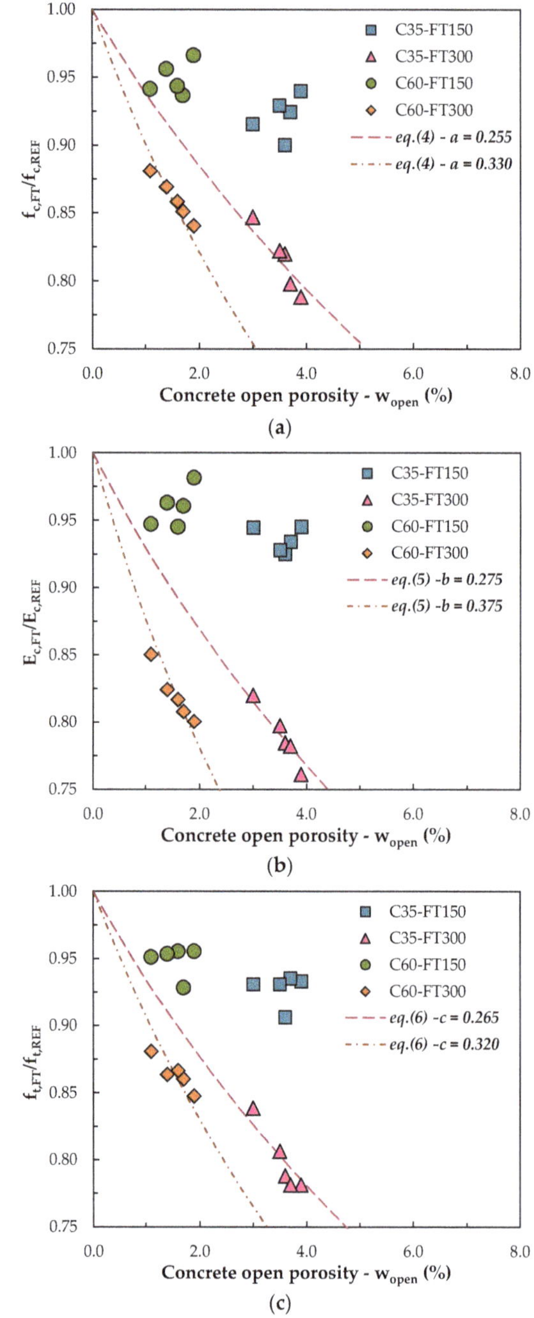

Figure 8. Relation between concrete open porosity and mechanical properties in the freeze-thaw durability study: (**a**) compressive strength; (**b**) elastic modulus; (**c**) tensile strength.

The relationship between the compressive strength after cycles divided by the reference compressive strength is a function of the initial open porosity of the concrete and, this relationship exists for compressive strength, as well as for elastic modulus and tensile strength. The proposed Equations (4)–(6) have a physical meaning being the w_{open} directly correlated to the concrete degradation. In fact, when a theoretical w_{open} is equal to zero (impermeable), no degradation is observed, since the water cannot enter within the matrix and create damages. Contrarily, when the w_{open} reach high values, the concrete degradation is significant up to a theoretical case, in which the w_{open} is ∞ and the corresponding residual strength/elastic modulus is zero.

By looking at the results in the graph, the points above refer to 150 cycles and indicate that at 150 cycles the freeze-thaw degradation is not yet so relevant, because the points are within a small range of variation and, at the same time, a clear trend cannot be detected. When 300 cycles are performed, there are two tendencies for normal- and high-performance classes, respectively. With the results of compressive strength, two numbers were calibrated for parameter a, one for each strength class. The same was done for the elastic modulus for parameter b and for the tensile strength for parameter c.

The results related to the reference mixtures are the points with lower porosity for each of the four groups of points (more to the left), because this points represent the concretes of lower total mortar volume (since the natural concretes have only the new mortar and the recycled concretes also have an aged attached mortar). Up to 150 cycles, it is possible to have recycled concrete decreases even smaller than natural concrete decreases. However, the presence of attached mortar to the concretes causes a "delta" of increase in degradation at 300 cycles. The explanation is that there is one ITZ in natural concrete (between aggregate and new paste), while recycled concrete has two ITZs (one between aggregate and new paste and another between old paste and original aggregate). By having more ITZs, with the increase in the number of cycles, more damage develops in RACs than in NACs.

On the x-axis, the range of variation in absorption when observing the high strength class is higher. When observing the normal strength class, the range of variation is smaller. Thus, the C60 class shows a decrease referring to 300 FTCs in the graph more vertical than the other class. This confirms that the presence of a more porous aggregate causes a greater impact on the absorption of C60 class. As a matter of fact, the data plotted in Figure 8 highlight that higher open concrete porosity (w_{open}) results in more "rapid" concrete degradation for C60. On the other hand, since w_{open} is directly correlated to the volume of AM present in the mixture (see Equations (1)–(3)), this means that the C60 mixtures are more "sensible" to the presence of AM. This is confirmed also by the fact the calibrated values for the parameter a, b and c (Equations (4)–(6)) are always higher for C60 mixtures.

It is worth noting that for the elastic modulus the influence of porosity on degradation is slightly different than for compressive and tensile strengths. The proposed relation is more horizontal in the compressive and tensile strengths graphs, while the relation is a slightly more vertical in the elastic modulus graph. Therefore, the difference of the use of a recycled aggregate for the elastic modulus is more prominent, because the drop in this property is more significant.

In summary, the proposed freeze-thaw degradation-law states that concrete degradation is directly related to the initial open porosity of the concrete (shown in Figure 8), which consequently is directly related (see Figure 2 and Equation (3)) to the total volume of mortar ($V_{M,tot}$). Moreover, as emerges from Equations (1) and (2), $V_{M,tot}$ depends on the amount of AM presents within the RCAs. Consequently, the proposed Equations (4)–(6) allow to unveil the existing relationship between the RCAs properties and the corresponding RAC degradation due to freeze-thaw cycles. Therefore, there are two main questions regarding the use of RCAs in concrete that will be subjected to freeze-thaw degradation processes: the first question would be whether the use of this type of aggregate influences the durability of the concrete exposed to this extreme condition; and the second question it would be if it is possible to use recycled concrete in structures subject to this type of freeze-thaw variation. First, the experimental results of this study proved that the presence of the recycled aggregate increases the total mortal volume, which consequently increases the open porosity of the concrete, and finally increases the impact caused by the freeze-thaw cycles on the mechanical properties of the concrete. However, if the degradation is governed by the open porosity of the concrete, the answer to make it possible to use RACs in structures subject to freeze-thaw would be to perform the mix-design of the concrete in order to control this physical property. That is, it would be possible to use RACs in cold regions by controlling the open porosity of this material.

Despite this, it is worth mentioning that the higher degradation registered for RAC in comparison with the reference natural mixtures is relatively low, and this variation is less significant especially for high-strength mixtures. As a matter of the fact, as also highlighted in Figure 7, when RCAs are employed in normal-strength concrete the degradation increase of around 5–6% (at 300 cycles); meanwhile, this variation is in the range of 3–4% for high-strength mixtures. This evidence confirms the fact, although the presence of RCA could affect the concrete durability, the overall concrete durability performances is more significantly governed by the properties of the matrix (i.e., the mortar) in which the coarse aggregates are employed.

The proposed formulation is a first step toward a more general model capable to predict the evolution of degradation processes in RAC as a function of the actual number of freeze-thaw cycles. Specifically, the proposed relationships, referred to 300 cycles (which corresponds to an extremely severe protocol) can be generalized to predict the effect to a lower number of cycles. Although the available experimental results (for 150 and 300 cycles) already identify possible interpolation rules, further data are available to achieve a more accurate and reliable formulation, which is among the future developments of the present research.

5. Conclusions

This study reported an analysis on the durability of natural and recycled aggregate concrete subjected to severe environmental conditions. Specifically, it summarizes the results of an experimental investigation aimed at analyzing the mechanical behavior and the degradation processes of normal- and high-strength concrete produced with four different types of RCAs, after being subjected to different levels of freeze-thaw cycles.

According to the obtained results, the following conclusions can be highlighted:

- The surface damage of C35 indicates that the degradation mainly affected the (new) concrete mortar, indicating that the mortar is less resistant to degradation processes than coarse aggregates. The superficial damage of C60, on the other hand, occurs with fragmentation of the RCAs when exposed to the degradation process, in which the attached mortar (more fragile) did not resist the internal efforts generated by the cycles;
- All concrete mixtures of classes C35 and C60 showed a reduction in the compressive and tensile strength, as well as for the concrete mass after being subjected to 150 and 300 freeze-thaw cycles;
- The degradation cycles have a higher impact on the mechanical and physical performances of normal strength concretes than the high strength mixtures;
- For both C30 and C60 strength class, the natural mixtures showed a greater capacity to resist against freeze-thaw degradation than RACs;
- The presence of RCAs increases the porosity of the concrete and, consequently, decreases the freeze-thaw durability of RACs;
- A freeze-thaw degradation-law for RAC is proposed in this study for the main mechanical properties after freeze-thaw degradation (compressive strength, elastic modulus and tensile strength) as a function of the initial open porosity of the concrete;

The data presented herein demonstrated that although the presence of RCAs could affect the concrete durability, the overall concrete durability performances is more significantly governed by the properties of the matrix (i.e., the mortar), in which the coarse aggregates are employed. However, it would be possible to use RACs in structures subjected to freeze-thaw situations by controlling the open porosity value of the concrete in the mix-design stage.

Author Contributions: Conceptualization, M.P., E.M. and R.D.T.F.; methodology, C.S.R. and M.A.; validation, C.S.R., M.A. and M.P.; formal analysis, C.S.R., M.A. and M.P.; investigation, C.S.R. and M.A.; resources, R.D.T.F.; data curation, C.S.R., M.A. and M.P.; writing—original draft preparation, C.S.R.; writing—review and editing, M.A., M.P., E.M. and R.D.T.F.; visualization, C.S.R. and M.P.; supervision, M.P., E.M. and R.D.T.F.; project administration, E.M. and R.D.T.F.; funding acquisition, E.M. and R.D.T.F. All authors have read and agreed to the published version of the manuscript.

Funding: This research was financed in part by the *Coordenação de Aperfeiçoamento de Pessoal de Nível Superior*—Brazil (CAPES)—Finance Code 001. This work was carried out with support from CNPq, *Conselho Nacional de Desenvolvimento Científico e Tecnológico*—Brazil.

Acknowledgments: The present research is also part of the activities carried out by the third Author (M.P.) within a post-doctoral project co-funded by the Department of Civil Engineering of University of Salerno (Italy) and COPPETEC foundation (UFRJ, Brazil). The present study is part of the activities carried out by the Authors within the "SUPERCONCRETE" Project (www.superconcrete-h2020.unisa.it) funded by the European Union within the Horizon 2020 Framework Programme (H2020-MSCA-RISE-2014 n°645704).

Conflicts of Interest: The authors declare no conflict of interest.

References

1. Gálvez-Martos, J.L.; Styles, D.; Schoenberger, H.; Zeschmar-Lahl, B. Construction and demolition waste best management practice in Europe. *Resour. Conserv. Recycl.* **2018**, *136*, 166–178. [CrossRef]
2. Amaral, R.E.; Brito, J.; Buckman, M.; Drake, E.; Ilatova, E.; Rice, P.; Sabbagh, C.; Voronkin, S.; Abraham, Y.S. Waste Management and Operational Energy for Sustainable Buildings: A Review. *Sustainability* **2020**, *12*, 5337. [CrossRef]
3. Pavlů, T.; Kočí, V.; Hájek, P. Environmental Assessment of Two Use Cycles of Recycled Aggregate Concrete. *Sustainability* **2019**, *11*, 6185. [CrossRef]
4. Taboada, G.L.; Seruca, I.; Sousa, C.; Pereira, Á. Exploratory Data Analysis and Data Envelopment Analysis of Construction and Demolition Waste Management in the European Economic Area. *Sustainability* **2020**, *12*, 4995. [CrossRef]
5. Sadati, S.; Arezoumandi, M.; Khayat, K.H.; Volz, J.S. Shear performance of reinforced concrete beams incorporating recycled concrete aggregate and high-volume fly ash. *J. Clean. Prod.* **2016**, *115*, 284–293. [CrossRef]
6. Arezoumandi, M.; Smith, A.; Volz, J.S.; Khayat, K.H. An experimental study on flexural strength of reinforced concrete beams with 100% recycled concrete aggregate. *Eng. Struct.* **2015**, *88*, 154–162. [CrossRef]
7. Francesconi, L.; Pani, L.; Stochino, F. Punching shear strength of reinforced recycled concrete slabs. *Constr. Build. Mater.* **2016**, *127*, 248–263. [CrossRef]
8. Xiao, J. Recycled Aggregate Concrete Structures. *Springer Tracts Civ. Eng.* **2018**, 623.
9. Kalinowska-Wichrowska, K.; Suescum-Morales, D. The Experimental Study of the Utilization of Recycling Aggregate from the Demolition of Elements of a Reinforced Concrete Hall. *Sustainability* **2020**, *12*, 5182. [CrossRef]
10. Richardson, A.; Coventry, K.; Bacon, J. Freeze/thaw durability of concrete with recycled demolition aggregate compared to virgin aggregate concrete. *J. Clean. Prod.* **2011**, *19*, 272–277. [CrossRef]
11. Guo, H.; Shi, C.; Guan, X.; Zhu, J.; Ding, Y.; Ling, T.; Zhang, H.; Wang, Y. Durability of recycled aggregate concrete—A review. *Cem. Concr. Compos.* **2018**, *89*, 251–259. [CrossRef]
12. Šavija, B. Smart Crack Control in Concrete through Use of Phase Change Materials (PCMs): A Review. *Materials* **2018**, *11*, 654. [CrossRef] [PubMed]
13. Pigeon, M. *Durability of Concrete in Cold Climates*; CRC Press: Boca Raton, FL, USA, 2014.
14. Li, Z.; Liu, L.; Yan, S.; Zhang, M.; Xia, J.; Xie, Y. Effect of freeze-thaw cycles on mechanical and porosity properties of recycled construction waste mixtures. *Constr. Build. Mater.* **2019**, *210*, 347–363. [CrossRef]
15. Shang, H.; Song, Y.; Ou, J. Behavior of air-entrained concrete after freeze-thaw cycles. *Acta Mech. Solida Sin.* **2009**, *22*, 261–266. [CrossRef]
16. Cheng, Y.; Wang, W.; Gong, Y.; Wang, S.; Yang, S.; Sun, X. Comparative study on the damage characteristics of asphalt mixtures reinforced with an eco-friendly basalt fiber under freeze-thaw cycles. *Materials* **2018**, *11*, 2488. [CrossRef] [PubMed]
17. Li, J.; Wu, Z.; Shi, C.; Yuan, Q.; Zhang, Z. Durability of ultra-high performance concrete—A review. *Constr. Build. Mater.* **2020**, *255*, 119296. [CrossRef]

18. Kim, N.; Kim, J.; Yang, S. Mechanical Strength Properties of RCA Concrete Made by a Modified EMV Method. *Sustainability* **2016**, *8*, 924. [CrossRef]
19. Tuyan, M.; Mardani-Aghabaglou, A.; Ramyar, K. Freeze–thaw resistance, mechanical and transport properties of self-consolidating concrete incorporating coarse recycled concrete aggregate. *Mater. Des.* **2014**, *53*, 983–991. [CrossRef]
20. Mardani-Aghabaglou, A.; Andiç-Çakir, Ö.; Ramyar, K. Freeze–thaw resistance and transport properties of high-volume fly ash roller compacted concrete designed by maximum density method. *Cem. Concr. Compos.* **2013**, *37*, 259–266. [CrossRef]
21. Diagne, M.; Tinjum, J.M.; Nokkaew, K. The effects of recycled clay brick content on the engineering properties, weathering durability, and resilient modulus of recycled concrete aggregate. *Transp. Geotech.* **2015**, *3*, 15–23. [CrossRef]
22. Wawrzeńczyk, J.; Molendowska, A.; Juszczak, T. Determining k-Value with Regard to Freeze-Thaw Resistance of Concretes Containing GGBS. *Materials* **2018**, *11*, 2349. [CrossRef] [PubMed]
23. Zhu, P.; Hao, Y.; Liu, H.; Wei, D.; Liu, S.; Gu, L. Durability evaluation of three generations of 100% repeatedly recycled coarse aggregate concrete. *Constr. Build. Mater.* **2019**, *210*, 442–450. [CrossRef]
24. Wu, J.; Jing, X.; Wang, Z. Uni-axial compressive stress-strain relation of recycled coarse aggregate concrete after freezing and thawing cycles. *Constr. Build. Mater.* **2017**, *134*, 210–219. [CrossRef]
25. Wang, Y.; Liu, Z.; Fu, K.; Li, Q.; Wang, Y. Experimental studies on the chloride ion permeability of concrete considering the effect of freeze–thaw damage. *Constr. Build. Mater.* **2020**, *236*, 117556. [CrossRef]
26. Yang, H.; Liu, C.; Jiang, J. Damage constitutive model of stirrup-confined recycled aggregate concrete after freezing and thawing cycles. *Constr. Build. Mater.* **2020**, *253*, 119100. [CrossRef]
27. Ren, G.; Shang, H.; Zhang, P.; Zhao, T. Bond behaviour of reinforced recycled concrete after rapid freezing-thawing cycles. *Cold Reg. Sci. Technol.* **2019**, *157*, 133–138. [CrossRef]
28. Rangel, C.S.; Toledo Filho, R.D.; Amario, M.; Pepe, M.; de Castro Polisseni, G.; de Andrade, G.P. Generalized quality control parameter for heterogenous recycled concrete aggregates: A pilot scale case study. *J. Clean. Prod.* **2019**, *208*, 589–601. [CrossRef]
29. Amario, M.; Rangel, C.S.; Pepe, M.; Toledo Filho, R.D. Optimization of normal and high strength recycled aggregate concrete mixtures by using packing model. *Cem. Concr. Compos.* **2017**, *84*, 83–92. [CrossRef]
30. Rangel, C.S.; Amario, M.; Pepe, M.; Yao, Y.; Mobasher, B.; Toledo Filho, R.D. Tension stiffening approach for interface characterization in recycled aggregate concrete. *Cem. Concr. Compos.* **2017**, *82*, 176–189. [CrossRef]
31. Pepe, M.; Toledo Filho, R.D.; Koenders, E.A.; Martinelli, E. Alternative processing procedures for recycled aggregates in structural concrete. *Constr. Build. Mater.* **2014**, *69*, 124–132. [CrossRef]
32. NBR 248. *Aggregates—Sieve Analysis of Fine and Coarse Aggregates*; ABNT: Rio de Janeiro, Brazil, 2003.
33. NBR NM 30. *Fine Aggregate—Test Method for Water Absorption*; ABNT: Rio de Janeiro, Brazil, 2001.
34. NBR NM 52. *Fine Aggregate—Determination of the Bulk Specific Gravity and Apparent Specific Gravity*; ABNT: Rio de Janeiro, Brazil, 2009.
35. NBR NM 53. *Coarse Aggregate—Determination of the Bulk Specific Gravity, Apparent Specific Gravity and Water Absorption*; ABNT: Rio de Janeiro, Brazil, 2009.
36. NBR NM 51. *Small-size Coarse Aggregate—Test Method for Resistance to Degradation by Los Angeles Machine*; ABNT: Rio de Janeiro, Brazil, 2001.
37. De Larrard, F. *Concrete Mixture Proportioning: A Scientific Approach*; Series: Modern Concrete Technology; CRC Press: Boca Raton, FL, USA, 1999.
38. Robu, I.; Mazilu, C.; Deju, R. Study concerning characterization of some recycled concrete aggregates. *Math. Model. Civ. Eng.* **2016**, *12*, 1–12. [CrossRef]
39. NBR 16697. *Portland Cement—Requirements*; ABNT: Rio de Janeiro, Brazil, 2018.
40. Pepe, M.; Toledo Filho, R.D.; Koenders, E.A.; Martinelli, E. A novel mix design methodology for Recycled Aggregate Concrete. *Constr. Build. Mater.* **2016**, *122*, 362–372. [CrossRef]
41. Tam, V.W.; Gao, X.F.; Tam, C.M. Microstructural analysis of recycled aggregate concrete produced from two-stage mixing approach. *Cem. Concr. Res.* **2005**, *35*, 1195–1203. [CrossRef]
42. Tam, V.W.; Tam, C.M. Assessment of durability of recycled aggregate concrete produced by two-stage mixing approach. *J. Mater. Sci.* **2007**, *42*, 3592–3602. [CrossRef]
43. Tam, V.W.; Tam, C.M. Diversifying two-stage mixing approach (TSMA) for recycled aggregate concrete: TSMAs and TSMAsc. *Constr. Build. Mater.* **2008**, *22*, 2068–2077. [CrossRef]

44. NBR NM 67. *Concrete—Slump Test for Determination of the Consistency*; ABNT: Rio de Janeiro, Brazil, 1998.
45. NBR 5739. *Concrete—Compression Test of Cylindric Specimens—Method of Test*; ABNT: Rio de Janeiro, Brazil, 2007.
46. NBR 7222. *Concrete and Mortar—Determination of the Tension Strength by Diametrical Compression of Cylindrical Test Specimens*; ABNT: Rio de Janeiro, Brazil, 2011.
47. ASTM C666. *Resistance of Concrete to Rapid Freezing and Thawing*; American Society for Testing and Materials: West Conshohocken, PA, USA, 2008.
48. NBR 9778. *Hardened Mortar and Concrete—Determination of Absorption, Voids and Specific Gravity*; ABNT: Rio de Janeiro, Brazil, 2009.

 *sustainability*

Article

Effect of Parent Concrete on the Performance of Recycled Aggregate Concrete

Luisa Pani, Lorena Francesconi, James Rombi, Fausto Mistretta, Mauro Sassu and Flavio Stochino *

Department of Civil, Environmental Engineering and Architecture, University of Cagliari, 09123 Cagliari, Italy; lpani@unica.it (L.P.); lorenafrancesconi@libero.it (L.F.); james.14@tiscali.it (J.R.); fmistret@unica.it (F.M.); msassu@unica.it (M.S.)

* Correspondence: fstochino@unica.it; Tel.: +39-070-675-5115

Received: 5 October 2020; Accepted: 10 November 2020; Published: 12 November 2020

Abstract: Recycling concrete construction waste is a promising way towards sustainable construction. Indeed, replacing natural aggregates with recycled aggregates obtained from concrete waste lowers the environmental impact of concrete constructions and improves natural resource conservation. This paper reports on an experimental study on mechanical and durability properties of concretes casted with recycled aggregates obtained from two different parent concretes, belonging to two structural elements of the old Cagliari stadium. The effects of parent concretes on coarse recycled aggregates and on new structural concretes produced with different replacement percentages of these recycled aggregates are investigated. Mechanical properties (compressive strength, modulus of elasticity, and splitting tensile strength) and durability properties (water absorption, freeze thaw, and chloride penetration resistance) are experimentally evaluated and analyzed as fundamental features to assess structural concrete behavior. The results show that the mechanical performance of recycled concrete is not related to the parent concrete characteristics. Furthermore, the resistance to pressured water penetration is not reduced by the presence of recycled aggregates, and instead, it happens for the chloride penetration resistance. The resistance to frost–thawing seems not related to the recycled aggregates replacement percentage, while an influence of the parent concrete has been assessed.

Keywords: concrete; recycled concrete; durability; recycled aggregate

---

1. Introduction

The environmental impact of concrete constructions is huge. For this reason, recycling concrete construction waste to obtain recycled concrete aggregate can lower the environmental impoverishment. Indeed, the use of Construction and Demolition Waste (C&DW) as alternative aggregates for new concrete production improves natural resource preservation, reduces landfill disposal, and promotes construction sustainability [1].

The physical properties of Recycled Aggregates (RA) depend on the quality and amount of the adhered cement mortar [1,2]. Actually, the quantity of adhered mortar increases with the decrease of the RA size [1,2]. Furthermore, the crushing procedure modifies the amount of adhered mortar. Due to this mortar, RA have higher water absorption and lower density in comparison to natural ones. In addition, the un-hydrated cement on the RA surface can modify the properties of concrete [3] and crack propagation [4,5].

It is observed that the mechanical properties (compressive strength, splitting tensile strength, and modulus of elasticity) of concrete with recycled concrete aggregates decrease with the increase of the replacement percentages of Natural Aggregates (NA) [6–8]. The different mechanical performances can be explained considering the different stress distribution and failure mechanisms caused by the different micro-structures of concrete with RA in comparison to the ones with NA. The failure mechanism of the concrete with RA is complex and it is influenced by the geometrical and mechanical properties of the aggregates but also by two different interfacial transition zones. Indeed, one is located between the original NA and the old mortar and the other one is between the old and the new mortar. Clearly, the situation is different in the case of normal concrete with NA where there is only one interfacial transition zone [9].

Often, RA have been used for concrete block pavements [10,11], but other research [9,12–14] has shown how it is possible to produce structural concrete with RA. Limbachiya et al. [14] found that flexural strength and modulus of elasticity of concrete containing recycled aggregates are similar to the ones of concrete made with NA. Recently, many researchers have investigated the influences of polymer additives on Self Compacting Concrete (SCC) cast with recycled and natural aggregates [15]—see [16] for a review—proving how it is possible to employ RA in the production of structural elements casted with SCC.

The durability properties of concrete with RA (chloride diffusion, freeze thaw resistance, and abrasion resistance) are still under investigation, since a wide variability in the results is reported [17]. The durability of concrete with recycled concrete aggregates is generally lower in comparison with traditional concrete [17–21]. Pereira et al. [22] suggest that the concrete containing recycled concrete aggregates should be avoided in aggressive environments. Actually, the adhered mortar that remains attached to the recycled concrete aggregates also influences the durability properties of concrete [19]. Saravanakumar and Dhinakaran [23] show that resistance to chloride ion penetration, water absorption, and acid attack resistance of concrete decrease with addition of recycled concrete aggregates. Kwan et al. [24] report that using recycled concrete aggregates as partial replacement of NA yields to low Water Absorption (WA) and low intrinsic permeability compared to the control concrete mix. Medina et al. [25] show that concrete with higher ratios of RA have higher freeze–thaw resistance. This can be explained considering the high mechanical quality of RA and the intrinsic properties of the new aggregates. Olorunsogo and Padayachee [26] reveal that the durability characteristics of concrete with RA are reduced by the increase in RA content. However, the durability of concrete with recycled concrete aggregates can be improved by the addition of pozzolanic materials, such as superfine phosphorus slag and ground granulated blast-furnace slag [27]. Xiao et al. [28], considering a Chinese experimental database, summarizes that the resistance of chloride penetration of Recycled Aggregate Concrete (RAC) is lower compared to that of Normal Concrete (NC), and that the resistance of chloride penetration of RAC decreases with the increase of RA replacement percentage. Similar studies confirm these conclusions; see [29,30]. Kurda et al. [31], considering both literature experimental data and their new experimental campaign, show that water absorption increases, and the electrical resistivity decreases with increasing replacement percentage of RA. An opposite result is obtained if fly ash is added to concrete for both tests. The reduction of water absorption is higher in mixes with both RA and fly ash in comparison to the mixes with only RA or fly ash. In addition, the benefit of incorporating fly ash and RA in concrete increases even more when superplasticizers are used. In addition, Lima et al. [32] prove that the presence of fly ash in the mixture improves the concrete workability, and compressive and tensile strengths.

In order to develop the marketing of recycled aggregates and the management of recycling plants, it is important to know whether their chemical, physical, and mechanical characteristics are influenced by parent concrete and also whether it influences the properties and performance of the concrete with RA. The experimental data representing the properties of RAC are characterized by high dispersion [33]. According to some authors [34,35], the quality of RA is mostly influenced by the quality of original demolished concrete. Even if more research is needed, some general statements can be drawn. For example, RAC with low to medium compressive strength can be produced independently from the characteristics of parent concrete [1,36–39]. On the other hand, Tabsh and Abdelfatah [40] state that the influence of the parent concrete is more significant in a weak concrete than in stronger one. Actually, this can be explained considering that the strength of concrete depends on both coarse aggregates and cement. Therefore, if more cement is used, then the effect of the coarse aggregate is reduced.

Given that non conclusive statements have been proved on this issue, in this paper, an extensive experimental campaign was carried out to evaluate the mechanical performance and durability of concrete with coarse recycled concrete aggregates obtained through the demolition of concrete with quite low compressive strength ($R_{ck} \leq 20$ MPa). In this case, the old football stadium located in Cagliari (Italy) has been used as an artificial "quarry". Indeed, in the future, the stadium will be demolished and rebuilt with a modern design. Thus, the RA are obtained from its concrete cantilever beams and foundations. Tests were carried out to evaluate the concrete mechanical performance of these concrete structures. Parts of cantilever beams and foundations have been separately demolished and crushed in order to obtain two types of coarse RA with a size range between 4 and 16 mm. Three different replacement percentages (30, 50, and 80%) of NA with RA have been used to produce different six concrete mixes. Three of them were casted using the RA obtained from the beams and the others were produced using the RA obtained from the foundations. An additional mix of NC with only NA was produced as a benchmark. Further tests were carried out to obtain a full description of physical and mechanical properties and durability of these concretes.

The aim of this work is twofold: to verify the feasibility of using concrete debris of the old Cagliari stadium for new structural concrete and to investigate the influence of the parent concrete on the new concrete obtained with RA.

After this brief introduction, Section 2 presents the experimental program, while Section 3 describes the characteristics of the RA. Section 4 deals with the mechanical and durability properties of the concrete with RA, discussing the influence of the parent concrete. Some discussions and conclusive remarks are presented in Section 5.

2. Experimental Program

The mechanical characteristics of concrete structures of the old Cagliari football stadium (built between 1965 and 1970, as shown in Figure 1) were investigated in the first step of the experimental program. Beams and foundation blocks (see blue elements in Figure 2) were chosen as the sources of the recycled concrete.

Figure 1. Aerial view of old Cagliari football stadium, cropped version of the File: "Stadio_Sant'Elia _Cagliari_Italy_23OOcto2008.jpg" posted to Flickr by Cristiano Cani, license CC 2.0.

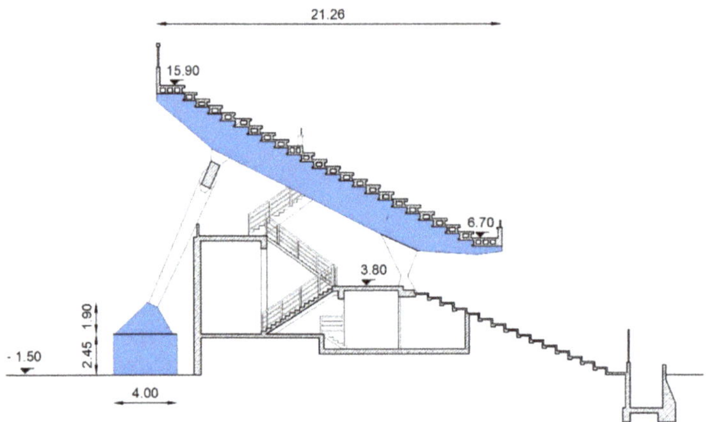

Figure 2. Cross section of reinforced concrete structures of old Cagliari football stadium. The analyzed structures are highlighted in blue (measures are in m).

Parent Concrete Quality

Twelve core samples were extracted from both the beams and the foundations, respectively labelled C. Beam and C. Found. Table 1 presents the average values of parent concrete mechanical characteristics and carbonation depth. These experimental results prove that concrete used to cast the beams is different from the ones used for the foundations. Indeed, different mechanical properties and carbonatation depth were assessed. The mechanical performances and the carbonatation depth of foundations concrete are better than those of the one used for beams.

Table 1. Properties of parent concrete.

Identification	Carbonatation Depth (mm)	Density (kg/m^3)	Compressive Strength (MPa)	Elasticity Modulus (MPa)	Tensile Strength (MPa)
C. Found. 1	30	2299	26.8	24,470	-
C. Found. 2	30	2334	32.2	27,751	-
C. Found. 3	0	2283	24.7	23,785	-
C. Found. 4	0	2345	-	-	2.04
C. Found. 5	0	2298	-	-	1.83
C. Found. 6	0	2327	-	-	2.28
C. Found. Average	10	2314	27.9	25,335	2.05
C. Beam 1	50	2271	22.2	19,744	-
C. Beam 2	0	2315	22.1	18,537	-
C. Beam 3	60	2233	18.7	15,845	-
C. Beam 4	0	2295	-	-	1.50
C. Beam 5	40	2248	-	-	1.58
C. Beam 6	40	2259	-	-	1.40
C. Beam: Average	32	2270	21.0	18,042	1.49

Petrographic analysis on thin sections of the cores highlights differences in the composition of the two materials C. Found and C. Beam. The polarizing microscope detects a fine cement matrix with different kinds of aggregates. Size distribution and mineralogical composition allow us to distinguish them. C. Found is characterized by centimetric fragments of micritic limestone. The presence of a varied siliciclastic, fine-grained, millimetric and sub-millimetric, fraction of metamorphic rock and granite fragments, with feldspar free crystals and quartz was also detected. All the fragments are characterized by sharp edges. C. Beam presents a quite homogeneous siliciclastic composition. Millimetric-centimetric fraction of granite rocks, angular fragments, and various types of metamorphic rocks can be seen with a fine-grained, sub-millimetric fraction of the same materials and free crystals of feldspars, biotite, and quartz.

3. Recycled Aggregates

Taking into account that two different parent concretes have been considered, two kinds of RA have been produced: Recycled Aggregates obtained from crushed Foundations (RA_F), and Recycled Aggregates obtained from crushed Beams (RA_B). In both cases, the aggregates size range is 4–16 mm.

The tests following the indications of UNI EN 12620: 2008 [41] and UNI 8520-1: 2015 [42] have been performed on both types of RA. Table 2 presents the main test results while Figure 3 depicts the RA size distribution. It is interesting to point out that both RA types have very similar characteristics even if they have been obtained by crushing two different concretes. Indeed, only four parameters (content of acid-soluble sulfate and water-soluble sulfates, percentage of fines, shape index) out of twenty-one are different.

The physical properties, workability, mechanical performances, and durability of concrete with RA is strongly influenced by the Residual Mortar Content (RMC) attached onto the original NA particles [2,29,43–47]. Indeed, previous studies have proved that the reduction in compressive strength of concrete with RA [43–48] and in modulus of elasticity [49] are related to the presence of RMC. Thus, in order to evaluate the properties of concrete with RA, the determination of the RMC is critical. However, currently no standard method is available. In this research, the authors follow the strategy proposed by Abbas et al. in [50]. RA samples were exposed to daily cycles of freezing and thawing in a sodium sulphate solution. Table 3 presents the RMC obtained in RA_F and RA_B considering two fraction sizes (retained by a 4 and 10 mm sieve) and it highlights that RMC is almost similar for RA_B and RA_F.

Table 2. Recycled aggregate properties.

Property	RA_F	RA_B
Size designation	4/16	4/16
Category grading	GC 90/15, GT 17.5	GC 90/15, GT 17.5
Flakiness Index	4	4
Shape Index	59	34
Saturated surface-dried particle density	2.39 Mg/m^3	2.38 Mg/m^3
Loose bulk density and voids	ρ_b = 1.23 Mg/m^3 v% = 45	ρ_b = 1.14 Mg/m^3 v% = 49
Percentage of fines	0.15%	0.59%
Percentage of shells	absent	absent
Resistance to fragmentation	39	39
Constituents of coarse RA	X = 0; Rc = 74%; Ru = 27%; Rb = 0; Ra = 0; Rg = 0	X = 0; Rc = 78%; Ru = 22%; Rb = 0; Ra = 0; Rg = 0
Content of water-soluble chloride salts	0.005%	0.005%
Content of acid-soluble chloride salts	0.325%	0.325%
Content of acid-soluble sulphate	0.43%	0.26%
Content of total sulfur	S < 0.1%	S < 0.1%
Content of water-soluble sulphates	SS = 0.148%	SS = 0.068%
Lightweight contaminator	absent	absent
Water absorption	WA24 = 7.0%	WA24 = 6.7%
Resistance to freezing and thawing	41%	42%
Resistance to magnesium sulphate	2.56%	0%
Presence of humus	absent	absent

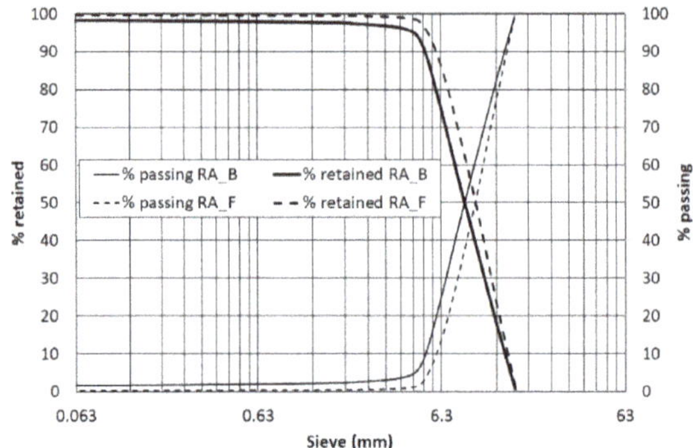

Figure 3. Recycled Aggregates (RA) size distribution.

Table 3. Residual mortar content.

Residual Mortar Content (%)	RA_F	RA_B
Sieve Retained 4 mm	55.81%	49.67%
Sieve Retained 10 mm	45.82%	45.65%

4. Concrete

Cement CEM II/A-LL 42,5 R was adopted for each concrete mix. Sand is the fine aggregate while the coarse aggregates are crushed granite and the two kinds of recycled aggregates (RA_F and RA_B). In addition, a super plasticizer based on polycarboxylate was also used. Different replacement percentages (30, 50, and 80%) of coarse RA belonging to RC_B (Reinforced Concrete of the Beams) and to RC_F (Reinforced Concrete of the Foundations) were considered. Thus, the label RC_B_X% represents a mix with X% replacement percentage using RC_B. In addition, a normal mix of concrete without RA and with only NA was produced and labelled NC. Table 4 presents the characteristics of each mix.

Table 4. Mix proportions of concretes per m^3.

Notation	w/c Ratio	Cement (kg/m^3)	Water (l/m^3)	Fine NA (kg/m^3)	Coarse NA (kg/m^3)	Coarse RA_F (kg/m^3)	Coarse RA_B (kg/m^3)	Additive (kg/m^3)	Density (kg/m^3)
NC	0.463	400	185	847.49	880.06	-	-	2.91	2322
RC_B 30%	0.463	400	185	821.8	616.04	-	263.69	3.31	2293
RC_F 30%	0.463	400	185	821.8	616.04	263.69	-	3.31	2287
RC_B 50%	0.463	400	185	802.97	440.03	-	440.27	3.31	2298
RC_F 50%	0.463	400	185	802.97	440.03	440.27	-	4.00	2283
RC_B 80%	0.463	400	185	778.15	176.01	-	703.96	4.00	2268
RC_F 80%	0.463	400	185	778.15	176.01	703.96	-	4.00	2229

4.1. Concrete Mechanical Properties

The standard slump test UNI EN 12350-2:2019 [51] was used to measure the fresh concrete workability. Two tests were performed for each mix at different times: immediately after the mixing process and after 30 min. Figure 4 presents the obtained values. It is interesting to highlight that slump values of the mixes with RA are very similar to NC.

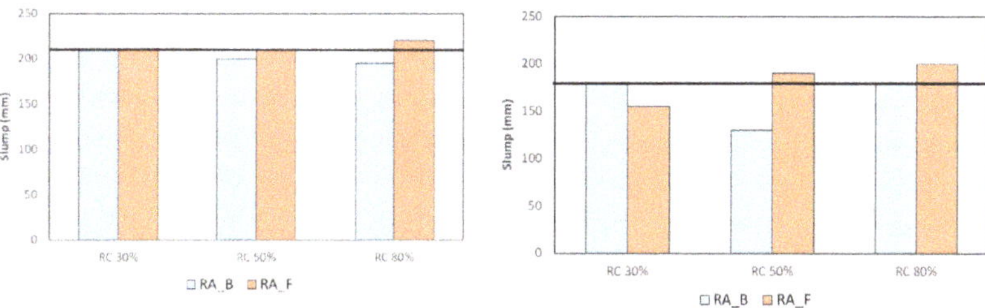

Figure 4. Slump test immediately (left) and 30 min after (right) mixing. The horizontal black line represents the slump value of Normal Concrete (NC).

Compressive strength and secant modulus of elasticity in compression tests were performed, respectively, according to UNI EN 12390-3: 2019 [52] and UNI EN 12390-13: 2013 [53], while splitting tensile strength was obtained following UNI EN 12390-6: 2010 [54].

After 14 and 28 days from the casting date, the compressive strength was measured, while modulus of elasticity and splitting tensile strength were obtained after 28 days. Table 5 presents the above-mentioned mechanical tests results. The average compressive strength at 14 and 28 days is quite high even when the percentage of coarse RA reaches 80%. Indeed, the compressive strength of concrete with RA seems not influenced by the parent concrete. Actually, some tests show how the compressive strength of concrete with RA is higher than NC. The splitting tensile strength of concrete with RA is almost equal or slightly higher than NC. Actually, the greater roughness of RA improves the aggregate interlocking, which produces an increase in tensile strength of concrete. As already shown in other research [49,55], the secant modulus of elasticity of concrete with RA is slightly lower than the one of NC.

The results shown in Table 5 prove that concrete with RA can be considered as a structural concrete, even when the replacement percentage reaches 80%. It is also important to point out that the performance characteristics of the parent concrete do not affect the performance of concrete with RA while the mix design plays a very important role [56].

4.2. Concrete Durability Properties

The durability of concrete is due to degradation phenomena that are produced by chemical and electro-chemical or physical causes [18]. The chemical and electro-chemical causes are related to reactions between aggressive fluids coming from the external environment and the ingredients or hydration products of the cement. The physical causes are determined by the temperature variations and relative humidity gradients, but they are also generated by static and dynamic loads acting on the structure and by abrasive actions. In this work, the durability properties related to the cementitious matrix characteristics have been analyzed in order to assess the concrete water permeability, the freeze–thaw resistance, and resistance to chloride penetration.

4.2.1. Permeability of Concretes

In general, concrete is not very permeable, and the higher the quality, the lower the permeability. Actually, permeability is an important parameter capable of assessing both the ability to avoid liquid loss, in the case of structures designed to contain liquids, and the material durability. The method currently used to estimate the permeability of concrete is based on the resistance to pressurized water penetration. The result of this test is the measurement of the water penetration depth in a cubic specimen (non-steady state, without water permeation), due to the effect of pressure acting on the specimen for the test time.

Table 5. Mechanical properties of concrete with RA and NC, $R_{c,14d}$ represents the cubic compressive strength at 14 days, $R_{c,28d}$ represents the cubic compressive strength at 28 days, f_{ct} is the splitting tensile strength, while E_c is the secant elastic modulus.

Notation	N.	$R_{c,14d}$ (MPa)	$R_{c,28d}$ (MPa)	f_{ct} (MPa)	E_c (MPa)
NC	1	37.4	41.7	3.53	26,601
	2	41.1	41.4	3.71	25,473
	3	40.3	45.2	3.75	26,037
	Average Value	39.6	42.8	3.66	26,037
RC_B 30%	1	44.4	45.5	3.46	24,138
	2	41.7	47.3	3.83	23,553
	3	41.5	44.8	4.06	22,846
	Average Value	42.5	45.9	3.78	23,512
RC_F 30%	1	43.1	44.2	3.87	25,081
	2	38.5	46.3	3.95	25,081
	3	42.0	43.1	3.87	24,543
	Average Value	41.2	44.5	3.89	24,902
RC_B 50%	1	45.5	43.9	3.70	23,383
	2	44.9	41.8	4.04	22,976
	3	43.9	47.5	3.95	22,675
	Average Value	44.8	44.4	3.90	23,011
RC_F 50%	1	45.5	48.6	3.19	25,796
	2	45.3	46.3	3.26	23,842
	3	44.0	48.9	4.60	26,889
	Average Value	44.9	47.9	3.68	25,509
RC_B 80%	1	43.4	45.6	4.10	25,314
	2	42.8	47.9	3.59	22,602
	3	43.1	48.1	3.87	22,541
	Average Value	43.1	47.2	3.85	23,486
RC_F 80%	1	39.4	42.6	3.77	25,398
	2	41.0	43.5	3.47	23,415
	3	40.6	47.3	3.84	23,315
	Average Value	40.3	44.5	3.69	24,043

The standard considered for this test is UNI EN 12390-8: 2009 [57]. Water pressure of 500 kPa is applied for 72 h, as shown in Figure 5a, to the specimen. During the test, the presence of water on the specimen surfaces not exposed to water pressure was periodically observed. The pressure was applied for 72 h and then the specimen was split in half. The cutting surface was perpendicular to the face on which the water pressure was applied, as shown in Figure 5b. The water penetration front could be clearly seen on the split face and then it was marked, as shown in Figure 5c, and measured.

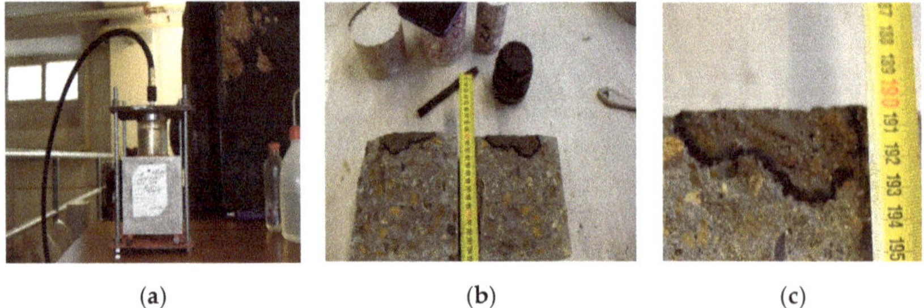

(**a**) (**b**) (**c**)

Figure 5. Test of penetration of water under pressure: (**a**) apparatus to apply water under pressure; (**b**) split specimen; (**c**) marked waterfront.

Figure 6 presents the maximum penetration depth measured on the marked waterfront. It was found that NC presents a greater permeability compared to concrete with RA. The RC with RA_B has higher permeability than RC with RA_F. The influence of parent concrete in recycled concrete with 30% and 50% replacement percentage was also highlighted. Concrete made with RA_F shows less permeability. When the percentage of substitution is 30%, the depth of penetration of pressurized water is greater (1.5 times) in RC with RA_B than in RC with RA_F. The difference in penetration depth between RC with RA_B and with RA_F tends to disappear as the percentage of substitution increases. Thus, the permeability of concrete with RA was lower than the one of NC.

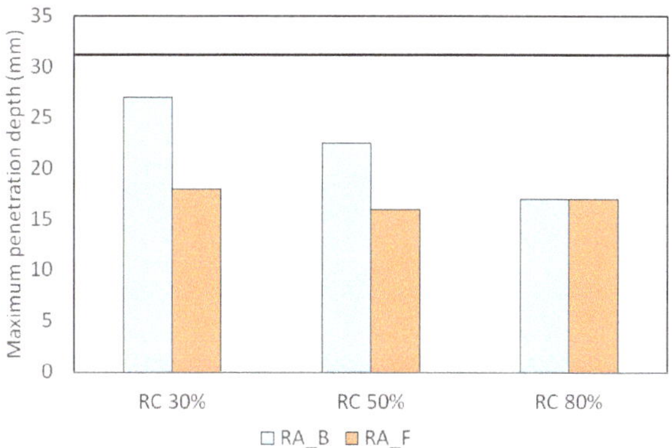

Figure 6. Maximum penetration depth of water under pressure. The black horizontal line represents the penetration depth of water under pressure for NC.

4.2.2. Resistance to Chloride Penetration

The resistance to chloride penetration has been measured following the international standard UNI EN 12390-11: 2015 [58]. Seven cube specimens have been casted and cured for a period of 28 days. Each specimen was divided into two sub-specimens: a "profile specimen" that was used to determine the chloride profile after exposure to unidirectional chloride ingress, and an initial chloride sub-specimen that was used to determine the initial chloride level, C_i. The profile specimen was vacuum saturated with demineralized water, coated on all sides but one, and then the uncoated face was exposed to a chloride solution (3% mass sodium chloride (NaCl) solution) by complete immersion. After 90 days of exposure, 8 layers parallel to the chloride exposed surface but with different depths were ground. The acid-soluble chloride content of each layer and the average depth of the layer from the surface of concrete exposed to the chloride solution were determined. The initial chloride content was also determined by grinding a sample from the other sub-specimen and the acid soluble chloride content determined.

By non-linear regression analysis using the least squares approach, the surface chloride constant (C_s) and the non-steady state chloride diffusion coefficients (D_{nss}) were determined. The regression is necessary to find the parameters C_s and D_{nss}, which minimize the differences between the measured experimental data and the solution to Fick's 2nd law:

$$C_x = C_i + (C_s - C_i)\left(1 - erf\left[\frac{x}{2\cdot\sqrt[2]{D_{nss}\cdot t}}\right]\right). \quad (1)$$

Table 6 presents the experimental data related to the initial chloride level C_i (% by mass of concrete), acid-soluble chloride content C_x (% by mass of concrete), and related average depth, x (mm), of the layer from the surface of concrete exposed for an exposure period of 90 days, and the parameters C_s (% by mass of concrete), D_{nss} (mm^2/days), and the coefficient of determination R^2.

Table 6. Experimental chloride penetration parameters.

	Concrete	NC	RC_F30%	RC_F50%	RC_F80%	RC_B30%	RC_B50%	RC_B80%
	C_i (%)	**0.007**	**0.016**	**0.012**	**0.010**	**0.011**	**0.006**	**0.007**
Layer 1	x (mm)	1.000	1.000	1.000	1.000	1.000	1.000	1.000
	C_x (%)	0.176	0.074	0.274	0.157	0.272	0.130	0.108
Layer 2	x (mm)	2.310	2.690	2.950	2.520	3.000	2.800	2.810
	C_x (%)	0.060	0.061	0.075	0.070	0.070	0.083	0.099
Layer 3	x (mm)	4.190	4.950	4.930	4.740	5.000	4.450	4.940
	C_x (%)	0.055	0.052	0.063	0.057	0.058	0.077	0.080
Layer 4	x (mm)	7.270	7.060	6.720	6.950	7.500	6.600	7.200
	C_x (%)	0.040	0.044	0.055	0.050	0.048	0.060	0.060
Layer 5	x (mm)	9.930	9.180	10.97	10.90	10.46	10.11	9.960
	C_x (%)	0.038	0.040	0.038	0.038	0.032	0.054	0.050
Layer 6	x (mm)	13.35	13.50	14.10	14.83	14.17	13.87	14.39
	C_x (%)	0.020	0.021	0.030	0.026	0.028	0.038	0.015
Layer 7	x (mm)	16.63	17.23	17.80	17.67	17.17	17.78	17.67
	C_x (%)	0.013	0.007	0.009	0.020	0.020	0.017	0.013
Layer 8	x (mm)	20.06	21.14	20.40	21.75	20.73	21.88	21.95
	C_x (%)	0.011	0.006	0.006	0.009	0.009	0.012	0.007
	C_s (%)	0.071	0.075	0.090	0.083	0.085	0.099	0.128
	D_{nss} (mm^2/days)	0.787	0.516	0.731	0.711	0.649	1.019	0.496
	R^2	0.953	0.961	0.998	0.993	0.980	0.970	0.970

The resistance of concrete to chloride penetration can be defined by three parameters: initial chloride content in concrete, C_i; surface chloride content in concrete after exposure to chlorides, C_s; and diffusion coefficient, D_{nss}. Although the adopted exposure solution does not simulate an actual condition produced by seawater or thawing salts, it is nevertheless useful to reduce the problem of durability to a single parameter. To this end, the time T necessary for the chlorides to destroy the protective film was obtained using Equation (1). The calculation was done with reference to a particular level of concentration of chlorides critical C_{cr} and a depth x equal to the reinforcement depth. In this case, C_{cr} = 0.05 (% by mass of concrete) and x = 40 mm have been considered. Figure 7 presents the theoretical service life T (in years) of concretes.

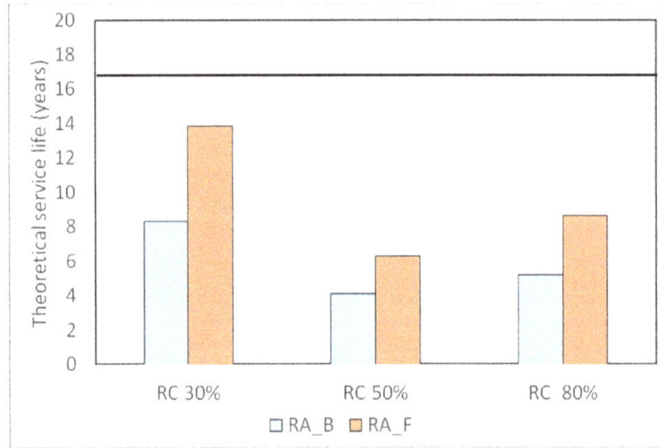

Figure 7. Theoretical service life of concretes. The horizontal black line represents the theoretical service life of NC.

The theoretical service life is not the actual service life of these concretes but is indicative for their chloride penetration resistance. The chloride penetration resistance of concrete with RA is lower than that of NC. When 30% of the aggregates were replaced by RA_F, a decrease of 20% in the theoretical service life was observed when compared to the control mix NC, decreasing up to 50% when 30%

of the aggregates were replaced by RA_B. The chloride penetration resistance of RC appears to be influenced by the parent concrete of RA. On average, the theoretical service life of RC with RA_F is 40% higher than that of RC with RA_B.

4.2.3. Freeze–Thaw Resistance of Concrete

Concrete elements frequently exposed to water, to high relative humidity (more than 75%), and to low environmental temperature (−5 °C or lower), can be subjected to deterioration caused by frost and thaw. This damage allows the penetration of aggressive external agents (such as sulphates and chlorides) and consequently the rebars corrosion is started. It consists mainly of micro- and macro-cracking of the cement matrix and also of spalling of the exposed surface [25,59,60]. In order to investigate this aspect, frost and thaw resistance tests were developed on the different concrete mixes, as shown in Table 4. The standard test procedure UNI CEN/TS 12390-9: 2017 [61] was adopted to assess the frost and thaw resistance of concrete in the presence of a sodium chloride solution. The test does not fully reproduce any possible real field condition, as that can be extremely variable and can be influenced by random parameters as the chemical composition of the environment surrounding the concrete. Indeed, real concretes can be exposed to different types of pollution, chemical aggression, and freezing and thawing cycles at the same moment. In addition, the porosity of the test specimens can be different from the porosity of real concrete elements that is influenced by the presence of reinforcing bars, different casting methods, etc. However, the test is useful to compare the behavior of different concrete mixes with and without RA in given conditions.

Four cubic specimens (150 mm side length), for each concrete mix, were packed and cured for 28 days under standard thermo-hygrometric conditions. From each cube, a prismatic specimen of size 150 × 150 × 50 mm was obtained using a water-cooled diamond saw. A rubber coating was glued to each face of the specimen except for the test surface, which coincides with the sawdust surface. A silicone cord was applied around the entire perimeter of the test surface between the concrete and the rubber coating. The individual specimens were placed in polystyrene honeycomb plastic boxes to ensure thermal insulation on all surfaces except for the test surface, as shown in Figure 8.

Figure 8. Samples exposed to freeze–thaw cycles.

All 28 samples (4 specimens for 7 concretes mixes, as shown in Table 4) were placed in the freezing chamber and subjected to repeated freezing and thawing cycles. The temperature of the freezing liquid in one specimen was monitored continuously and Figure 9 shows the temperatures measured over a 24 h cycle of freezing and thawing for a representative sample.

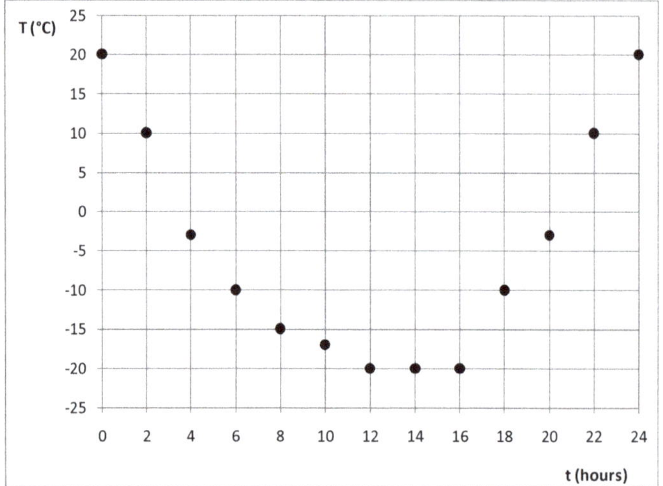

Figure 9. Temperature time history for a 24 h freezing and thawing cycle.

After 14, 28, 42, and 56 freezing and thawing cycles, the specimens were subjected to flaking. The flakes of material detached from the test surface were collected by rinsing and brushing. The collected material was subjected to 110 °C in an oven in order to be dried and then weighed.

Figure 10 shows the cumulative quantity of dried flakes per unit area (S_n) for the different concrete mixes. It is interesting to point out that the concretes with RA_B have the same qualitative and quantitative trend regardless of the percentage of recycled aggregates replacement percentage. The mass value of the cumulated flakes is always higher than the one of the NC for all the monitored freezing and thawing cycles. Instead the concrete with RA_F has a lower, or at most the same, cumulative flake mass values of the NC. Looking at Figure 10 there is no obvious link between the percentage of substitution and S_n.

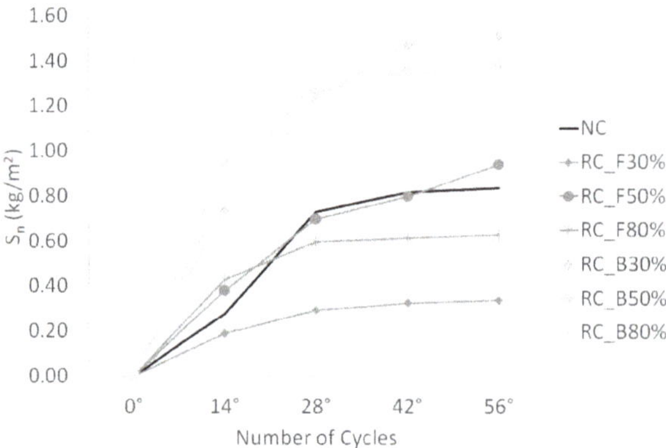

Figure 10. S_n versus number of freeze–thaw cycles.

After 56 freeze-thaw cycles, the cumulative quantity of dried material flakes per unit area of S_n (kg/m^2) and the average value of the four samples for each concrete mix was evaluated and reported in Table 7. It can be stated that the resistance to the frost–thawing cycle, measured using the S_n parameter, is higher in concrete with RA_B. However, the S_n value of the concrete RA_F is lower or approximately equal to the value of the NC. The obtained results do not show any relationship between the replacement percentage of recycled aggregates and the resistance to frost and thaw.

Table 7. Cumulative quantity of flakes of dried material per unit area (S_n) after 56 cycles of frost and thaw.

Concrete	Sample	S_n (kg/m^2)	$S_{n, average}$ (kg/m^2)
RC_B30%	1	1.39	1.52
	2	0.76	
	3	1.71	
	4	2.20	
RC_B50%	1	1.97	1.39
	2	1.00	
	3	0.66	
	4	1.93	
RC_B80%	1	1.03	1.39
	2	2.60	
	3	0.79	
	4	1.16	
NC	1	1.12	0.88
	2	0.21	
	3	1.17	
	4	1.00	
RC_F30%	1	0.20	0.34
	2	0.21	
	3	0.63	
	4	0.31	
RC_F50%	1	0.97	0.94
	2	0.94	
	3	1.27	
	4	0.57	
RC_F80%	1	0.36	0.63
	2	0.70	
	3	0.86	
	4	0.60	

5. Discussion and Conclusions

In this paper, an experimental campaign has been developed in order to assess the mechanical and durability properties of concrete with recycled concrete aggregates. Two different parent concretes have been used to produce the recycled aggregates. In this way, it was possible to investigate what is the influence of the parent concrete on the performance of recycled concrete. RC_F and RC_B denote the concrete with recycled concrete aggregates respectively obtained from the foundation and the beam of the old Cagliari Stadium. The foundation concrete showed better mechanical performance in comparison to the beams one. The following conclusions can be drawn from the results:

- Recycled concrete produced with coarse recycled aggregates has shown similar mechanical performances to normal concrete produced with natural aggregate, even when the natural aggregates replacement percentage reaches 80%.
- The mechanical performance of recycled concrete is not related to the parent concrete mechanical characteristics.
- Concerning the durability, experimental results show that:
- The resistance to pressured water penetration is not reduced by the presence of recycled aggregates.
- The chloride penetration resistance of concrete with RA is lower than that of normal concrete (NC). In addition, it appears to be influenced by the parent concrete. Indeed, the theoretical

service life of RC_F is 40% higher than that of RC_B, regardless of the percentage of recycled aggregate replacement.

- The resistance to the frost–thawing cycle is higher in concrete with RA_B. Instead the S_n value of the concrete with RA_F is lower or approximately equal to the value of the normal concrete NC. The results obtained do not show a relationship between the replacement percentage of recycled aggregates and the resistance to frost and thaw.

These results highlight the importance of the mix design that can allow the obtaining of structural concrete even with concrete demolition waste with different mechanical characteristics.

Recycled aggregates can represent an efficient way to lower the buildings' impact on the environment, improving their sustainability. At the same time, RA can create new opportunities for the companies that re-design their production workflow. For instance, the processing scraps of precast concrete elements should be used to create recycled aggregates, reducing losses and maximizing earnings with a beneficial effect on the environment.

Actually, the transportation costs of construction materials have a paramount relevance in the economic analysis. Thus, recycled aggregates can be very effective when the source of the parent concrete is near the location of the construction, as happens in the case of demolition and re-building, or in the case of retrofitting of existing structures and infrastructures (see [62,63]). Finally, it should be considered that if the environmental impact of the retrofitting intervention is taken into account (see [64,65]), the equivalent CO_2 cost is reduced by the use of RA. Furthermore, the combined use of RA and alternative bio-natural aggregate [66] and structures [67] represent an effective approach to lower the environmental impact of constructions.

Further developments of this work are expected considering whole structural elements like those presented in [68–71].

Author Contributions: Conceptualization, validation, and formal analysis, L.P. and F.S.; methodology, investigation, data curation, L.F. and J.R.; writing—original draft preparation, L.P. and F.S.; writing—review and editing, F.M. and M.S.; supervision, project administration, and funding acquisition, L.P. All authors have read and agreed to the published version of the manuscript.

Funding: This research was funded by Sardegna Ricerche for the financial support (POR FESR 2014/2020-ASSE PRIORITARIO I "RICERCA SCIENTIFICA, SVILUPPO TECNOLOGICO E INNOVAZIONE). The financial support by Autonomous Region of Sardinia under grant PO-FSE 2014–2020, CCI: 2014-IT05SFOP021, through the project "Retrofitting, rehabilitation and requalification of the historical cultural architectural heritage (R3-PAS)" is acknowledged by Flavio Stochino.

Conflicts of Interest: The authors declare no conflict of interest.

References

1. Etxeberria, M.; Mari, A.R.; Vazquez, E. Recycled aggregate concrete as structural material. *Mater. Struct.* **2007**, *40*, 529–541. [CrossRef]
2. De Juan, M.S.; Gutiérrez, P.A. Study on the influence of attached mortar content on the properties of recycled concrete aggregate. *Constr. Build. Mater.* **2009**, *23*, 872–877. [CrossRef]
3. Katz, A. Properties of concrete made with recycled aggregate from partially hydrated old concrete. *Cem. Concr. Res.* **2003**, *33*, 703–711. [CrossRef]
4. Li, W.; Long, C.; Tam, V.W.Y.; Poon, C.-S.; Duan, W.H. Effects of nano-particles on failure process and microstructural properties of recycled aggregate concrete. *Constr. Build. Mater.* **2017**, *142*, 42–50. [CrossRef]
5. Li, W.; Luo, Z.; Sun, Z.; Hu, Y.; Duan, W.H. Numerical modelling of plastic–damage response and crack propagation in RAC under uniaxial loading. *Mag. Concr. Res.* **2018**, *70*, 459–472. [CrossRef]
6. Silva, R.V.; de Brito, J.; Dhir, R.K. The influence of the use of recycled aggregates on the compressive strength of concrete: A review. *Eur. J. Environ. Civ. Eng.* **2015**, *19*, 825–849. [CrossRef]
7. Silva, R.V.; de Brito, J.; Dhir, R.K. Establishing a relationship between modulus of elasticity and compressive strength of recycled aggregate concrete. *J. Clean. Prod.* **2016**, *112*, 2171–2186. [CrossRef]
8. Silva, R.V.; de Brito, J.; Dhir, R.K. Tensile strength behaviour of recycled aggregate concrete. *Constr. Build. Mater.* **2015**, *83*, 108–118. [CrossRef]

9. Francesconi, L.; Pani, L.; Stochino, F. Punching shear strength of reinforced recycled concrete slabs. *Constr. Build. Mater.* **2016**, *127*, 248–263. [CrossRef]
10. Rahman, M.M.; Beecham, S.; Iqbal, A.; Karim, M.R.; Rabbi, A.T.Z. Sustainability assessment of using recycled aggregates in concrete block pavements. *Sustainability* **2020**, *12*, 4313. [CrossRef]
11. Shi, X.; Grasley, Z.; Mukhopadhyay, A.; Zollinger, D. Use of recycled aggregates in concrete pavement: Pavement design and life cycle assessment. In Proceedings of the International Symposium on Pavement. Roadway, and Bridge Life Cycle Assessment 2020 (LCA 2020), Sacramento, CA, USA, 3–6 June 2020; CRC Press: Boca Raton, FL, USA, 2020; p. 324.
12. Silva, R.V.; de Brito, J. Reinforced recycled aggregate concrete slabs: Structural design based on Eurocode 2. *Eng. Struct.* **2020**, *204*, 110047. [CrossRef]
13. Cantero, B.; Bravo, M.; de Brito, J.; del Bosque, I.S.; Medina, C. Mechanical behaviour of structural concrete with ground recycled concrete cement and mixed recycled aggregate. *J. Clean. Prod.* **2020**, *275*, 122913. [CrossRef]
14. Limbachiya, M.C.; Leelawat, T.; Dhir, R.K. Use of recycled concrete aggregate in high-strength concrete. *Mater. Struct.* **2000**, *33*, 574–580. [CrossRef]
15. Ardalan, R.B.; Emamzadeh, Z.N.; Rasekh, H.; Joshaghani, A.; Samali, B. Physical and mechanical properties of polymer modified self-compacting concrete (SCC) using natural and recycled aggregates. *J. Sustain. Cem. Based Mater.* **2020**, *9*, 1–16. [CrossRef]
16. Singh, A.; Duan, Z.; Xiao, J.; Liu, Q. Incorporating recycled aggregates in self-compacting concrete: A review. *J. Sustain. Cem. Based Mater.* **2020**, *9*, 165–189. [CrossRef]
17. ACI Committee 555. *Removal and Reuse of Hardened Concrete*; American Concrete Institute: Farmington Hills, MI, USA, 2001.
18. Kisku, N.; Joshi, H.; Ansari, M.; Panda, S.K.; Nayak, S.; Dutta, S.C. A critical review and assessment for usage of recycled aggregate as sustainable construction material. *Constr. Build. Mater.* **2017**, *131*, 721–740. [CrossRef]
19. Guo, H.; Shi, C.; Guan, X.; Zhu, J.; Ding, Y.; Ling, T.-C.; Zhang, H.; Wang, Y. Durability of recycled aggregate concrete—A review. *Cem. Concr. Compos.* **2018**, *89*, 251–259. [CrossRef]
20. Silva, R.V.; de Brito, J.; Dhir, R.K. Prediction of chloride ion penetration of recycled aggregate concrete. *Mater. Res.* **2015**, *18*, 427–440. [CrossRef]
21. Sasanipour, H.; Aslani, F. Durability properties evaluation of self-compacting concrete prepared with waste fine and coarse recycled concrete aggregates. *Constr. Build. Mater.* **2020**, *236*, 117540. [CrossRef]
22. Pereira, P.; Evangelista, L.; de Brito, J. The effect of superplasticizers on the mechanical performance of concrete made with fine recycled concrete aggregates. *Cem. Concr. Compos.* **2012**, *34*, 1044–1052. [CrossRef]
23. Saravanakumar, P.; Dhinakaran, G. Durability characteristics of recycled aggregate concrete. *Struct. Eng. Mech.* **2013**, *47*, 701–711. [CrossRef]
24. Kwan, W.H.; Ramli, M.; Kam, K.J.; Sulieman, M.Z. Influence of the amount of recycled coarse aggregate in concrete design and durability properties. *Constr. Build. Mater.* **2012**, *26*, 565–573. [CrossRef]
25. Medina, C.; de Rojas, M.I.S.; Frías, M. Freeze-thaw durability of recycled concrete containing ceramic aggregate. *J. Clean. Prod.* **2013**, *40*, 151–160. [CrossRef]
26. Olorunsogo, F.; Padayachee, N. Performance of recycled aggregate concrete monitored by durability indexes. *Cem. Concr. Res.* **2002**, *32*, 179–185. [CrossRef]
27. Wang, H.; Sun, X.; Wang, J.; Monteiro, P.J.M. Permeability of concrete with recycled concrete aggregate and pozzolanic materials under stress. *Materials* **2016**, *9*, 252. [CrossRef]
28. Xiao, J.; Li, W.; Fan, Y.; Huang, X. An overview of study on recycled aggregate concrete in China (1996–2011). *Constr. Build. Mater.* **2012**, *31*, 364–383. [CrossRef]
29. Otsuki, N.; Miyazato, S.S.; Yodsudjai, W. Influence of recycled aggregate on interfacial transition zone, strength, chloride penetration and carbonation of concrete. *J. Mater. Civil. Eng.* **2003**, *15*, 443–551. [CrossRef]
30. Sim, J.; Park, C. Compressive strength and resistance to chloride ion penetration and carbonation of recycled aggregate concrete with varying amount of fly ash and fine recycled aggregate. *Waste Manag.* **2011**, *31*, 2352–2360. [CrossRef]
31. Kurda, R.; de Brito, J.; Silvestre, J.D. Water absorption and electrical resistivity of concrete with recycled concrete aggregates and fly ash. *Cem. Concr. Compos.* **2019**, *95*, 169–182. [CrossRef]

32. Lima, C.; Caggiano, A.; Faella, C.; Martinelli, E.; Pepe, M.; Realfonzo, R. Physical properties and mechanical behaviour of concrete made with recycled aggregates and fly ash. *Constr. Build. Mater.* **2013**, *47*, 547–559. [CrossRef]
33. Pacheco, J.; de Brito, J.; Chastre, C.; Evangelista, L. Experimental investigation on the variability of the main mechanical properties of concrete produced with coarse recycled concrete aggregates. *Constr. Build. Mater.* **2019**, *201*, 110–120. [CrossRef]
34. Padmini, A.K.; Ramamurthy, K.; Mathews, M.S. Influence of parent concrete on the properties of recycled aggregate concrete. *Constr. Build. Mater.* **2009**, *23*, 829–836. [CrossRef]
35. Kou, S.-C.; Poon, C.-S. Effect of the quality of parent concrete on the properties of high performance recycled aggregate concrete. *Constr. Build. Mater.* **2015**, *77*, 501–508. [CrossRef]
36. Ajdukiewicz, A.; Kliszczewicz, A. Influence of recycled aggregates on mechanical properties of HS/HPC. *Cem. Concr. Compos.* **2002**, *24*, 269–279. [CrossRef]
37. Rahal, K.N. Mechanical properties of concrete with recycled coarse aggregate. *Build. Environ.* **2007**, *42*, 407–415. [CrossRef]
38. González-Fonteboa, B.; Martínez-Abella, F. Concretes with aggregates from demolition waste and silica fume. Materials and mechanical properties. *Build. Environ.* **2008**, *43*, 429–437. [CrossRef]
39. Pani, L.; Francesconi, L.; Rombi, J.; Naitza, S.; Balletto, G.; Mei, G. Recycled aggregates, mechanical properties and environmental sustainability. In *Planning, Nature and Ecosystem Services, INPUT aCAdemy;* FedOAPress—Federico II Open ACCESS University Press: Cagliari, Italy, 2019; pp. 431–442.
40. Tabsh, S.W.; Abdelfatah, A.S. Influence of recycled concrete aggregates on strength properties of concrete. *Constr. Build. Mater.* **2009**, *23*, 1163–1167. [CrossRef]
41. *Aggregates for Concrete;* UNI EN 12620: 2008; Ente Nazionale Italiano di Unificazione (UNI): Milan, Italy, 2008.
42. *Aggregates for Concrete. Additional Provisions for the Application of EN 12620 Part 1: Designation and Conformity Criteria;* UNI 8520-1: 2015; Ente Nazionale Italiano di Unificazione (UNI): Milan, Italy, 2015.
43. Dimitriou, G.; Savva, P.; Petrou, M.F. Enhancing mechanical and durability properties of recycled aggregate concrete. *Constr. Build. Mater.* **2018**, *158*, 228–235. [CrossRef]
44. Omary, S.; Ghorbel, E.; Wardeh, G.; Nguyen, M.D. Mix design and recycled aggregates effects on the concrete's properties. *J. Civ. Eng.* **2018**, *16*, 973–992. [CrossRef]
45. Pani, L.; Francesconi, L.; Concu, G. Influence of replacement percentage of recycled aggregates on recycled aggregate concrete properties. In Proceedings of the 2011 FIB Symposium: Concrete Engineering for Excellence and Efficiency, Prague, Czech Republic, 8–10 June 2011.
46. Pani, L.; Francesconi, L.; Concu, G. Relation between static and dynamic moduli of elasticity for recycled aggregate concrete. In Proceedings of the First International Conference on Concrete Sustainability, Tokyo, Japan, 27–29 May 2013; pp. 676–681.
47. Pani, L.; Balletto, G.; Naitza, S.; Francesconi, L.; Trulli, N.; Mei, G.; Furcas, C. Evaluation of mechanical, physical and chemical properties of recycled aggregates for structural concrete. In Proceedings of the Sardinia 2013, Fourteenth International Waste Management and Landfill Symposium, Cagliari, Italy, 30 September–4 October 2013.
48. Tavakoli, M.; Soroushian, P. Strength of recycled aggregate concrete made using field-demolished concrete as aggregate. *ACI Mat. J.* **1996**, *93*, 182–190.
49. Salem, R.M.; Burdette, E.G. Role of chemical and mineral admixtures on physical properties and frost-resistance of recycled aggregate concrete. *ACI Mat. J.* **1998**, *95*, 558–563.
50. Abbas, A.; Fathifazl, G.; Isgor, O.B.; Razaqpur, A.G.; Fournier, B.; Foo, S. Proposed method for determining the residual mortar content of recycled concrete aggregates. *J. ASTM Intern.* **2007**, *5*, 1–12.
51. *Testing Fresh Concrete—Part 2: Slump Test;* UNI EN 12350-2:2019; Ente Nazionale Italiano di Unificazione (UNI): Milan, Italy, 2019.
52. *Testing Hardened Concrete—Part 3: Compressive Strength of Test Specimens;* UNI EN 12390-3: 2019; Ente Nazionale Italiano di Unificazione (UNI): Milan, Italy, 2019.
53. *Testing Hardened Concrete—Part 13: Determination of Secant Modulus of Elasticity in Compression;* UNI EN 12390-13: 2013; Ente Nazionale Italiano di Unificazione (UNI): Milan, Italy, 2013.
54. *Testing Hardened Concrete—Part 6: Tensile Splitting Strength of the Specimens;* UNI EN 12390-6: 2010; Ente Nazionale Italiano di Unificazione (UNI): Milan, Italy, 2010.

55. Stochino, F.; Pani, L.; Francesconi, L.; Mistretta, F. Cracking of reinforced recycled concrete slabs. *Int. J. Struct. Glass Adv. Mater. Res.* **2017**, *1*, 3–9. [CrossRef]
56. Pepe, M.; Filho, R.D.T.; Koenders, E.A.; Martinelli, E. A novel mix design methodology for recycled aggregate concrete. *Constr. Build. Mater.* **2016**, *122*, 362–372. [CrossRef]
57. *Testing Hardened Concrete. Part. 8 Depth of Penetration of Water Under Pressure;* UNI EN 12390-8: 2009; Ente Nazionale Italiano di Unificazione (UNI): Milan, Italy, 2009.
58. *Testing Hardened Concrete. Part. 11: Determination of the Chloride Resistance of Concrete, Unidirectional Diffusion;* UNI EN 12390-11: 2015; Ente Nazionale Italiano di Unificazione (UNI): Milan, Italy, 2015.
59. Hanjari, K.Z.; Utgenannt, P.; Lundgren, K. Experimental study of the material and bond properties of frost-damaged concrete. *Cem. Concr. Res.* **2011**, *41*, 244–254. [CrossRef]
60. Harrison, T.A.; Dewar, J.D.; Brown, B.V. *Freeze-Thaw Resisting Concrete: Its Achievement in the UK;* CIRIA: London, UK, 2001.
61. *Testing Hardened Concrete. Part 9. Freeze-Thaw Resistance with Deicing Salts. Scaling;* UNI CEN/TS 12390-9: 2017; Ente Nazionale Italiano di Unificazione (UNI): Milan, Italy, 2017.
62. Stochino, F.; Fadda, M.L.; Mistretta, F. Assessment of RC Bridges integrity by means of low-cost investigations. *Frat. Integrita Strutt.* **2018**, *46*, 216–225. [CrossRef]
63. Stochino, F.; Fadda, M.L.; Mistretta, F. Low cost condition assessment method for existing RC bridges. *Eng. Fail. Anal.* **2018**, *86*, 56–71. [CrossRef]
64. Sassu, M.; Stochino, F.; Mistretta, F. Assessment method for combined structural and energy retrofitting in masonry buildings. *Buildings* **2017**, *7*, 71. [CrossRef]
65. Mistretta, F.; Stochino, F.; Sassu, M. Structural and thermal retrofitting of masonry walls: An integrated cost-analysis approach for the Italian context. *Build. Environ.* **2019**, *155*, 127–136. [CrossRef]
66. Sassu, M.; Giresini, L.; Bonannini, E.; Puppio, M.L. On the use of vibro-compressed units with bio-natural aggregate. *Buildings* **2016**, *6*, 40. [CrossRef]
67. Sassu, M.; de Falco, A.; Giresini, L.; Puppio, M.L. Structural solutions for low-cost bamboo frames: Experimental tests and constructive assessments. *Materials* **2016**, *9*, 346. [CrossRef] [PubMed]
68. Xu, J.J.; Chen, Z.P.; Zhao, X.Y.; Demartino, C.; Ozbakkaloglu, T.; Xue, J.Y. Seismic performance of circular recycled aggregate concrete-filled steel tubular columns: FEM modelling and sensitivity analysis. *Thin Walled Struct.* **2019**, *141*, 509–525. [CrossRef]
69. Xu, J.J.; Chen, Z.P.; Ozbakkaloglu, T.; Zhao, X.-Y.; Demartino, C. A critical assessment of the compressive behavior of reinforced recycled aggregate concrete columns. *Eng. Struct.* **2018**, *161*, 161–175. [CrossRef]
70. Xu, J.J.; Chen, Z.P.; Xiao, Y.; Demartino, C.; Wang, J.H. Recycled aggregate concrete in FRP-confined columns: A review of experimental results. *Comp. Struct.* **2017**, *174*, 277–291. [CrossRef]
71. Deresa, S.T.; Xu, J.; Demartino, C.; Heo, Y.; Li, Z.; Xiao, Y. A review of experimental results on structural performance of reinforced recycled aggregate concrete beams and columns. *Adv. Struct. Eng.* **2020**, *23*, 3351–3369. [CrossRef]

Publisher's Note: MDPI stays neutral with regard to jurisdictional claims in published maps and institutional affiliations.

MDPI

Article

Influence of Steel Slag Type on Concrete Shrinkage

Maria Dolores Rubio-Cintas [1], Maria Eugenia Parron-Rubio [1,*], Francisca Perez-Garcia [2], António Bettencourt Ribeiro [3] and Miguel José Oliveira [4]

[1] Department of Industrial and Civil Engineering, Universidad de Cádiz, 11003 Cádiz, Spain; mariadolores.rubio@uca.es
[2] Departamento de Ingeniería Civil, Universidad de Málaga, 29016 Málaga, Spain; perez@uma.es
[3] Laboratório Nacional de Engenharia Civil, 1700-075 Lisboa, Portugal; bribeiro@lnec.pt
[4] Departamento de Engenharia Civil (DEC), Universidade do Algarve, 8005-139 Faro, Portugal; mjolivei@ualg.pt
* Correspondence: mariaeugenia.parron@uca.es; Tel.: +34-639-030-147

Abstract: Building construction and building operations have a massive direct and indirect effect on the environment. Cement-based materials will remain essential to supply the growth of our built environment. Without preventive measures, this necessary demand in cement production will imply a substantial increase in CO_2 generation. Reductions in global CO_2 emissions due to cement consumption may be achieved by improvements on two main areas: increased use of low CO_2 supplementary cementitious materials and a more efficient use of Portland cement clinker in mortars and concretes. The use of ground granulated blast furnace slag in concrete, as cement constituent or as latent hydraulic binder, is a current practice, but information of concrete with ladle furnace slag is more limited. Specific knowledge of the behavior of mixtures with steel slag in relation to certain properties needs to be improved. This paper presents the results of the shrinkage (total and autogenous) of five concrete mixtures, produced with different percentages of two different slags in substitution of cement. The results show that shrinkage of concrete with the two different slags diverges. These different characteristics of the two materials suggest that their use in combination can be useful in optimizing the performance of concrete.

Keywords: shrinkage; slags; cement replacement; concrete

Citation: Rubio-Cintas, M.D.; Parron-Rubio, M.E.; Perez-Garcia, F.; Bettencourt Ribeiro, A.; Oliveira, M.J. Influence of Steel Slag Type on Concrete Shrinkage. *Sustainability* **2021**, *13*, 214. https://doi.org/10.3390/su13010214

Received: 12 November 2020
Accepted: 19 December 2020
Published: 28 December 2020

Publisher's Note: MDPI stays neutral with regard to jurisdictional claims in published maps and institutional affiliations.

1. Introduction

The building and construction sector is a key player that contributes to meeting the needs of housing, hospitals, schools, among others developments, but it is also a large consumer of materials and natural resources. Building construction and building operations have a massive direct and indirect effect on the environment [1–6]. Not considering water, concrete may be considered the most used construction material in the world. The most important constituent of concrete is cement. It is produced in a dynamic process including high temperature treatment and a release of approximately 710 kg CO_2 per cement ton cement [7].

Cement-based materials will remain essential to supply the growth of our built environment, mainly by those located in the developing world. Without preventive measures, this needed demand in cement production will imply a substantial increase in CO_2 generation, further aggravating the environmental aspects and global warming. According to United Nations [8] there are two main areas that can deliver very considerable additional reductions in global CO_2 emissions related to cement and concrete production and use (not including carbon capture and storage (CCS)):

1. Increased use of low-CO_2 supplements (supplementary cementitious materials) as partial replacements for Portland cement clinker.
2. More efficient use of Portland cement clinker in mortars and concretes.

In fact, the use of industrial and construction waste as raw materials for the production of cement and concrete can be considered not only to reduce CO_2 emissions and the amount of embodied energy, but also to contribute to a circular economy and to diminish the environmental threats associated with industrial waste materials. This partial substitution of Portland cement clinker by appropriate composite materials need to feature a certain chemical activity in order to assure cement quality [9]. Moreover, the main problem associated to the iron and steel industry is waste generation and by-products. To promote environmental sustainability and circular economy it is essential to promote the reuse of these by-products. One of these by-products is steel slag. The use of slag in the concrete has been going on for some time all over the world, especially in cement as an addition to the mix of cement manufactured in plant, under certain conditions, to make up the cements CEM II, CEM III and cement V [10]. This subject is relatively well studied and documented for main properties, however, for more specific properties such as shrinkage further studies are needed [11–15].

Depending on different factors, mainly due to the manufacturing process, four types of steel slag can be identified: blast furnace slag (BFS), ladle furnace slag (LFS), electric arc furnace (EAF) and basic oxygen furnace slag (BOFS) [16]. On the other hand, one of the essential aspects in promoting the most efficient use of cement materials in the production of concrete is the maximization of its durability. For good long-term behavior, it is essential to reduce the cracking of reinforced concrete elements. In this sense shrinkage evaluation and shrinkage crack reduction is an important factor. The assessment of time-dependent behavior is still one of the most difficult features in designing a concrete structure [17]. The structural concrete codes dealing with time dependent behavior provide common rules for regular concrete and the confirmation of some established stress-strain-relations have to be confirmed through laboratory tests when special mixtures are used [18–20].

Shrinkage is the diminution in either length or volume of a material, after suffering changes in chemical properties or moisture content, and occurs in the absence of external actions applied to the concrete [21]. If moisture transfer with the adjacent environment is not permitted, and temperature is kept constant, this volume variation is called autogenous shrinkage and is attributed to self-desiccation due to binder hydration [22].

Standardization in the field of construction products, and in particular in the field of concrete with hydraulic binders, has been providing an increase in the quality and durability requirements of concrete structures. Indeed, in order to achieve sustainable development in construction, while continuing to use concrete, it is necessary, among other aspects, to maximize the durability of the structures. According to Mehta [23], this solution represents a major step in the optimization of resources in the construction industry. In addition, it is necessary to ensure that the capacities of the materials and elements of the structures that are designed and built are fully exploited and maximized in service, in the respective useful life period [24]. The reduction of shrinkage cracking is very important from the point of view of durability, as well as resistance and behavior in service.

This paper presents the results of the shrinkage (total and autogenous) of five concrete mixtures, produced with different percentages of two different slags in substitution of cement. This information can be useful for minimizing the cracking of concrete elements, due to shrinkage, and improving its performance in service, contributing to the greater use of slag and to a more sustainable construction.

2. Materials and Methods

In this work, five concrete mixtures were prepared, namely, a reference mixture without cement replacement and four mixtures in which 25% and 40% cement replacement (mass), using separately ground granulated blast furnace slag (GGBFS) or ladle furnace slag (LFS). Portland cement, class strength 52.5 R, was used because this cement contains 95–100% Portland cement clinker. Since this research is part of a larger project that is studying the use of landfill slag as a concrete constituent, for the production of precast concrete, the concrete dosage has been designed in order to meet a dry consistency.

The percentages of cement replacement by slag were defined based on the results obtained by Rubio [25].

The mixtures were prepared using the following materials:

- Cement: Portland Cement CEM I 52.5 R.
- Slag 1: Ground granulated blast furnace slag (GGBFS), particle size <0.063 µm provided by the company (ESTABISOL S.A.).
- Slag 2: Unprocessed ladle furnace slag (LFS). This slag was sieved in laboratory in order to increase the particle size <0.063 µm fraction (23%).
- Aggregates: crushed limestone. Fine sand (0/2), medium size sand (0/4) and gravel (4/12).
- Chemical admixture: Superplasticizer.
- Water: Tap water.

Physical and chemical characteristics of the cement and slag used are presented in Table 1.

Table 1. Cement and slag chemical composition, density and specific surface area. (Data provided by the suppliers).

	Chemical Composition						Density	Specific Surface Area
Material	**SiO_2**	**AL_2O_3**	**Fe_2O_3**	**CaO**	**MgO**	**Σ Others**		
	%	%	%	%	%	%	**gr/cm³**	**cm²/gr**
CEM	20–22	4–8	4–6	55–60	2–3	2–15	3.01	>2800
GGBFS	32–36	11–12	0–2	40–42	7–8	1–10	2.91	4500–4700
LFS	20–23	8–10	0–2	55–60	7–10	5–9	2.65	2000–2500

The grading curves of the used aggregates are presented in Figure 1.

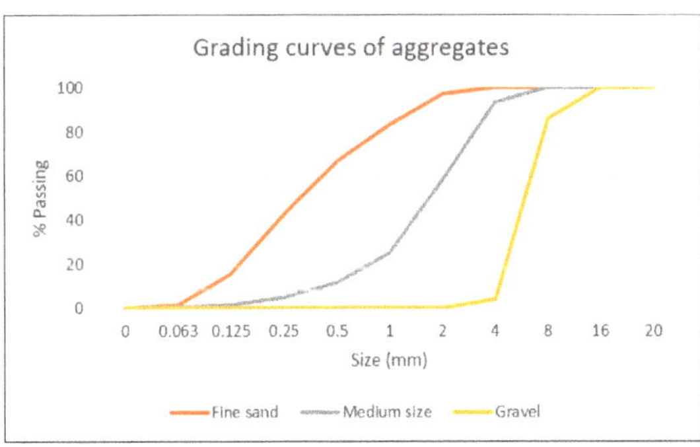

Figure 1. Grading curves of used aggregates.

Table 2 presents the concrete mixture proportions. They were named as follow:

Mixture 1 (REF): Ordinary concrete without slag;
Mixture 2 (25GGBFS): Concrete with 25% cement replaced with GGBFS;
Mixture 3 (40GGBFS): Concrete with 40% cement replaced with GGBFS;
Mixture 4 (25LFS): Concrete with 25% cement replaced with LFS;
Mixture 5 (40LFS): Concrete with 40% cement replaced with LFS

Table 2. Concrete mixture proportions.

Materials	REF [kg/m^3]	25GGBFS [kg/m^3]	40GGBFS [kg/m^3]	25LFS [kg/m^3]	40LFS [kg/m^3]
CEM I 52.5 R	300	225	180	225	180
GGBFS	—	75	120	—	—
LFS	—	—	—	75	120
Sand 0–2	306	306	306	306	306
Sand 0–4	712	712	712	712	712
Gravel 4–12	1017	1017	1017	1017	1017
Water	150	150	150	150	150
Superplasticizer	3.9	3.9	3.9	3.9	3.9
Water/Powder	0.5	0.5	0.5	0.5	0.5

The preparation of the prismatic specimens (160 mm × 40 mm × 40 mm) was performed according to EN 196-1 [26], in a room with a temperature of 20 ± 2 Celsius degrees and a relative humidity of 55% ± 5. It was decided to use small specimens since the ratio between smallest size of the specimen and largest aggregate size is about 3. Nevertheless, the mixture proportions and the aggregates used in this research are different from those indicated in [26].

Aitcin [27] recommends that the shrinkage measurement should be started as soon as possible; otherwise the measurement performed may underestimate the real shrinkage. In this sense the compressive strength of the mixtures in the first hours was monitored, through the procedure described in the EN 196-1, although adapted to concrete. The results are presented on Table 3.

Table 3. Concrete mixtures: early age compressive strength.

Mixtures	2 h	3 h	4 h	5 h	6 h
REF	0.8 MPa	2.2 MPa	3.1 MPa	6 MPa	7.6 MPa
25GGBFS	0.9 MPa	1.8 MPa	3.5 MPa	5.9 MPa	6.3 MPa
40GGBFS	—	—	2.2 MPa	3.4 MPa	4.0 MPa
25LFS	0.9 MPa	1.9 MPa	3.2 MPa	5.8 MPa	6.4 MPa
40LFS	—	—	2.1 MPa	3.0 MPa	3.6 MPa

The removal of molds took place about 4 h after mixing. This time delay was defined as the minimum required time to ensure concrete strength between 2 MPa and 4 MPa, in order to avoid specimens damage. At 28 days the compression strength of the REF mixture was 50 MPa [13]. The tests performed on the prismatic specimens were: drying shrinkage, autogenous shrinkage and expansion under water immersion. The samples used for the autogenous shrinkage test were sealed with a plastic film (Figure 2). Samples for autogenous and drying shrinkage tests were placed on two thin supports and the others samples were immersed in water to achieve saturated conditions (Figure 3). Linear deformations of each specimen were measured using a length comparator with the sensitivity of 1 µm (Figure 4). Gage studs were located at the end sections of the concrete prisms (Figure 5). At the ages of 1, 3, 7, 14 and 28 days, and 2, 3, 4, 5, 6, 7, 8 and 9 months, the specimens were weighed and the length variation was measured, using the procedures described in Draft prEN 12390-16 [28]. For shrinkage, the measuring positions were along the principal axis of the specimen.

Figure 2. Sample for measurement of autogenous shrinkage.

Figure 3. Samples immersed in water.

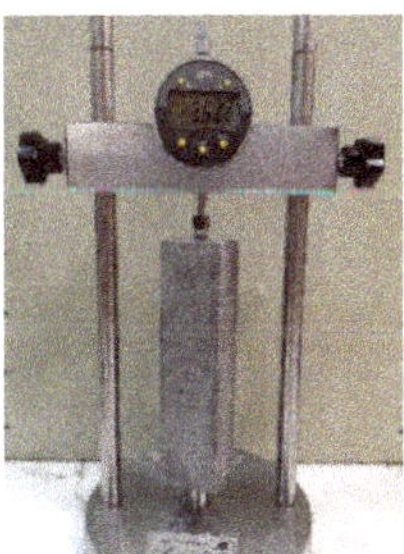

Figure 4. Length comparator.

Figure 5. Gage studs at the end sections of the specimens.

3. Results

The test results obtained on the five series of specimens allowed comparing their relative performance regarding mass change and shrinkage. Sections 3.1 and 3.2 are dedicated to the presentation of the results of mass change and shrinkage, respectively, using figures but without discussion. The analysis of the results is presented in Section 4.

3.1. Mass Change

The charts in the following figures (Figures 6–10) show the mass variation for the mixtures: REF; 25GGBFS, 40GGBFS, 25LFS, 40LFS, recorded up to 9 months (each value presented is the average of three specimens). The individual results deviation were very small (SD < 0.14%).

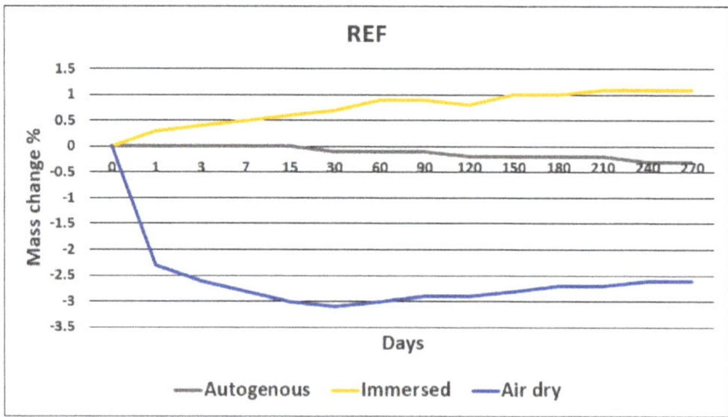

Figure 6. Mass change of mixture REF.

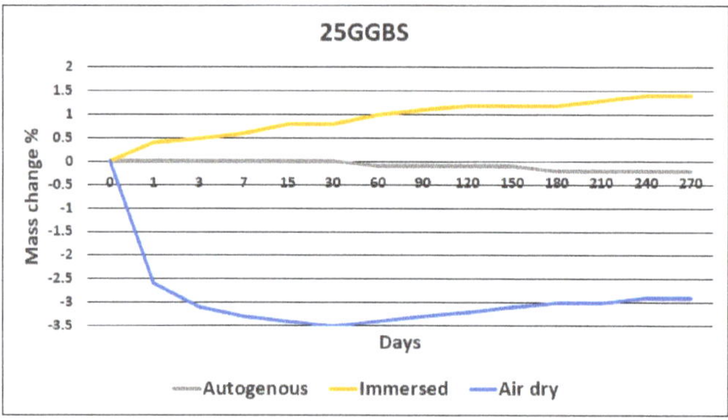

Figure 7. Mass change of mixture 25GGBS.

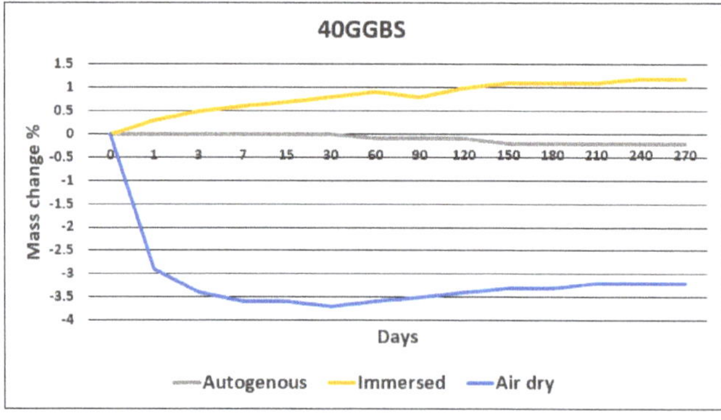

Figure 8. Mass change of mixture 40GGBS.

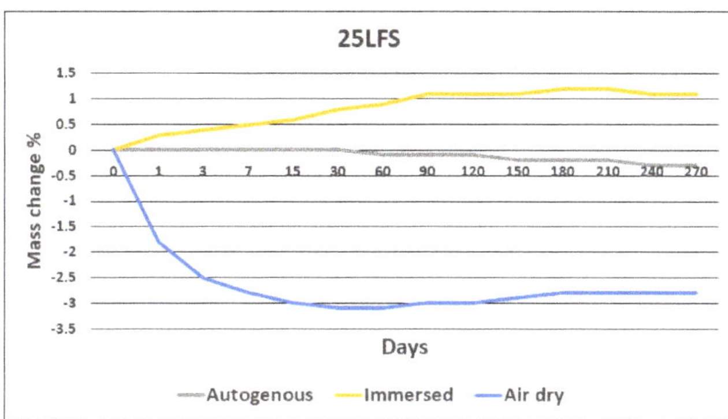

Figure 9. Mass change of mixture 25LFS.

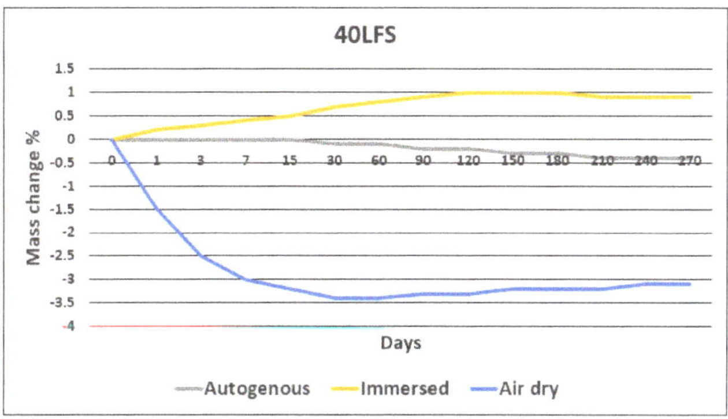

Figure 10. Mass change of mixture 40LFS.

3.2. Shrinkage

The following results, presented in the form of graphs (Figures 11–15), were obtained using the shrinkage measurements of nine specimens per mixture (three for each condition: air dry, sealed and immersed). For each mixture, the solid curves present the average and the dashed curves present the average plus or minus one standard deviation (SD), recorded up to 9 months of age.

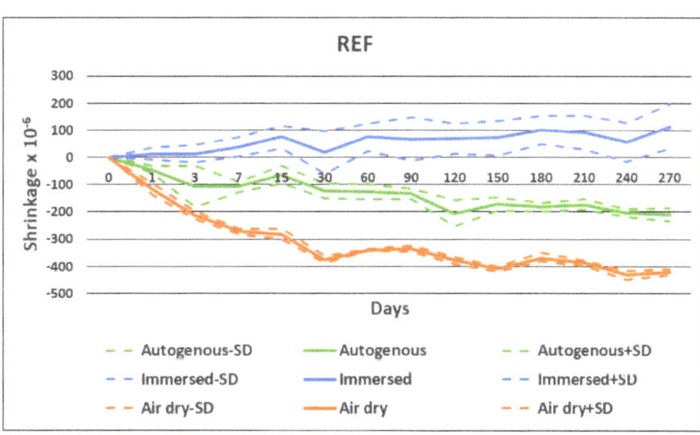

Figure 11. Strain of mixture REF.

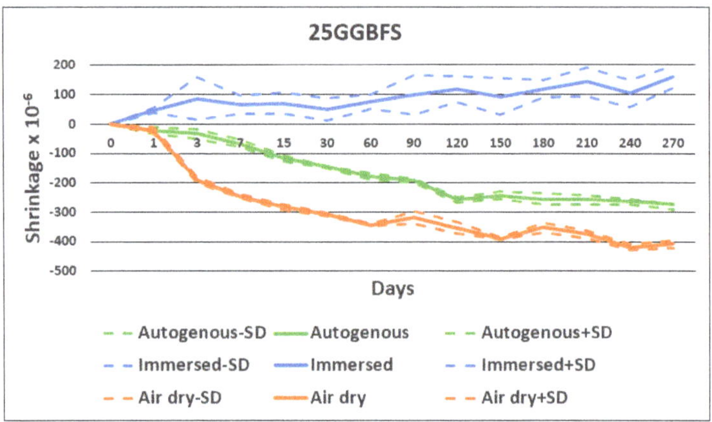

Figure 12. Strain of mixture 25GGBS.

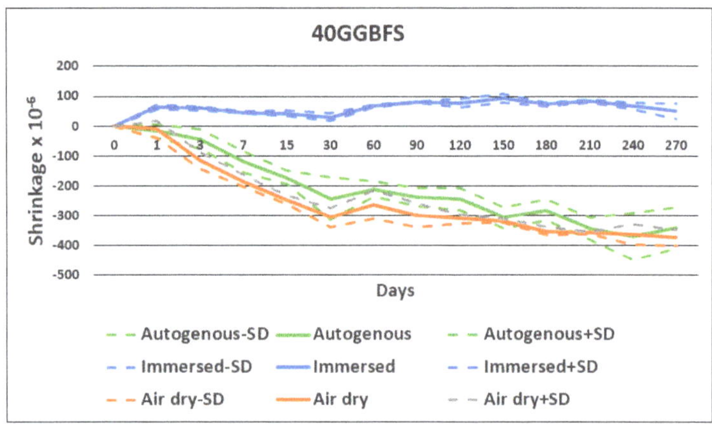

Figure 13. Strain of mixture 40GGBS.

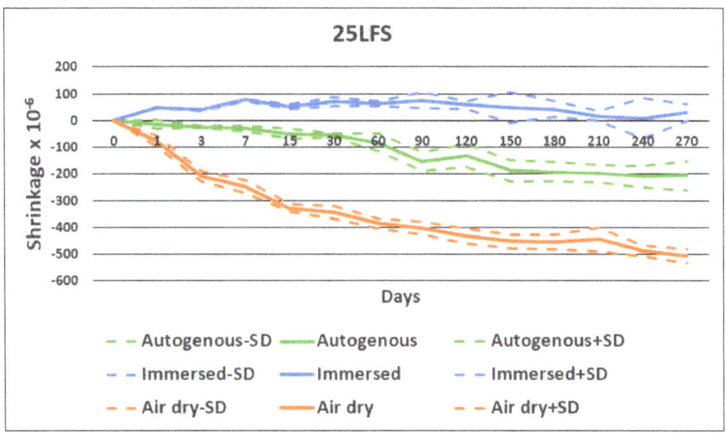

Figure 14. Strain of mixture 25LFS.

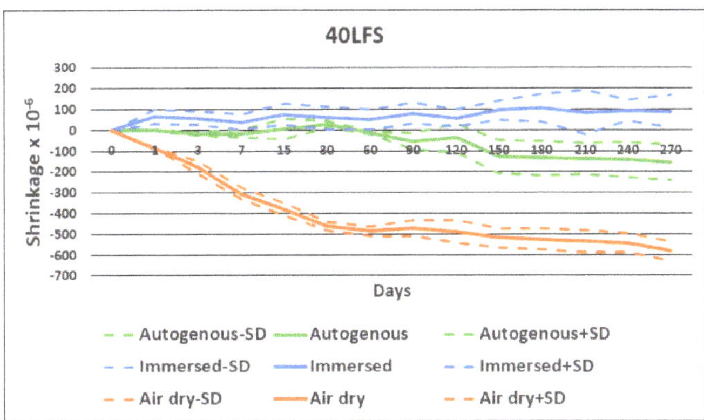

Figure 15. Strain of mixture 40LFS.

4. Discussions

Regarding mass variation of sealed specimens, in the first month there was almost no mass change. After 30 days, there was a very small loss of mass, but it did not reach even 0.5% at 270 days. The immersed specimens presented mass gain, as expected.

The mass loss in the first 30 days of specimens exposed to air drying is presented in Figure 16. In first 3 days, this figure shows an increase of the mass loss with increasing GGBFS. The lower early age strength of mixtures with increasing GGBFS contents (Table 3) is a consequence of the slower growth of the structural network, leaving the solid body more exposed to drying. However, despite the lower strength, at 3 days, mixtures with LFS did not show higher mass loss than the reference mixture. As the LFS density is lower than the density of cement or GGBFS, the LFS paste has higher solids content, decreasing the porosity when expressed in volume. Thus, the volume of liquid available for evaporation, in relative terms, is smaller, which may be the cause of the initial low loss of mass. In addition, LFS has higher content of CaO than GGBFS. If some of this CaO is in the form of CaCO3, the fine particles of calcium carbonate accelerate the initial hydration reaction and influence the hydrate assemblage of the hydrating cement pastes [29]. At later ages, the influence of the porous structure becomes dominant, and the mass loss of the LFS specimens progressively becomes higher than the mass loss of the reference specimens.

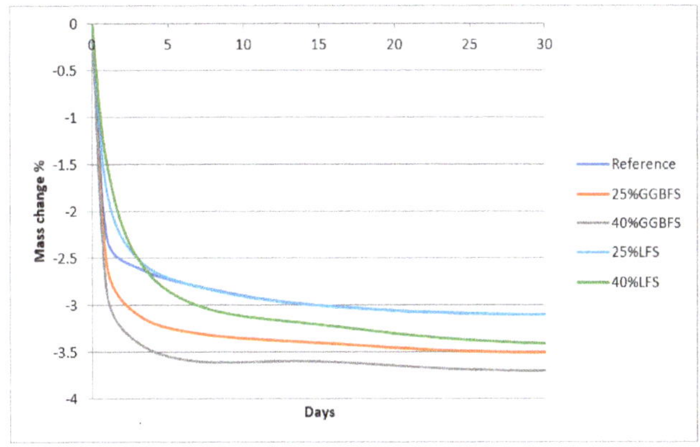

Figure 16. Mass loss in the first 30 days of specimens exposed to air drying.

Figure 17 shows the results of the autogenous shrinkage tests. The reference mixture presents a shrinkage of about 100 microstrains at 30 days, which is in agreement with *w/c* of 0.5, that is, poor conditions to develop desiccation. In fact, autogenous shrinkage depends on internal relative humidity [30], and autogenous shrinkage decreases as *w/c* increases [31,32]. A paste with *w/c* = 0.6 may present a swelling behavior in the first month, since the amount of available internal water largely surpasses the strictly required water to full hydration [33]. Additionally, as aggregates act as restrictions to the free paste movements [34], a concrete with *w/c* = 0.5 is expected to present low autogenous shrinkage values in the first days, in accordance with reference ones. At later ages, the reference shrinkage reaches 200 microstrains, due to further hydration and marginal drying. In the first days, mixtures with GGBFS had lower shrinkage than the reference mixture, due to the delay in the formation of the structural network [35,36], but after 2 weeks the autogenous shrinkage of mixtures with GGBFS is higher than the reference one. This is in agreement with other works [35], indicating that with the increase in the content of GGBFS, the autogenous shrinkage increases. This is not the case of LFS. The results obtained with LFS at later ages suggest absence of more compact microstructure that is expected when using slag [36]. It is well-known that the change of microstructure due to the formation of hydration products usually leads to an increase of the autogenous shrinkage. Indeed, autogenous shrinkage is used as an indirect measure of the degree of hydration [37]. With age, there is a shrinkage increase due to grow of the degree of hydration, which is seen in LFS mixtures. However, the autogenous shrinkage of LFS mixtures is not higher than the autogenous shrinkage of the reference mixture, and the shrinkage values did not increase with increasing the LFS dosage. This indicates that, at later ages, the influence of a large content of LFS in the rate of hydration is small or absent.

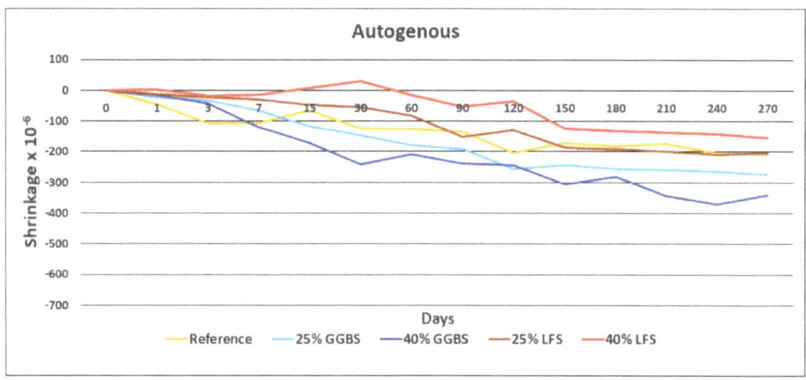

Figure 17. Autogenous shrinkage—evolution of the five mixtures.

Figure 18 presents the expansion of the five mixtures for immersed conditions. The results do not show evidence of presence of deleterious materials. Expansion of 40–160 microstrain at nine months in those conditions may be considered usual values for normal concrete. Considering the total deformation in autogenous shrinkage tests is actually the sum of the "pure" autogenous shrinkage and the swelling [38], specimens underwater with *w/c* = 0.5 should present very low "pure" autogenous shrinkage, due to the large source of water, prevailing the swelling component. According to [38] hydration always causes expansion, and consequently, higher swelling may indicate higher hydration. However, considering the expansion is caused by growing diameters of C-S-H shells surrounding the remnants of anhydrous cement grains, the swelling depends on contact pressure which is related with solid part configuration differing from paste to paste. Taking into account the accuracy of the test and the low values of expansion, the results does not allow detecting significant variations due to the presence of slag in the mixtures.

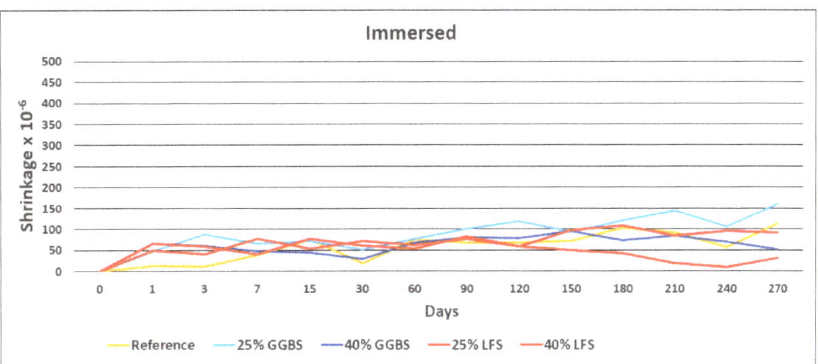

Figure 18. Shrinkage of the immersed specimens—evolution of the five mixtures.

Figure 19 presents the results of shrinkage in air-drying conditions. Reference and GGBFS specimens present similar air-drying shrinkage, which is in line with the indication of [39] which suggests that drying shrinkage is similar in Portland-cement concrete and concrete containing slag cement. However, the mixtures containing LFS show higher drying shrinkage, with increasing shrinkage as the LFS dosage increases. This also indicates that LFS contribution to a more compact microstructure is smaller than the contribution of GGBFS.

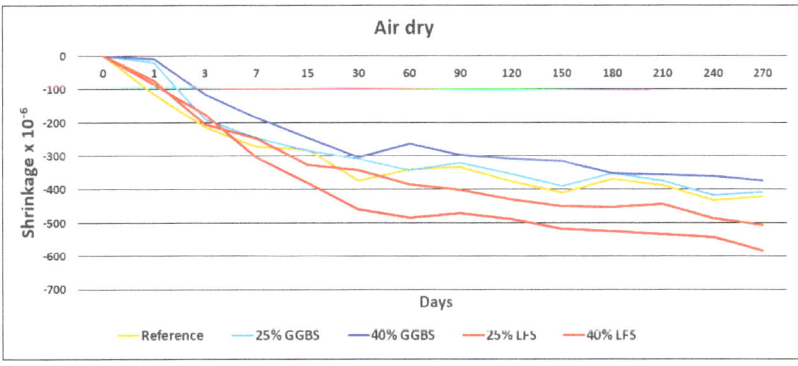

Figure 19. Shrinkage of the air dry specimens—evolution of the five mixtures.

5. Conclusions

The work presented is dedicated to the evaluation of the influence of steel slag type on concrete shrinkage. For comparison purposes a reference mixture without slag and mixtures with the same content of GGBFS and LFS were also prepared. The results show that shrinkage of concrete with LFS is different from concrete with GGBFS. Concrete with LFS show smaller autogenous shrinkage and higher drying shrinkage than concrete with GGBFS. The swelling underwater of GGBS and LFS mixtures was similar. At very early age, concrete with LFS shows lower mass loss and higher strength than concrete with GGBFS. The tests performed are not enough to support a full rational for this behavior; however, they show that LFS is not improving the hydration at later ages when compared with GGBS. It is speculated that the higher early strength of LFS mixtures may be related with higher CaO content of LFS, when compared to GGBS. As the two materials present different characteristics, their use in combination may be useful in optimizing the performance of concrete, that is, when shrinkage at early age is more relevant, partial replacement of GGBFS by LFS may reduce early age cracking, but the subject deserves further studies.

Author Contributions: Conceptualization, M.J.O., A.B.R., M.E.P.-R., F.P.-G. and M.D.R.-C.; methodology, M.J.O., M.E.P.-R.; software, M.J.O., A.B.R.; validation, M.J.O., A.B.R., M.E.P.-R., F.P.-G., M.D.R.-C.; formal analysis, M.J.O., A.B.R.; investigation, M.J.O., A.B.R., M.E.P.-R., F.P.-G., M.D.R.-C., writing—original draft preparation, M.J.O., A.B.R., M.E.P.-R., F.P.-G., M.D.R.-C. All authors have read and agreed to the published version of the manuscript.

Funding: This research received no external funding.

Institutional Review Board Statement: Not applicable.

Informed Consent Statement: Not applicable.

Data Availability Statement: Not applicable.

Acknowledgments: The authors express thanks to technical teams of: "Laboratorio de Materiales de Construcción" of "Escuela Politécnica Superior de Algeciras"—Cádiz University and "Laboratório de Materiais de Construção (LMC)" from Engineering Institute—Algarve University.

Conflicts of Interest: The authors declare no conflict of interest.

References

1. Levin, H. Systematic evaluation and assessment of building environmental performance (SEABEP). In Proceedings of the Second International Conference on Buildings and the Environment, CSTB and CIB, Paris, France, 6–7 June 1997; pp. 9–12.
2. Hu, M.; Milner, D. Visualizing the research of embodied energy and environmental impact research in the building and construction field: A bibliometric analysis. *Dev. Built Environ.* **2020**, *3*, 100010. [CrossRef]
3. Junnila, S.; Horvath, A. Life-cycle environmental effects of an office building. *J. Infrastruct. Syst.* **2003**, *9*, 157–166. [CrossRef]
4. Zabalza Bribián, I.; Aranda Usón, A.; Scarpellini, S. Life cycle assessment in buildings: State-of-the-art and simplified LCA methodology as a complement for building certification. *Buil. Environ.* **2009**, *12*, 2510–2520. [CrossRef]
5. Green Building Council. Buildings and Climate Change, United Green Building Council. 2020. Available online: https://cdn01.rockwool.com/siteassets/rw-na-ranson/sustainability/usgbc-buildings-and-climate-change-report.pdf?f=20181205094253 (accessed on 20 January 2020).
6. United National Environment Programme, 2018, Global Status Report 2018, United National Environment Programme. Available online: https://www.unenvironment.org/resources/report/global-status-report-2018 (accessed on 20 January 2020).
7. Verein Deutscher Zementwerke e.V. (VDZ). *Monitoring-Abschlussbericht 1990–2012—Verminderung der CO_2-Emissionen*, Germany, 2013. Available online: https://www.vdz-online.de/wissensportal/publikationen/monitoring-abschluss-bericht-1990-2012-verminderung-der-co2-emissionen (accessed on 26 December 2020).
8. United Nations Environment Programme. Eco-Efficient Cements: Potential Economically Viable Solutions for a Low-CO_2 Cement Based Materials Industry. Available online: www.unep.org (accessed on 20 October 2020).
9. Scrivener, K.L.; John, V.M.; Gartner, E.M. *Eco-Efficient Cements: Potential, Economically Viable Solutions for a Low-CO_2, Cement-Based Materials Industry*; UNEP: Paris, France, 2016.
10. CEN, EN 197-1. *Cement—Part 1: Composition, Specifications and Conformity criteria for Common Cements*; CEN, Central Secretariat, Rue de Stassart: Brussels, Belgium, 2000.
11. Yuksel, I. *Blast-Furnace Slag, Waste and Supplementary Cementitious Materials in Concrete, Characterization, Properties and Applications.* Woodhead Publishings Series in Civil and Structural Enginnering. 2018, pp. 361–415. Available online: https://www.elsevier.com/books/book-series/woodhead-publishing-series-in-civil-and-structural-engineering (accessed on 26 December 2020).
12. Parron-Rubio, M.E.; Perez-Garcia, F.; Gonzalez-Herrera, A.; Oliveira, M.J.; Rubio-Cintas, M.D. Slag Substitution as a Cementing Material in Concrete: Mechanical, Physical and Environmental Properties. *Materials* **2019**, *12*, 2845. [CrossRef] [PubMed]
13. Parron-Rubio, M.E.; Perez-García, F.; Gonzalez-Herrera, A.; Rubio-Cintas, M.D. Concrete Properties Comparison When Substituting a 25% Cement with Slag from Different Provenances. *Materials* **2018**, *11*, 1029. [CrossRef] [PubMed]
14. Perez-Garcia, F.; Parron-Rubio, M.E.; Garcia-Manrique, J.M.; Rubio-Cintas, M.D. Study of the suitability of different types of slag and its influence on the quality of green grouts obtained by partial replacement of cement. *Materials* **2019**, *12*, 1166. [CrossRef] [PubMed]
15. Perez-Garcia, F.; Rubio-Cintas, M.D.; Parron-Rubio, M.E.; Garcia-Manrique, J.M. Advances in the Analysis of Properties Behaviour of Cement-Based Grouts with High Substitution of Cement with Blast Furnace Slags. *Materials* **2020**, *13*, 561. [CrossRef] [PubMed]
16. Setién, J.; Hernández, D.; González, J.J. Characterization of ladle furnace basic slag for use as a construction material. *Constr. Build. Mater.* **2009**, *23*, 1788–1794. [CrossRef]
17. Bisschop, J. Evolution of Solid Behaviour. In Early Age Cracking in Cementitious Systems, RILEM TC 181-EAS. July 2009; 27–34.
18. Oliveira, M.J.; Ribeiro, A.B.; Branco, F.G. Shrinkage of self-compacting concrete. A comparative analysis. *J. Build. Eng.* **2017**, *9*, 117–124. [CrossRef]
19. CEN. *Eurocode 2, Design of Concrete Structures, Part1: General Rules for Building. Europen Standard*; European Committee for Standardization: Brussels, Belgium, 1992; pp. 28–31.

20. ACI 318-11: Building Code Requirements for Structural Concrete and Commentary. Available online: https://www.pecivilexam.com/Study_Documents/Struc-Materials-Online/ACI_318-Building-Code-2011.pdf (accessed on 26 December 2020).
21. ACI CT-13, ACI Concrete Terminology, American Concrete Institute. 2013. Available online: https://www.concrete.org/portals/0/files/pdf/aci_concrete_terminology.pdf (accessed on 26 December 2020).
22. Tazawa, E. Autogenous Shrinkage of Concrete. In Proceedings of the International Workshop organized by JCI (Japan Concrete Institute), Hiroshima, Japan, 3–8 June 1998; pp. 13–14.
23. Mehta, K. Reducing the environment impact of concrete—Concrete can be durable and environmentally friendly. *Concr. Int.* **2001**, *23*, 61–66.
24. Swamy, R.N. Sustainable concrete for infrastructure regeneration and reconstruction, in Sustainable Construction into the next Millennium: Environmentally friendly and innovative cement based materials. In Proceedings of the International Conference, João Pessoa, Brazil, 24–26 November 2000; pp. 15–43.
25. Rubio, M.E.P. Industrialización de Procesos Y Valoración de Residuos Siderúrgicos. Ph.D. Thesis, Universidad de Malaga, Málaga, Spain, December 2018.
26. CEN. NP EN 196-1:2006—Métodos de ensaio de cimentos. Parte 1: Determinação das resistências mecânicas, Versão portuguesa da norma europeia EN 196-1:2005, IPQ. 2006. Available online: http://repositorio.furg.br/handle/1/1477 (accessed on 26 December 2020).
27. Aitcin, P. Autogenous shrinkage measurement. In *AUTOSHRINK'98, Proceedings of the International Workshop on Autogenous Shrinkage of Concrete, June 13-14 1998*; Tazawa, E., Ed.; Japan Concrete Institute: Hiroshima, Japan, 1998; pp. 245–256.
28. CEN. *DRAFT prEN 12390-16—Testing Hardened Concrete—Part 16: Determination of the Shrinkage of Concrete*; European Committee for Standardization: Brussels, Belgium, April 2018.
29. Lothenbach, B.; Le Saout, G.; Gallucci, E.; Scrivener, K. Influence of limestone on the hydration of Portland cements. *Cem. Concr. Res.* **2008**, *38*, 848–860. [CrossRef]
30. Lura, P.; Jensen, O.M.; Van Breugel, K. Autogenous shrinkage in high-performance cement paste: An evaluation of basic mechanisms. *Cem. Concr. Res.* **2003**, *33*, 223–232. [CrossRef]
31. Zhang, M.H.; Tam, C.T.; Leow, M.P. Effect of water-to-cementitious materials ratio and silica fume on the autogenous shrinkage of concrete. *Cem. Concr. Res.* **2003**, *33*, 1687–1694. [CrossRef]
32. Baroghel-Bouny, V.; Mounanga, P.; Khelidj, A.; Loukili, A.; Rafai, N. Autogenous deformations of cement pastes Part II. W/C effects, micro–macro correlations, and threshold values. *Cem. Concr. Res.* **2006**, *36*, 123–136. [CrossRef]
33. Powers, T.C.; Brownyard, T.L. Studies of the physical properties of hardened portland cement paste. Research Laboratories of the Portland Cement Association, Bulletin 22, Journal of the American Concrete Institute, March, 1948, Chicago, 992p. Available online: https://www.concrete.org/publications/internationalconcreteabstractsportal.aspx?m=details&i=15301 (accessed on 26 December 2020).
34. Holt, E. Contribution of mixture design to chemical and autogenous shrinkage of concrete at early ages. *Cem. Concr. Res.* **2005**, *35*, 464–472. [CrossRef]
35. Matthes, W.; Vollpracht, A.; Villagrán, Y.; Kamali-Bernard, S.; Hooton, D.; Gruyaert, E. Ground granulated blast-furnace slag. In *Properties of Fresh and Hardened Concrete Containing Supplementary Cementitious Materials. RILEM State-of-the-Art Reports*; De Belie, N., Soutsos, M., Gruyaert, E., Eds.; Springer: Cham, Switzerland, 2018; Volume 25. [CrossRef]
36. Giergiczny, Z. Fly ash and slag. *Cem. Concr. Res.* **2019**, *124*, 105826. [CrossRef]
37. Morina, V.; Cohen-Tenoudjia, F.; Feylessoufib, A.; Richard, P. Evolution of the capillary network in a reactive powder concrete during hydration process. *Cem. Concr. Res.* **2002**, *32*, 1907–1914. [CrossRef]
38. Rasoolinejad, M.; Rahimi-Aghdam, S.; Bažant, Z.P. Prediction of autogenous shrinkage in concrete from material composition or strength calibrated by a large database, as update to model B4. *Mater. Struct. Constr.* **2019**, *52*, 33. [CrossRef]
39. ACI-233R-03 (ACI 233R-03 (2011)). *Slag Cement in Concrete and Mortar*; American Concrete Institute: Farmington Hills, MI, USA, 2011; Available online: https://www.concrete.org/publications/internationalconcreteabstractsportal/m/details/id/12648 (accessed on 26 December 2020).

MDPI

Article

Mechanical and Environmental Proprieties of UHP-FRCC Panels Bonded to Existing Concrete Beams

Tomoya Nishiwaki [1], Oscar Mancinelli [2], Alessandro Pasquale Fantilli [2,*] and Yuka Adachi [1]

[1] Department of Architecture and Building Science, School of Engineering, Tohoku University, Sendai 980-8579, Japan; tomoya.nishiwaki.e8@tohoku.ac.jp (T.N.); yuka.adachi.t8@dc.tohoku.ac.jp (Y.A.)
[2] Department of Structural, Building and Geotechnical Engineering (DISEG), Politecnico di Torino, 10129 Torino, Italy; oscar.mancinelli@polito.it
* Correspondence: alessandro.fantilli@polito.it; Tel.: +39-011-0904900

Abstract: Among the techniques used to retrofit existing reinforced concrete structures, methods involving Ultra High Performance Fiber Reinforced Cementitious Composites (UHP-FRCC) are widely regarded. However, current practices make the use of this material for in-situ application expensive and complicated to perform. Accordingly, a new method to strengthen existing concrete beams by applying a precast UHP-FRCC layer on the bottom side are introduced and described herein. Two test campaigns are performed with the aim of defining the best conditions at the interface between the reinforcing layer and the existing beam and to reducing the environmental impact of UHP-FRCC mixtures. As a result, the eco-mechanical analysis reveals that the best performances are attained when the adhesion at interface is enhanced by means of steel nails on the upper surface of the UHP-FRCC layer, in which 20% of the cement is replaced by fly ash.

Keywords: existing beams; retrofitting method; environmental assessment; fly ash; moment–curvature relationship; precast elements

Citation: Nishiwaki, T.; Mancinelli, O.; Fantilli, A.P.; Adachi, Y. Mechanical and Environmental Proprieties of UHP-FRCC Panels Bonded to Existing Concrete Beams. *Sustainability* **2021**, *13*, 3085. https://doi.org/10.3390/su13063085

Academic Editor: Fausto Minelli

Received: 5 February 2021
Accepted: 5 March 2021
Published: 11 March 2021

1. Introduction

Several studies aim at finding the best way to strengthen concrete columns and beams. Among the possible solutions, reinforced concrete (RC) jackets and steel cages are the most used [1,2]. Additionally, jacketing using innovative materials, such as Fiber Reinforced Polymer (FRP) [3–5] and Ultra High-Performance Fiber-Reinforced Cementitious Composites (UHP-FRCC) [6–8], have been successfully introduced and applied in the last decades. In particular, UHP-FRCCs have also brought the interest of many researchers, who assert that for both new and existing structure a new construction era has started [9]. In addition to the high tensile and compressive strength, UHP-FRCC has remarkable waterproofing properties, and therefore can protect structures not only from water, but also from aggressive agents.

Cast-in-situ coating layers, made of UHP-FRCC and cured at ambient conditions, are used to enhance the bearing capacity and stiffness of exiting RC beams [10], and to repair those damaged as well [11]. Nevertheless, this strengthening procedure requires laborious formworks and long casting procedures compared to the use of FRP. In addition, it is not easy to apply cast-in-situ layers on the bottom of beams because of the gravity action. For these reasons, some studies have been devoted to the mechanical performance of precast HP-FRCC slabs [12] used for strengthening RC structures. For instance, Jongvivatsakul et al. significantly increased the shear capacity of RC beams when Steel Fiber-Reinforced Precast Panels are applied on the faces [13].

However, when a cast-in-situ or precast panel overlays an existing structure, the effectiveness of the strengthening depends on the bond condition at the interface between new and old structures. To avoid the delamination phenomena produced by a weak adhesion, roughening treatments of the existing concrete surfaces ensure better performances than using bonding agents, such as epoxy resin [14]. In particular, sandblasting and chipping enable a higher surface roughening than grooves and drill holes [15]. Moreover, the environmental impact of the reinforcing layers made with UHP-FRCC is also an important aspect that has been taken into consideration. In fact, recent reports indicate that the clinker burning accounts for approximately 4% of the global CO_2 emissions [16]; hence, the cement manufacturing process emits huge amounts of greenhouse gases [17]. As a consequence, the massive content of cement and the presence of large volume of steel fibers make UHP-FRCC a high carbon footprint material. Thus, the Material Substitution Strategy (MSS), which consists of replacing a large part of clinker with mineral additives, can be an effective way to reduce the embodied CO_2 [18].

Accordingly, a new approach for retrofitting concrete structures is proposed herein. It consists of enhancing the resisting bending moment, and therefore the lifespan, of existing concrete beams by adding UHP-FRCC in the tensile zone. The best stratigraphy of the applied materials, made both with cast-in-situ and precast layers, is selected by means of the eco-mechanical analyses. It is a comparative study in which not only the mechanical performances (bearing capacity, bond conditions, etc.), but also the environmental impact of different UHP-FRCC layers are taken into consideration.

2. Materials and Test Procedure

As shown in Figure 1, the new retrofitting method consists of a precast flat layer applied on the tensile zone of an existing RC beam. The layer is hung to the existing structure by means of plugs and screws, leaving an empty gap between the new and the old parts.

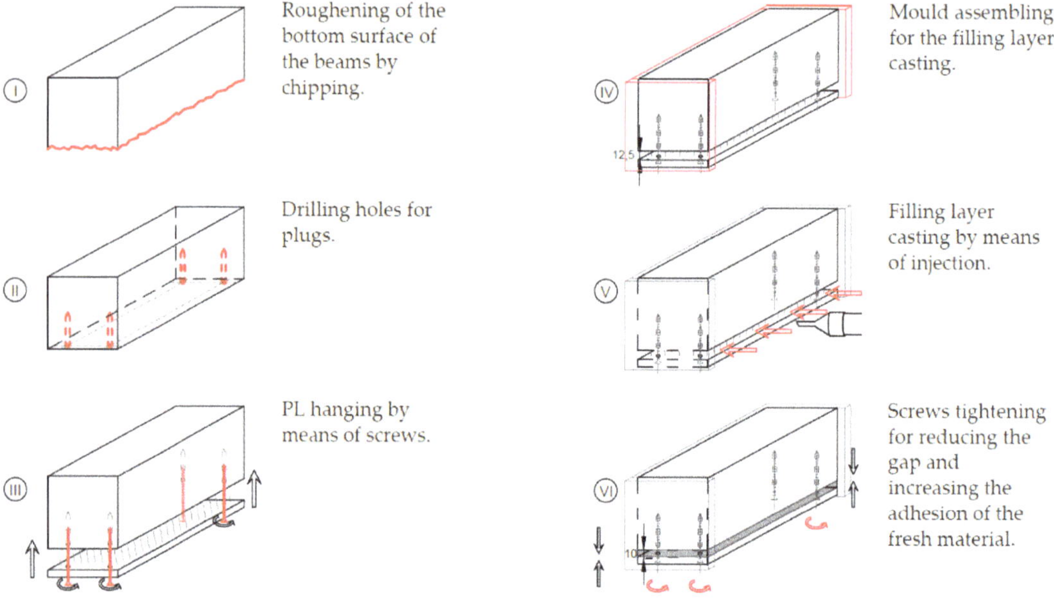

Figure 1. Procedure of strengthening existing beams by applying UHP-FRCC layers.

This gap is then filled by injecting a cement-based mortar. After the injection, a further tightening of the screws tends to reduce the thickness of the gap, and to improve the adhesion between old concrete and the new precast panel.

To have the highest performances of the panel, UHP-FRCC is used. As known, in such a cement-based composite, compressive strength is generally higher than 150 MPa and, in the pre-softening stage, the energy absorption capacity is larger than 50 kJ/m^3 [19]. These performances are achieved not only by adding fibers, but also with a dense microstructure, which is in turn tailored with an extremely low water–binder ratio (lower than 0.2) and by using ultra-fine additive, such as silica fume, wollastonite, and fine sand. Table 1 reports the density of all the materials used to tailor three series of mortar.

Table 1. Materials used to cast the samples (catalog data of producers).

Material	Symbol	Density [kg/m^3]
High Early Strength Portland Cement	HESPC	3140
Low Heat Cement	LHC	3240
Crushed Sand	S1	2610
Land Sand	S2	2580
Silica Fume	SF	2200
Silica Sand	Ss	2600
Wollastonite	Wo	2900
Water	W	1000
Superplasticizer	SP	1050
Defoaming Agent	DA	1010
Macro-fibers (30 mm long)	HDR	7850
Micro-fibers (6 mm long)	OL	7850

Existing beams have been cast with a normal strength mortar, according to mix proportion shown in Table 2. Three UHP-FRCC (namely, FA0, FA20, and FA70) have been used to strengthen the existing beams by means of precast panels. As reported in Table 3, with respect to the reference FA0, containing only cement and silica fume as a binder, in the mix proportions of FA20 and FA70, 20% and 70% of cement have been replaced by fly ash, respectively. Finally, the mix design of the filler layer is shown in Table 4. To reduce the viscosity and facilitate the injection, this filler is obtained from FA0 by removing only macro-fibers.

Table 2. Composition of the normal strength mortar simulating the existing beams (kg per m^3 of concrete).

HESPC	S [1]	W
485.6	1456.7	291.4

[1] S = S1 (50% weight) + S2 (50% weight).

Table 3. Composition of the retrofitting UHP-FRCC layers.

Series	W [1]	LHC [1]	SF [1]	FA [1]	Ss [1]	Wo [1]	SP [1]	DA [1]	HDR [2]	OL [2]
FA0	201	1197	263	0	511	190	32.1	0.3	1.5	1
FA20	195	928	255	232	495	184	31.0	0.3	1.5	1
FA70	181	323	236	753	459	170	31.5	0.3	1.5	1

[1] kg per m^3 of concrete., [2] % Vol of concrete.

Table 4. Composition of the filler layer.

W [1]	LHC [1]	SF [1]	FA [1]	Ss [1]	Wo [1]	SP [1]	DA [1]	HDR [2]	OL [2]
201	1197	263	0	511	190	32.1	0.3	0	1

[1] kg per m^3 of concrete., [2] % Vol of concrete.

The mechanical properties of all the mixtures are reported in Table 5. Compressive strength and Young's modulus have been determined by testing cylindrical samples in uniaxial compression, whereas uniaxial tensile tests on dumbbell shaped specimens have been performed to measure tensile strength [20].

Table 5. Mechanical proprieties of the concrete (MPa).

Parameter	Normal Strength Mortar	FA0	FA20	FA70
Compressive strength	45.1	204.7	193.4	150.7
Young's modulus	24.9×10^3	46.4×10^3	45.5×10^3	40.1×10^3
Tensile strength	2.6	16.5	17.8	7.4

The use of UHP-FRCC panels is particularly effective in the refurbishment of existing buildings, due to their dual function: structural strengthening and protection against aggressive agents. Nevertheless, due to the high cost and to the environmental impact, UHP-FRCC panels are not used to cover the entire perimeter of the cross-section, or through a three-side jacket [21], but rather they are located only the bottom part of a RC beam. Accordingly, in this research project, small-size beams with a UHP-FRCC layer in the tensile zone (Figure 2a) are investigated with two different tests of Campaign 1 and Campaign 2. Campaign 1 aimed at measuring the effects of four precast reinforcing UHP-FRCC layers (i.e., Type 1—flat slab; Type 2—flat slab with steel nails; Type 3 and Type 4—ribbed slabs), as shown in Figure 2b.

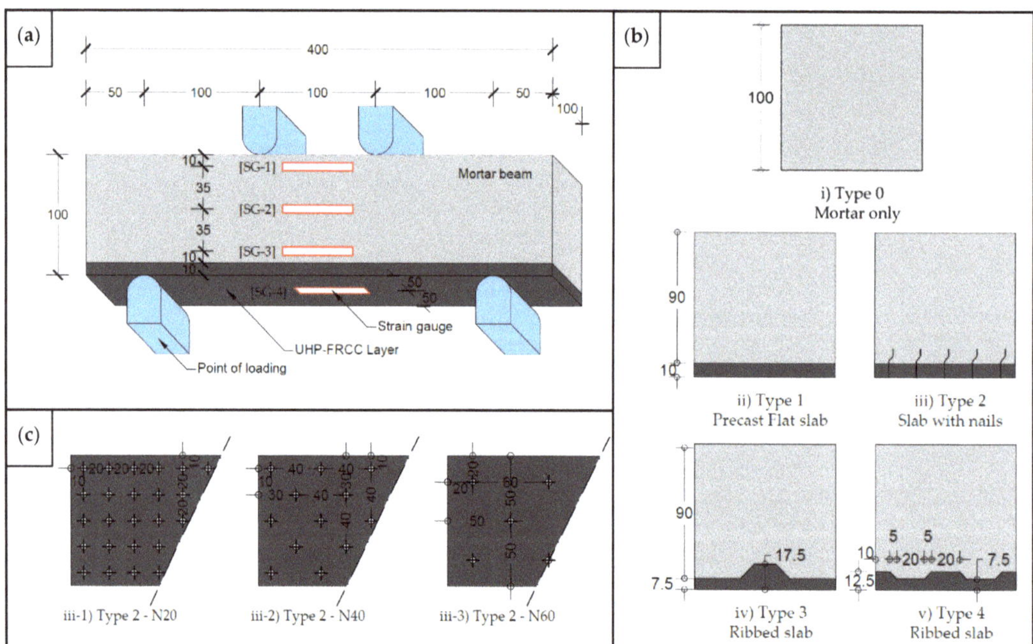

Figure 2. Mortar beams retrofitted by precast UHP-FRCC layer in Test Campaign 1. (**a**) Specimens shape, strain gauges arrangement, and test configuration. (**b**) Types of precast layer; (**c**) Neil density on the surface of the Type 2 precast layer for improving the connection with the mortar beam.

In the case of Type 2 layer, three different bond conditions (namely, N20, N40, and N60 in Figure 2c) are created with the aim of changing the bond conditions between the precast layer and the existing beam. Indeed, in the first phase of this research project, the performance provided by precast layers of different shape and bond condition has to be investigated. Thus, the assembling technique illustrated in Figure 1 is not used. On the contrary, normal strength mortar (see Table 2), reproducing the existing beam, has been directly cast on precast UHP-FRCC layer made with FA0 (see Table 3). For these specimens, the injection between the existing beams and the additional UHP-FRCC is not necessary.

In Campaign 2, the existing beams are reinforced with screws and plugs, following the procedure illustrated in Figure 1. Only Type 2 layer of Figure 2b, which performed better than the other panels, is used. Such layers are cast with three different series of UHP-FRCC (FA0, FA20, and FA70) to reduce the embodied CO_2. The results obtained from these specimens are benchmarked with those of unreinforced beams (Type 0 in Figure 3b) and with the samples representative of the current strengthening method (i.e., Type 5 in Figure 3b). It consists of a UHP-FRCC layer cast-in-situ on the surface of the beam, without any screws and plugs.

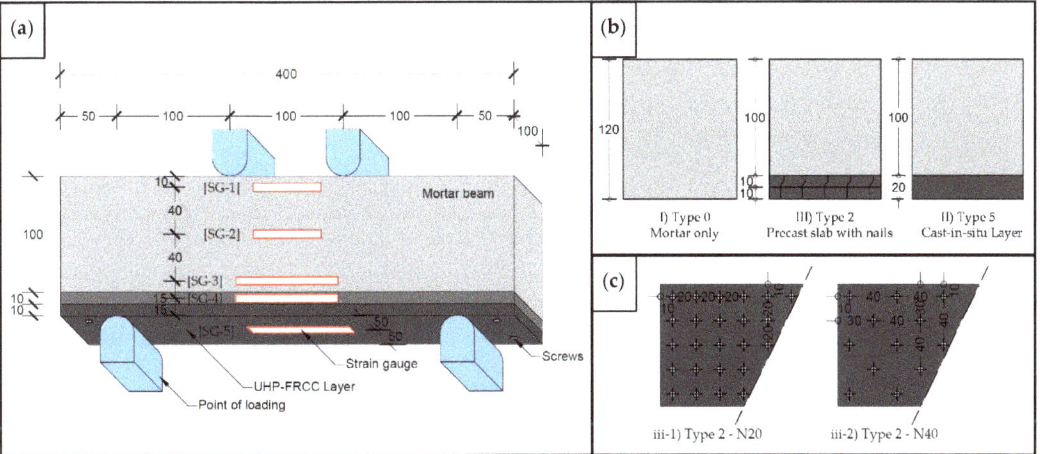

Figure 3. Mortar beams retrofitting by UHP-FRCC layer. (**a**) Specimens shape, strain gauges arrangement, and test configuration regarding the second test campaign; (**b**) Precast UHP-FRCC layers used to reinforce the beams; (**c**) Neil density on the surface of the type 2 precast layers for improving the connection between the parts.

3. Test Campaign 1

3.1. Specimens and Test Setup

Figure 2a illustrates the specimens tested in Campaign 1, which consist of mortar (Table 2) beams retrofitted with a precast UHP-FRCC layer. The thickness of the layer is 10 mm, whereas 100 mm is the depth of the beam, and both the layer and the beam are 100 mm wide. The length of the beam is 400 mm. The nails of the Type 2 series consist of hook-end steel fibers, arranged as in Figure 2c (i.e., at a distance of 20 mm—N20 series, of 40 mm—N40 series, and of 60 mm—N60 series) and embedded within the precast layer. Therefore, the parameters measured in this test campaign are the shape of the precast UHP-FRCC layers (Types 1, 2, 3, and 4) and the density of nails in the Type 2 series (N20, N40, and N60). The samples tested in the Campaign 1 are summarized in Table 6, where all the layers have been made with FA0 (see Table 3).

Table 6. Details of the specimens investigated in the Test Campaign 1.

Symbol	Layer Type	Bonding Surface	Number of Samples
B_I_T0	Type 0	[-]	2
B_I_T1	Type 1	Flat	2
B_I_T2_N20	Type 2	Steel fibers—Spacing: 20 mm	2
B_I_T2_N40	Type 2	Steel fibers—Spacing: 40 mm	2
B_I_T2_N60	Type 2	Steel fibers—Spacing: 60 mm	2
B_I_T3	Type 3	Ribbed	2
B_I_T4	Type 4	Ribbed	2

After casting the UHP_FRCC layers, they are steam cured for 48 h at 90 °C (95% RH) [22], and then used to cast the mortar beams. Finally, the composite beams are stored for 28 days in the curing room (at 20 °C and 95% RH).

The mechanical performances of the specimens are measured through four-point bending tests performed by using a 1000 kN Universal Testing Machine. A constant speed of 0.3 mm/min of the ram stroke is employed. As shown in Figure 2a, four strain gauges (with a length of 60 mm) are glued on the constant moment zone of the beam. Three of them measured the strains on a single face of the existing beam, whereas the fourth is located on the bottom side (i.e., on the UHP-FRCC layer). Through these strain gauges, the average curvature χ in the constant moment zone is calculated by means of the following equation:

$$\chi = (\varepsilon_4 - \varepsilon_1)/d_0 \quad (1)$$

where ε_4 and ε_1 are the strains measured by the strain gauge [SG-4] (on the bottom side) and [SG-1], respectively, and d_0 = 90 mm is the distance between the two gauges.

3.2. Results and Discussion

In Figure 4, the moment–curvature curves of the mortar beam without strengthening (B_I_T0) is put into comparison with those of the beams retrofitted with the different types of precast layers (see Figure 2 and Table 6).

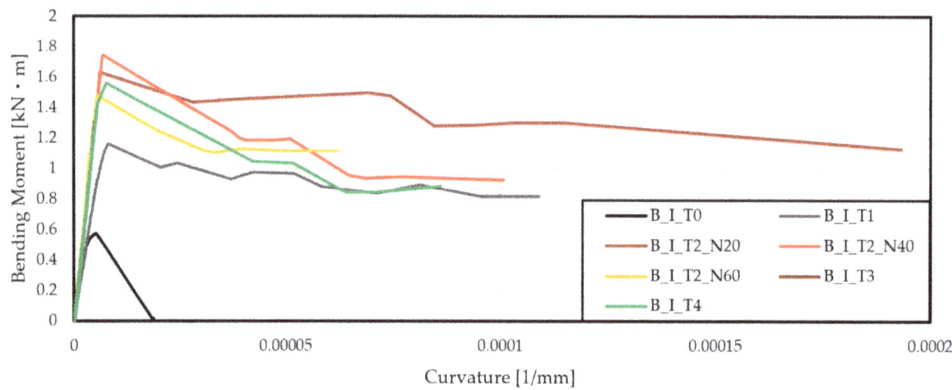

Figure 4. Moment–curvature curves measured in the beams of the Test Campaign 1.

Obviously, all beams reinforced with the UHP-FRCC precast layer show a maximum bending moment greater than that of the unreinforced mortar beam. Among them, the beam retrofitted with the precast layer without nails (i.e., B_I_T1) is less strong, whereas the B_I_T2_N20 and B_I_T2_N40 show the highest load-bearing capacity. In fact, the failure of the beams B_I_T2_N20, B_I_T2_N40, and B_I_T3 is due to the failure of the precast layer, whereas B_I_T1 and B_I_T2_N60 beams collapsed after the detachment of the reinforcing layer, which delaminates before exploiting the reinforcing capability. The density of the nails on the interface between the precast UHP-FRCC layer and mortar beam affects both the load-bearing capacity and the post peak behavior. Indeed, beams strengthened with N40 and N60 Type 2 layers (see Figure 2c) show a more brittle behavior than those reinforced by N20 precast UHP-FRCC.

Figure 5 depicts the strain profiles of the cross-sections, when in the constant moment zone of the beams B_I_T0 (Figure 5a), B_I_T3 (Figure 5b), and B_I_T2_N20 (Figure 5c), two different loads are acting. In all the specimens, plane cross-sections remain plane when 1/3 of the maximum load is applied. Conversely, nonlinear strain profiles can be observed at the ultimate bending moment. Nevertheless, in the bottom of the beam B_I_T2_N20, the strains measured during the tests are larger than those of the linear theoretical trend. The exact opposite occurs in the beams B_I_T0, without reinforcement, and B_I_T3, where the lower real strains indicate a weak transmission of the shear stresses due to the delamination of the UHP-FRCC layer from the mortar beam.

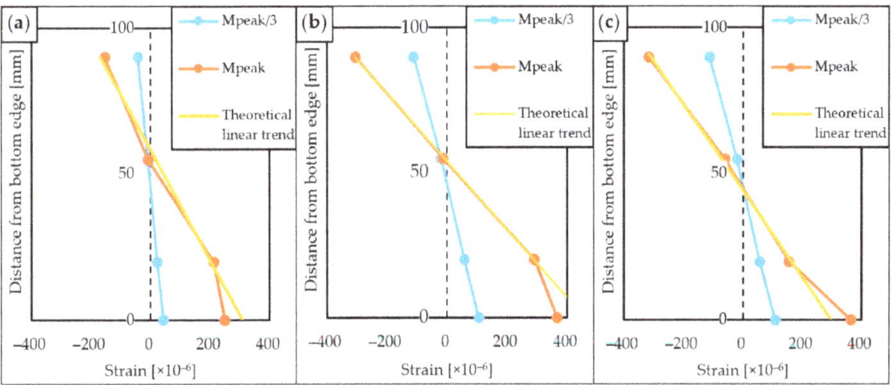

Figure 5. Strain profiles measured in the beams (**a**) B_I_T0, (**b**) B_I_T3, and (**c**) B_I_T2_N20.

This experimental result emphasizes the fundamental role that nails play in the transmission of stresses between the existing beam and the reinforcing layer. As illustrated in Figure 6 where the strain profiles of the beams B_I_T2_N20 (Figure 6a), B_I_T2_N40 (Figure 6b), and B_I_T2_N60 (Figure 6c) are reported, delamination does not occur when nails are present. Accordingly, in the Test Campaign 2, only Type 2 N20 and Type 2 N40 layers were used to reinforce the existing beams.

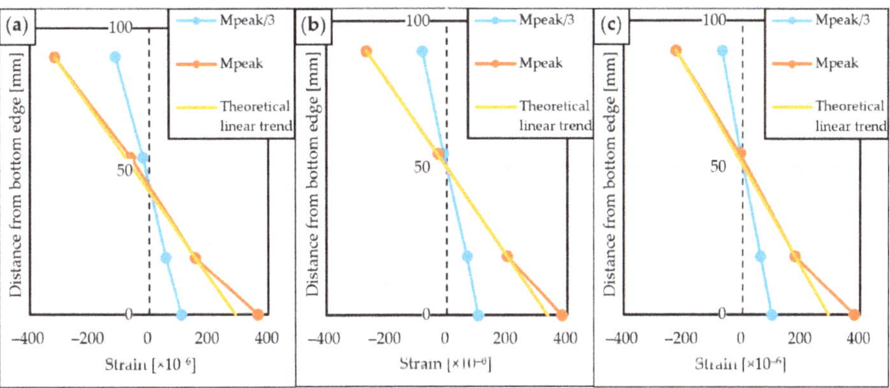

Figure 6. Strain profiles measured in the beams (**a**) B_I_T2_N20, (**b**) B_I_T2_N40, and (**c**) B_I_T2_N60.

4. Test Campaign 2

4.1. Specimens and Test Setup

In this second campaign, mortar beams and precast layers are cast, and the general scheme of Figure 1 is adopted to create the composite structure. Not only the different types of the layers shown in Figure 3b are tested, but also the environmental performances of the UHP-FRCC are analyzed. Such layers have been made by FA0, FA20, and FA70 mixtures shown in Table 3.

The mortar beams have been stored in the curing room for 28 days, whereas the precast layers have been steam cured and then stored in the curing room for the same lapse of time. To simulate the current reinforcing approach, the bottom surfaces of some mortar beams are chipped and subsequently reinforced with a layer of UHP-FRCC cast on the bottom surface, as shown in Figure 3b (Type 5 layer). The remaining mortar beams are strengthened by applying the precast UHP-FRCC layers (Type 2 in Figure 3b,c) by means of chipping, plugs, screws, and the filling layer, as illustrated in Figure 1. Figure 3a shows the general scheme of the 120 mm high specimens tested by means of a 1000 kN Universal Testing Machine, whereas Table 7 summarizes the composite beams investigated in Test Campaign 2. As every type of specimen counted four samples, a total of 40 beams were realized and tested.

Table 7. Details of the specimens tested in the Campaign 2.

Symbol	Layer Type	Bonding Surface	UHP-FRCC Series	Number of Samples
B_II_T0	None	[-]	[-]	4
B_II_T2_FA0_N20	Type 2	Steel fibers—Spacing: 20 mm	FA0	4
B_II_T2_FA0_N40	Type 2	Steel fibers—Spacing: 40 mm	FA0	4
B_II_T2_FA20_N20	Type 2	Steel fibers—Spacing: 20 mm	FA20	4
B_II_T2_FA20_N40	Type 2	Steel fibers—Spacing: 40 mm	FA20	4
B_II_T2_FA70_N20	Type 2	Steel fibers—Spacing: 20 mm	FA70	4
B_II_T2_FA70_N40	Type 2	Steel fibers—Spacing: 40 mm	FA70	4
B_II_T5_FA0	Type 5	Chipping only	FA0	4
B_II_T5_FA20	Type 5	Chipping only	FA20	4
B_II_T5_FA70	Type 5	Chipping only	FA70	4

In all the tests, five strain gauges were glued in the composite beams, one in the lower surface and four on a side face. As shown in Figure 3a, the strain gauges [SG-3], [SG-4] and [SG-5], closer to the bottom, are 90 mm long, and the rest are 60 mm long. By means of these instruments, the strain profile is measured in the cross-section of the constant moment zone, and the debonding phenomena among the layers are also detected. In addition, the measure of the curvature is carried out with the following equation:

$$\chi = (\varepsilon_5 - \varepsilon_1)/d_0 \quad (2)$$

where ε_5 and ε_1 are the strains measured by the strain gauge [SG-5] (on the bottom side) and strain gauge [SG-1], respectively; and d_0 = 110 mm is the distance between the two gauges.

4.2. Results and Discussion

Figure 7 shows the average moment–curvature relationships of the reinforced beams, whose strength capacity (i.e., the maximum bending moment) is summarized in the histogram of Figure 8.

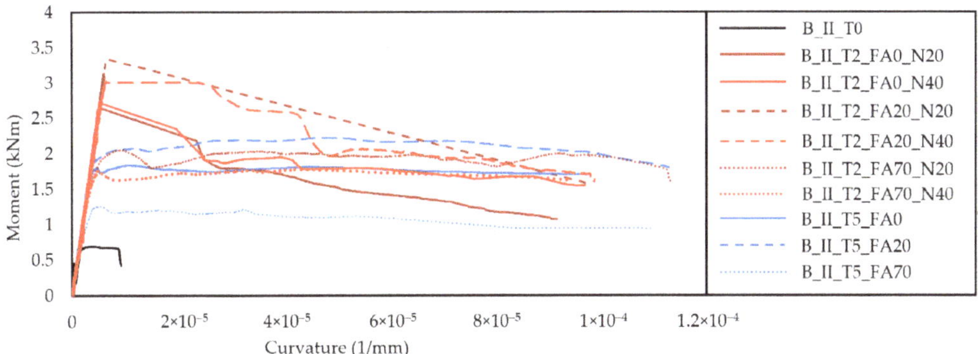

Figure 7. Bending moment–curvature relationship measured in test campaign 2.

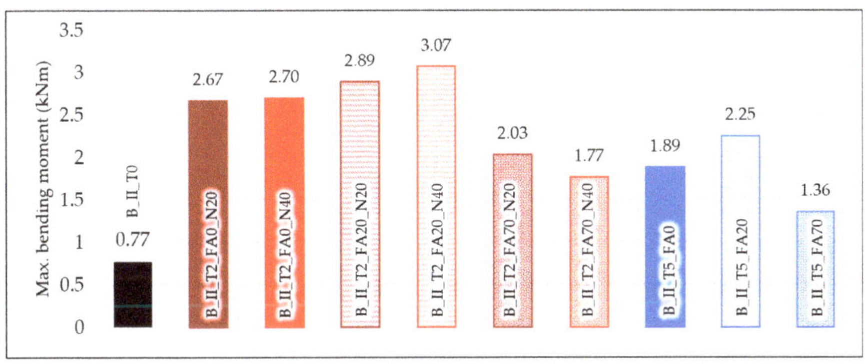

Figure 8. Average values of the maximum bending moment measured in test campaign 2.

Compared to the un-reinforced mortar beam (B_II_T0), a substantial increment of strength and toughness are observed when UHP-FRCC layers are used. However, the maximum bending moment of the beams B_II_T2 is higher than that of B_II_T5, regardless of the class of UHP-FRCC.

Three collapse modes have been identified by linking the stress–strain relationship of UHP-FRCC (obtained from the tensile tests performed on dumbbell shaped specimens) with the moment curvature–relationship (Figure 9):

- Failure in the tensile zone is illustrated in the beams B_II_T2_FA70 and B_II_T5 (Figure 9a). The value of the strain $\varepsilon_{D,y}$ corresponding to the first cracking of the UHP-FRCC substantially coincides with that measured on the bottom of the beam by [SG-5] (see Figure 3a) when first crack occurs. Afterwards, strain hardening appears both in the stress–strain relationship of the reinforcing layer and in the moment curvature relationship. At the peak of bending moment $M_{B,u}$, the strain gauge [SG-5] measured a value ε_5 equal to $\varepsilon_{D,u}$, which is the strain at the tensile strength $\sigma_{D,u}$ of UHP-FRCC. In other words, the ductile behavior of these beams strictly depends on the mechanical performances of the precast layer.
- Figure 9b illustrates the failure due to the crushing of mortar in the compressed zone of the beam. Indeed, during the strain hardening behavior of the UHP-FRCC layer, the moment–curvature relationship shows a softening branch. The resisting area in compression reduces due to the crushing, whereas in the precast layer, wide cracks are visible. In this case, the bending moment corresponding to the first crack, $M_{B,y}$, coincides with that at the peak $M_{B,u}$. This brittle behavior, which generally occurs

in over-reinforced concrete beams, can be observed in the beams B_II_T2_FA0 and B_II_T2_FA20_N40.

- Shear failure (Figure 9c) with a sudden drop in resistance. This brittle behavior is evident in the moment curvature diagram of the beams B_II_T2_FA20_N20, where a diagonal crack appears without crossing the strain gauges of the constant moment zone. As strain localizes in this crack, a reduction of the strain is measured by the gauges before reaching the cracking stress (and strain) in the reinforcing UHP-FRCC layer.

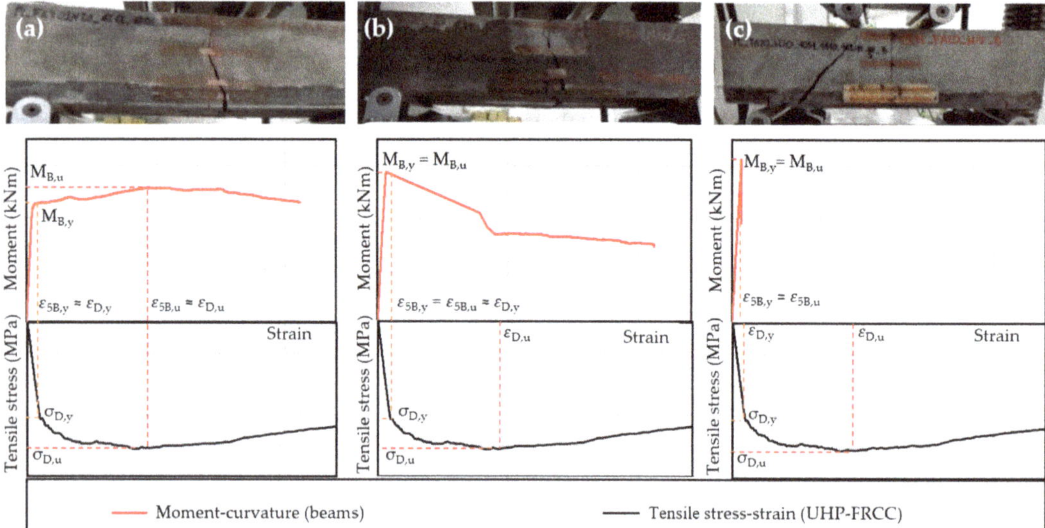

Figure 9. Failures of the beams in the test campaign 2: (**a**) failure in tension; (**b**) crushing in the compression zone; (**c**) shear failure.

Accordingly, the use of high-strength precast layers (such as FA0 and FA20) leads to the brittle behavior of the composite beams, either due to the crushing of mortar in the compression zone, or to the shear failure, especially in the presence of high bond strength (i.e., N20 series). Concerning the content of fly ash, low percentages of this industrial waste (i.e., FA20) produce an increment of strength in the UHP-FRCC layer, but a brittle behavior of the composite beam. On the contrary, the load-bearing capacity of the layer significantly reduces if the substitution rate of cement as fly ash increases (i.e., FA70), but the composite beams show a greater ductility.

To check the effectiveness of the three different types of bond between the reinforcing layer and the existing structure (see Figure 3), the strain profiles of the composite beams B_II_T2_N20, B_II_T2_N40, and B_II_T5 are illustrated in Figure 10. Except for the beam B_II_T5, plane sections remain plane up to the maximum bending moment. As a matter of fact, Figure 10c shows delamination between the mortar and the strengthening layer, which is prevented by combining screws and nails on the surface of the precast UHP-FRCC layer (i.e., Type 2—N20 and Type 2—N40 in Figure 3b,c).

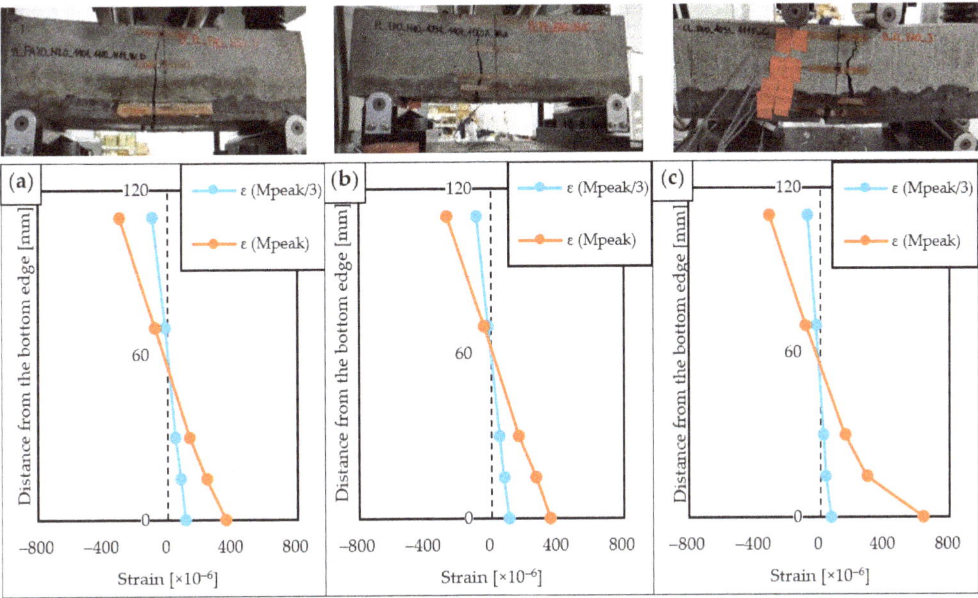

Figure 10. Strain profiles measured in the beams: (**a**) B_II_T2_N20, (**b**) B_II_T2_N40, and (**c**) B_II_T5.

5. Eco-Mechanical Analysis

A further study regarding Test Campaign 2 is herein performed by measuring the environmental performances of the beams, through the so-called eco-mechanical analysis [23]. Using the non-dimensional diagram of Figure 11, a comparative analysis among the beams is carried out in order to select the best reinforcing system, which contemporarily satisfies the environmental and mechanical performances.

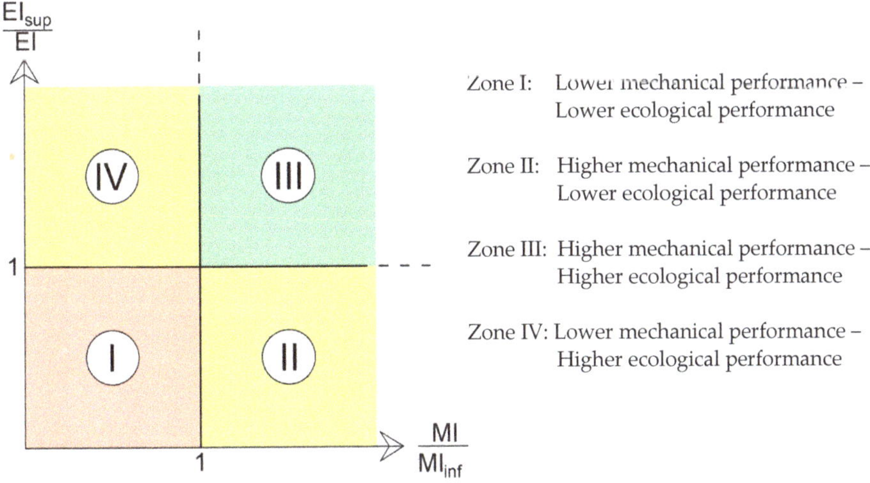

Figure 11. The non-dimensional diagram for evaluating the eco-mechanical performance of specimens.

The horizontal axis of Figure 11 reports the ratio between mechanical indexes (MI/MI_{inf}), whereas the vertical axis represents the ratio between ecological indexes (EI_{sup}/EI). Specifically, MI_{inf} is the lower bound value of mechanical performance, which is the so-called functional unit. In this study, MI_{inf} corresponds to the mechanical performance of a reference series. Similarly, EI_{sup} is the upper bound value of the environmental performance, corresponding to that of the layers of the reference series. In particular, the environmental impact was computed by multiplying the amount of materials used for each type of layer by the relevant unit carbon footprint, as given by the inventory data issued by the Japanese Concrete Institute (JCI) [24]. This computation is consistent with fib [25], where only the CO_2 released in the atmosphere has been taken into account.

The mechanical index, or the functional unit, could be the maximum bending moment in the moment–curvature relationship [26]. On the other hand, according to Fib [25], the mechanical index should also consider the overall behavior of the structure, including the ductility. Therefore, two different parameters are considered herein. The first parameter is the peak of bending moment, whereas in the second parameter (i.e., the ductility) is correlated to the work of deformation per unit length (J/m). It is the area D, defined by the moment–curvature diagram up to the maximum bending moment, as illustrated in Figure 12. Accordingly, it vanishes in the case of brittle behavior (see Figure 9b,c).

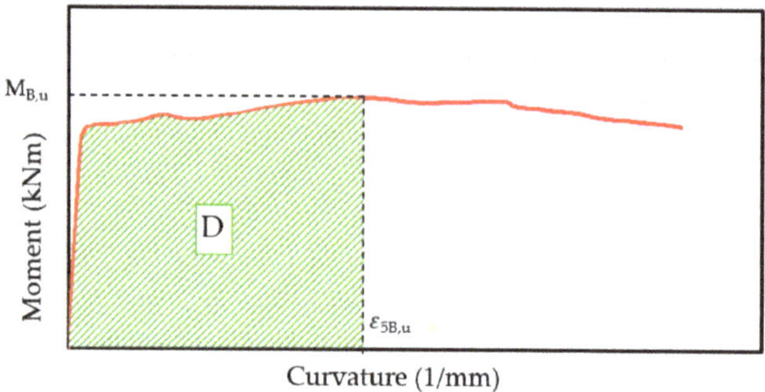

Figure 12. Evaluation of the mechanical indexes.

To calculate MI_{inf} and EI_{sup}, the mechanical and ecological performances of the beam B_II_T5_FA0 are considered as the references, because it represents the current method of retrofitting the existing RC beams. In this beam, UHP-FRCC does not include any supplementary cementitious materials to reduce the carbon footprint. Table 8 summarizes the values of the parameters of each retrofitting layer used to reinforce the beams. The embodied CO_2 computed for the B_II_T2 series also takes into account nails, screws, and the filling layers made with UHP-FRCC without fly ash (FA0). As a result, the environmental indicators are fairly high, even for the Type 2—FA20 and FA70 series. An additional reduction in CO_2 emission for the FA20 and FA70 series could be achieved by replacing the fly ash in the filling layer.

Table 8. Parameters used for computing the environmental and mechanical indexes.

Parameter	B_II_T2_FA0_		B_II_T2_FA20_		B_II_T2_FA70_		B_II_T5_		
	N20	N40	N20	N40	N20	N40	FA0	FA20	FA70
M_{peak} (kNm)	2.67	2.70	2.89	3.07	2.03	1.77	1.89	2.25	1.36
D. Work(J/m)	0.00	0.00	0.00	0.00	58.87	45.20	125.29	70.23	22.45
CO_2 (kg)	0.9719	0.9634	0.8882	0.8797	0.7045	0.6960	0.9679	0.8006	0.4331

All the values are reported within the non-dimensional diagram of Figure 13. In particular, the maximum bending moment is the functional unit in Figure 13a, whereas in Figure 13b, the functional unit is D.

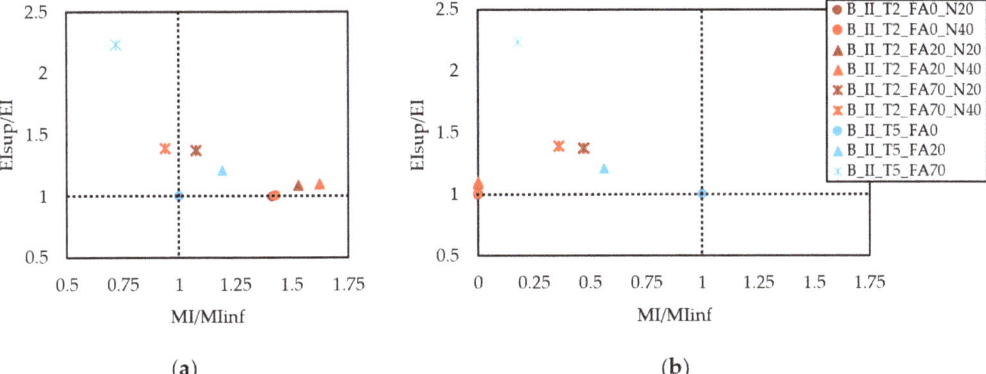

Figure 13. Mechanical and environmental assessment of the precast UHP-FRCC layers by considering (**a**) MI = maximum bending moment and (**b**) MI = work of deformation.

Figure 13a points out that the best result in terms of mechanical performances is achieved by the beam B_II_T2_FA20. On the other hand, B_II_T2_FA70 attains a fair reduction in emissions by preserving, approximately, the same resistance of the beam B_II_T5_FA0. When the work of deformation per unit length is the functional unit (see Figure 13b), the beams do not show any increment of ductility with respect to the reference beam B_II_T5_FA0. In particular, the beams B_II_T2_FA0_N40 and B_II_T2_FA20_N40 exhibited brittle failures, as shown in Figure 9b,c. Hence, their mechanical index MI is zero in Figure 13b.

As this brittle behavior generally affects the over reinforced beams under bending actions, it could be ascribed to the large thickness of the reinforcing UHP-FRCC layer. Thus, further numerical and experimental analyses have to be performed in order to also define the optimal geometry of the precast retrofitting layer, ensuring the strengthening in terms of both resisting bending moment and ductility.

6. Conclusions

The experimental results previously described lead to the following conclusions:

- The delamination failure of the strengthened beam can be avoided by introducing steel nails at the interface between the existing beam and the UHP-FRCC reinforcing layer.
- In addition to nails, the existing beams reinforced with screws and plugs show an increment of the resisting bending moment, regardless of the mixture used to cast the UHP-FRCC layer.
- The eco-mechanical analysis of the composite beams reveals that part of the cement used to cast UHP-FRCC layers can be effectively substituted by fly ash. When the rate of substitution is about 20%, both the environmental impact and the strength of the beams improve. Thus, with the strengthening procedure illustrated in Figure 1, a significant reduction of greenhouse emissions can be obtained while maintaining the mechanical performance provided by the current retrofitting method.
- On the other hand, when the resistance of the reinforcing layer is much higher than that of the existing beam, composite cross-section fails in a brittle manner. To avoid this undesired behavior, which generally occurs in over-reinforced concrete beams in bending, a suitable thickness of the UHP-FRCC layer has to be designed.

Author Contributions: Conceptualization: T.N. and A.P.F.; methodology: T.N., O.M., and A.P.F.; investigation: O.M. and Y.A.; resources: T.N.; data curation: O.M. and Y.A.; writing—original draft preparation: Y.A. and O.M.; writing—review and editing: T.N. and A.P.F.; visualization: O.M. and Y.A.; project administration: T.N., and A.P.F.; funding acquisition: T.N. and A.P.F. All authors have read and agreed to the published version of the manuscript.

Funding: This research received no external funding.

Institutional Review Board Statement: Not applicable.

Informed Consent Statement: Not applicable.

Data Availability Statement: The data presented in this study are available on request to the first author or corresponding author.

Acknowledgments: The authors would like to thank the Italian Civil Protection for the administrative and technical support given within the Research Program DPC Reluis—Politecnico di Torino (2019-2021)—WP11: Code rules contributions in the field of existing reinforced concrete structures.

Conflicts of Interest: The authors declare no conflict of interest.

References

1. Júlio, E.S.; Branco, F.; Silva, V.D. Structural rehabilitation of columns with reinforced concrete jacketing. *Prog. Struct. Eng. Mater.* **2003**, *5*, 29–37. [CrossRef]
2. Adam, J.M.; Ivorra, S.; Pallarés, F.J.; Giménez, E.; Calderón, P.A. Axially loaded RC columns strengthened by steel caging. Finite element modelling. *Constr. Build. Mater.* **2009**, *23*, 2265–2276. [CrossRef]
3. Olivova, K.; Bilcikm, J. Strengthening of concrete columns with CFRP. *Slovak J. Civ. Eng.* **2009**, *17*, 1–9.
4. Dasgupta, A. Retrofitting of concrete structure with fiber reinforced polymer. *Int. J. Innov. Res. Sci. Tech.* **2018**, *4*, 42–49.
5. Irshidat, M.; Al-Saleh, M.H. Strengthening RC columns using carbon fiber reinforced epoxy composites modified with carbon nanotubes. *Int. J. Civ. Struct. Constr. Arch. Eng. Dubai UAE* **2015**, *9*, 19–22.
6. Tayeh, B.A.; Bakar, B.H.A.; Johari, M.A.M.; Voo, Y.L. Utilization of ultra-high performance fibre concrete (Uhpfc) For rehabilitation—A review. *Proc. Eng.* **2013**, *54*, 525–538. [CrossRef]
7. Fantilli, A.P.; Meloni, L.P.; Nishiwaki, T.; Igarashi, G. Tailoring confining jacket for concrete column using ultra high performance-fiber reinforced cementitious composites (UHP-FRCC) With high volume fly ash (HVFA). *Materials* **2019**, *12*, 4010. [CrossRef] [PubMed]
8. Gholampour, A.; Hassanli, R.; Mills, J.E.; Vincent, T.; Kunieda, M. Experimental investigation of the performance of concrete col-umns strengthened with fiber reinforced concrete jacket. *Constr. Build Mater.* **2019**, *194*, 51–61. [CrossRef]
9. Brühwiler, E. "Structural UHPFRC": Welcome to the post-concrete era! In Proceedings of the First International Interactive Symposium on UHPC, Iowa State University, Ames, IA, USA, 18–20 July 2016.
10. Ruano, G.; Isla, F.; Pedraza, R.I.; Sfer, D.; Luccioni, B. Shear retrofitting of reinforced concrete beams with steel fiber reinforced concrete. *Constr. Build. Mater.* **2014**, *54*, 646–658. [CrossRef]
11. Martinola, G.; Meda, A.; Plizzari, G.A.; Rinaldi, Z. Strengthening and repair of RC beams with fiber reinforced concrete. *Cem. Concr. Compos.* **2010**, *32*, 731–739. [CrossRef]
12. Junior, P.R.R.S.; Maciel, P.S.; Barreto, R.R.; Neto, J.T.D.S.; Corrêa, E.C.S.; Bezerra, A.C.D.S. Thin slabs made of high-performance steel fibre-reinforced cementitious composite: Mechanical behaviour, statistical analysis and microstructural investigation. *Materials* **2019**, *12*, 3297. [CrossRef] [PubMed]
13. Jongvivatsakul, P.; Bui, L.V.H.; Koyekaewphring, T.; Kunawisarut, A.; Hemstapat, N.; Stitmannaithum, B. Using steel fiber-reinforced concrete precast panels for strengthening in shear of beams: An experimental and analytical investigation. *Adv. Int. Civ. Eng.* **2019**, *2019*, 4098505. [CrossRef]
14. Júlio, E.N.B.S.; Branco, F.A.B.; Silva, V.D. Concrete-to-concrete bond strength: Influence of an epoxy-based bonding agent on a roughened substrate surface. *Mag. Concr. Res.* **2005**, *57*, 463–468. [CrossRef]
15. Malek, N.A.A.A.; Muhamad, K.; Zahid, M.Z.A.M.; Hamiruddin, N.A.; Zainol, N.Z.; Yahya, N.; Bahaman, N.; Ramli, N.M. Evaluation of Bond Strength Between Normal Concrete and High Performance Fiber Reinforced Concrete (HPFRC). In Proceedings of the MATEC Web of Conferences 195, Solo Baru, Indonesia, 11–12 July 2018.
16. Oliver, J.G.J.; Peters, J.A.H.W. *Trends in Global CO_2 and Total Greenhouse Gas Emissions*; PBL Netherlands, Environmental Assessment Agency: The Hague, The Netherlands, 2020.
17. Damineli, B.L.; Kemeid, F.M.; Aguiar, P.S.; John, V.M. Measuring the eco-efficiency of cement use. *Cem. Concr. Compos.* **2010**, *32*, 555–562. [CrossRef]
18. Habert, G.; Roussel, N. Study of two concrete mix-design strategies to reach carbon mitigation objectives. *Cem. Concr. Compos.* **2009**, *31*, 397–402. [CrossRef]
19. Wille, K.; El-Tawil, S.; Naaman, A. Properties of strain hardening ultra high performance fiber reinforced concrete (UHP-FRC) under direct tensile loading. *Cem. Concr. Compos.* **2014**, *48*, 53–66. [CrossRef]
20. Japan Society of Civil Engineers (JSCE). Recommendations for Design and Construction of HPFRCC with Multiple Fine Cracks. 2018. Available online: http://www.jsce.or.jp/committee/concrete/e/hpfrcc_JSCE.pdf (accessed on 9 March 2021).
21. Lampropoulos, A.P.; Paschalis, S.A.; Tsioulou, O.T.; Dritsos, S.E. Strengthening of reinforced concrete beams using ultra high per-formance fibre reinforced concrete (UHPFRC). *Eng. Struct.* **2015**, *106*, 370–384. [CrossRef]

22. Kwon, S.; Nishiwaki, T.; Kikuta, T.; Mihashi, H. Development of ultra-high-performance hybrid fiber-reinforced cement-based composites. *ACI Mater. J.* **2014**, *111*. [CrossRef]
23. Fantilli, A.P.; Chiaia, B. The work of fracture in the eco-mechanical performances of structural concrete. *J. Adv. Concr. Technol.* **2013**, *11*, 282–290. [CrossRef]
24. Japan Concrete Institute. *A Textbook of Environmental Aspects for Concrete, Outline of Lci Data of Cement*; Japan Concrete Institute: Tokyo, Japan, 2013.
25. International Federation for Structural Concrete. *Guidelines for Green Concrete Structures*; FIB Bulletin No. 67; FIB Bulletin: Lausanne, Switzerland, 2012.
26. Flower, D.J.M.; Sanjayan, J.G. Green house gas emissions due to concrete manufacture. *Int. J. Life Cycle Ass.* **2007**, *12*, 282–288. [CrossRef]

MDPI

Article

Strength and Water Absorption of Sustainable Concrete Produced with Recycled Basaltic Concrete Aggregates and Powder

Ibrahim Sharaky [1,*], Usama Issa [1], Mamdooh Alwetaishi [1], Ahmed Abdelhafiz [1], Amal Shamseldin [1], Mohammed Al-Surf [2], Mosleh Al-Harthi [3] and Ashraf Balabel [4]

1 Civil Engineering Department, College of Engineering, Taif University, P.O. Box 11099, Taif 21099, Saudi Arabia; u.issa@tu.edu.sa (U.I.); m.alwetaishi@tu.edu.sa (M.A.); a.abdelhafiz@tu.edu.sa (A.A.); ashamseldin@tu.edu.sa (A.S.)

2 U.S. Green Building Council and Green Business Certification Inc., Jeddah 23525, Saudi Arabia; malsurf@gbci.org

3 Electrical Engineering Department, College of Engineering, Taif University, P.O. Box 11099, Taif 21099, Saudi Arabia; m.harthi@tu.edu.sa

4 Mechanical Engineering Department, College of Engineering, Taif University, P.O. Box 11099, Taif 21099, Saudi Arabia; a.balabel@tu.edu.sa

* Correspondence: i.sharaky@tu.edu.sa

Citation: Sharaky, I.; Issa, U.; Alwetaishi, M.; Abdelhafiz, A.; Shamseldin, A.; Al-Surf, M.; Al-Harthi, M.; Balabel, A. Strength and Water Absorption of Sustainable Concrete Produced with Recycled Basaltic Concrete Aggregates and Powder. *Sustainability* **2021**, *13*, 6277. https://doi.org/10.3390/su13116277

Academic Editors: Fausto Minelli, Enzo Martinelli and Luca Facconi

Received: 14 April 2021
Accepted: 25 May 2021
Published: 2 June 2021

Abstract: In this study, the recycled concrete aggregates and powder (RCA and RCP) prepared from basaltic concrete waste were used to replace the natural aggregate (NA) and cement, respectively. The NA (coarse and fine) was replaced by the recycled aggregates with five percentages (0%, 20%, 40%, 60% and 80%). Consequently, the cement was replaced by the RCP with four percentages (0%, 5%, 10% and 20%). Cubes with 100 mm edge length were prepared for all tests. The compressive and tensile strengths (f_{cu} and f_{tu}) and water absorption (WA) were investigated for all mixes at different ages. Partial substitution of NA with recycled aggregate reduced the compressive strength with different percentages depending on the type and source of recycled aggregate. After 28 days, the maximum reduction in f_{cu} value was 9.8% and 9.4% for mixtures with coarse RCA and fine RCA (FRCA), respectively. After 56 days, the mixes with 40% FRCA reached almost the same f_{cu} value as the control mix (M0, 99.5%). Consequently, the compressive strengths of the mixes with 10% RCA at 28 and 56 days were 99.3 and 95.2%, respectively, compared to those of M0. The mixes integrated FRCA and RCP showed higher tensile strengths than the M0 at 56 d with a very small reduction at 28 d (max = 3.4%). Moreover, the f_{cu} and f_{tu} values increased for the late test ages, while the WA decreased.

Keywords: basalt; recycled concrete aggregate; concrete properties; recycled natural basalt; recycled concrete powder

1. Introduction

Nowadays, a lot of attention is paid to the sustainable disposal of construction waste and its management, as it decreases the cost of disposal and reduces the environmental impact (EI) [1]. The construction industry has caused high environmental impacts worldwide as it demands excessive extraction of raw materials. Sand and gravel are the highest materials that are used on Earth next to water in contrast, and their natural regeneration rates are significantly lower than their usages as was specified by the United Nations Environment Programme. Natural aggregates of about 45 billion tons were extracted in 2017 and they are estimated to rise to 66 billion tons in 2025. Moreover, about 40 billion tons/year from aggregates were consumed in the cement product industries worldwide [2–4]. Furthermore, the European Statistical Office reported that about 923 billion tons/year of industrial wastes are produced from the construction works. The wastes generated from the construction industry are more than 33% of all wastes produced in the EU, of which about 90% could be reused as recycled materials. In 2018, about 10.93% of the consumed aggregate in Europe came from secondary sources (artificial, recycled, filled). More countries tried to increase their recycling rates of construction and demolition waste (C and DW) to replace the natural aggregate [2,5]. The extraction of recycled aggregates should be an essential part of the economy. The benefits of using recycled aggregates are reducing landfill space and energy of extracting raw materials, reducing greenhouse gases, preserving the natural resources, and achieving environmental sustainability [6].

Green concrete can meet structural function and service life with higher durability and strength than normal concrete (NC). It can be used to construct comfortable and suitable residential buildings for people [7]. Green concrete reduces the biological impact on the environment (environmentally friendly concrete material) either during the manufacturing process or use [8]. There are several modes of production, such as recycled aggregate concrete (RAC), fly ash concrete (FAC), and circular economy concrete (CEC) [9]. In the previous ways of making green concrete, urban construction waste is used as recycled aggregate while the fly ash (FA) as industrial waste is completely consumed. Green concrete can save resources of natural material, reduce spaces, save energy consumption, and reduce soil pollution [9]. Waste clay bricks (WCBF) are also used as a fine aggregate with various replacement ratios to produce recycled brick concrete (RBC) [10]. Increasing the WCBF replacement ratios decreased the density and compressive strength (f_{cu}) of RBC mainly with increasing water amount. Up to 50% WCBF replacement and without additional water, the properties of RBC (density, split tensile strength (f_{tu}), elastic modulus (E), and f_{cu}) were comparable to NC. Moreover, the use of untreated coal waste particles as concrete aggregate with an appropriate replacement ratio reduces the environmental impact of untreated coal waste and improves the mechanical properties of concrete by about (3–8%) [11].

The effects of ground granulated blast furnace slag (GGBFS) on the properties of recycled concrete (RCA) are also summarized [12–14]. Poured concrete with 50% RCA and 30% GGBFS showed comparable mechanical properties to the concrete with natural aggregate (NAC) [12]. The RCA concretes are usually used in road substructure because of the lower mechanical properties, higher WA and porosity f RCA compared to the NAC [15–17]. Consequently, the reduction in the mechanical performance of RCA concretes compared to NC is also reported in [18–20]. The reduction in the mechanical performance of RCA compared to NC is also presented in [15–17]. Concrete with 100% RCA showed 24% lower compressive strength compared to NAC [12]. In contrast, concrete containing up to 30% RCA as NA replacement can achieve the targeted NAC strength [17]. To achieve the sustainability of construction materials and international consensus, RCA should be used to produce new concrete, however, they show higher WA and weakness of interfacial transition zones (ITZs) that could reduce their mechanical properties compared to NAC [21–28]. The attached cement mortar to RCA surfaces affected their physical properties [15,29]. Consequently, the attached or adhered mortar quantity is also modified with the crushing procedure. Moreover, RCA has a lower density and higher WA than NA because of the adhered mortar. Furthermore, the adhered mortar with un-hydrated cement could modify crack propagation and concrete properties [30–32]. Thus, the concrete mechanical properties (compressive and tensile strengths) reduced as the RCA% in the mix increased [33–35].

Fly ash, cenosphere fly ash (CFA) and sintered fly ash are used to produce sustainable concrete as substitutes for cement, fine aggregate and coarse aggregate, respectively [36,37]. Moreover, several studies confirmed that the use of RCA in concrete has a positive effect on EI and cost [38–40]. In [38], the use of 30% and 100% RCA reduced the EI by up to 8% and 23%, respectively. Similarly, in [41], using 30% RCA and 100% RCA instead of NAC in concrete resulted in net cost benefits of 9 and 28%, respectively. Moreover, the environmental and cost impacts were reduced by 50.8 and 68.1% when waste concrete was used to produce RAC concrete [42]. Conversely, the use of 50%RCA was found to be the optimum percentage in terms of EI and cost efficiency [43], while in [44], the appropriate RCA percentage was 80%.

The effect of fine RCA (FRCA) on concrete strengths has been studied [2,38,45–50]. For concrete with 25% FRCA and 100% FRCA, the compressive strength decreased by 15% and 30% respectively, compared to NAC, while the shrinkage increased [45]. Consequently, an admixture of FRCA in concrete decreases its compressive strength and increases shrinkage compared to NAC [46]. Moreover, increasing the proportion of foundry sand (FS) and recycled fine aggregate (RFA) up to 100% resulted in a decrease in compressive strength [51]. In contrast, RFCA had no effect on concrete strength when the FRCA was added by less than 30% [38]. The method of FRCA production affected the properties and integrated concrete [48]. Conversely, the WA, and chloride penetration of concrete with integrated FRCA increased, while its carburization resistance decreased when FRCA was added [47]. The cement mortars grouted with FRCA exhibited a reduction in their compressive and flexural strengths with increasing FRCA content [49,50].

Concrete strength cast with 20% recycled glass sand (RGS) gained its design strength after 7 days. Concrete strength decreased slightly with increasing RGS content. The same previous effects of RGS on flexural and tensile strength were also reported, with f_{tu}/f_{cu}% ranging from 8% to 11% [52]. In addition, the use of individual plastic (PA), rubber (RA) and glass (GA) to partially replace the fine NA showed different effects on concrete strengths [53]. Except for 15% GA, all mixes showed a reduction in compressive strength compared to NAC. The PA concrete showed the highest strength reduction (about 50% for 30% PA) compared to NAC [53]. The same trends were observed for tensile strength as the mixes integrated 30% PA and 30% RA (f_{tu} reduction was 27% and 35%, respectively) [53]. The basalt powder can reduce the EI and improve the concrete properties [54,55]. Consequently, in [56,57], the basalt aggregate and basalt powder had an almost similar composition to Ordinary Portland Cement (OPC) and silica fume (SF). Although the waste glass had pozzolanic or cementitious properties, only part of those wastes can be used as cement replacement SCM as it can be reused to produce new glass [58–60]. Among the several types of glasses, soda-lime glasses are the most public type [61]. It also contains around 73% SiO_2, 10% CaO, and 13% Na_2O that makes it pozzolanic SCM that can be used in concrete [61]. The basalt powder can reduce the IE during the basalt extraction and can also be used to improve the concrete properties [54]. Additionally, in [56,57], the basalt aggregate and basalt powder had a near-similar composition as OPC and SF.

Several theoretical models have been used to evaluate the RAC mechanical properties. In [62], most of these available theoretical models were reported. The reported available models are not rather accurate [62]. For this reason, new mathematical approaches were developed to evaluate the RCA mechanical properties [62]. The proposed model could predict the RAC mechanical properties with higher accuracy and simpler formulae than the available existing models [62].

The above review shows that the use of recycled components obtained from concrete previously cast with basalt is still limited, especially for FRCA and recycled concrete powder (RCP). In this study, the coarse RCA (CRCA) produced from the basaltic concrete waste (CRCA) was used to replace the coarse NA with five percentages (0%, 20%, 40%, 60% and 80%). Consequently, the FRNA was used to partially replace the fine NA with five percentages (0%, 20%, 40%, 60% and 80%). Conversely, the RCP was used to replace the cement with four percentages (0%, 5%, 10% and 20%). Cubes with 100 mm edge length were cast, cured, and tested to obtain the concrete strengths and WA after different curing times (7, 28 and 56 days). The results obtained were also discussed and compared with those of NAC.

2. Research Significance

The use of concrete waste as a recycled aggregate is popular to produce new concrete. In contrast, the use of concrete waste powder as a cement substitute is still limited. Although the basaltic concrete waste has a pozzolanic effect, the use of this waste as a substitute for aggregate and cement is still limited. In this study, the use of basaltic concrete waste as a substitute for natural aggregates with low and high content was investigated. In addition, the use of basaltic concrete powder, that had pozzolanic activity, to replace cement and produce sustainable/green concrete was also investigated.

3. Experimental Work

3.1. Material Properties

3.1.1. Natural Aggregates

The crushed basalt (Figure 1a) and natural sand (Figure 1b) were the coarse and fine aggregates used for the control mixtures. The maximum nominal size (MNS) of the basalt gravel used in this experimental program was 12.5 mm. The natural sand had a fineness modulus of 3.2. The grading curves of the sand and basalt were in accordance with the limits of ASTM C33 [63] (Figure 2a). The physical properties of crushed basalt (coarse NA) and natural sand are listed in Table 1.

Figure 1. The aggregate and waste materials used in this study. (**a**) Natural basalt aggregate; (**b**) Natural sand; (**c**) CRCA; (**d**) FRCA; (**e**) RCP.

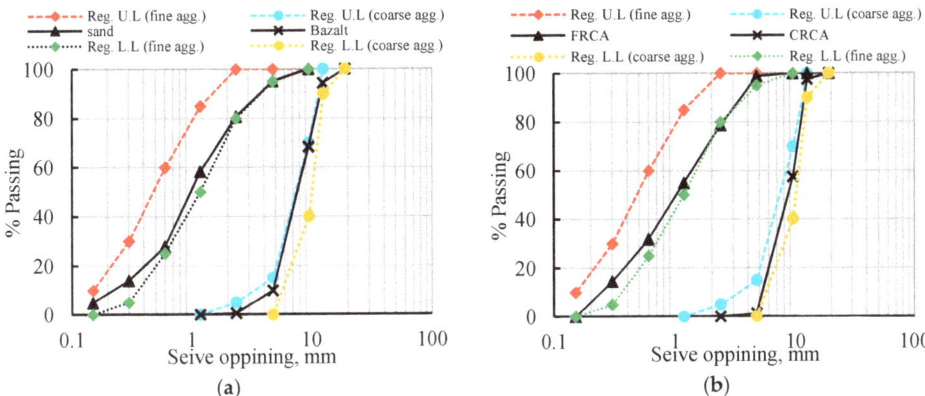

Figure 2. Sieve analysis for the natural and recycled aggregates. (**a**) NA; (**b**) RCA and RNA.

Table 1. Physical properties of natural and recycled aggregate.

Physical Properties	Coarse NA	CRCA	Natural Sand	FRCA
Apparent specific gravity (kg/m^3)	2.95	2.76	2.71	2.64
Bulk specific gravity (SSD) (gr/cm^3)	2.86	2.57	2.28	2.15
Bulk specific gravity (GD) (gr/cm^3)	2.81	2.46	2.03	1.86
Water absorption (%)	1.77	4.32	10.96	13.82
Moisture content (%)	0.93	1.25	2.73	2.37

3.1.2. Recycled Concrete Aggregates

The recycled basaltic concrete is taken from old concrete cubes and cylinders that were tested in the lab five to six years ago for previous research projects and kept outside the lab as wastes. The old cubes and cylinders were manually crushed then put in a crushing machine to obtain RCA, FRCA and powder. The concrete strength of the old concrete ranged from 33 MPa to 39 MPa. The CRCA and FRCA obtained from the basalt concrete are shown in Figure 1c,d, respectively. The standard sieves were used to separate the basaltic concrete wastes (BCWs) into CRCA, FRCA and RCP. The CRCA remaining on sieve No. 4 (sieve opening = 4.75 mm) was used as a replacement for the coarse NA. The MNS of the CRCA used in this experimental program was adjusted to correspond to 12.5 mm. The selected MNS (12.5 mm) was adopted to minimize the negative effects of the FRCA compared to other larger sizes (19 and 25 mm) on the durability and strength of the concrete [64]. Consequently, the FRCA that passed the No. 4 sieve and remained on the No. 100 sieve (sieve opening = 150 microns) were used to replace the fine NA (sand). The fineness modulus of the FRCA was also equal to 3.2. The particle size distribution curves of the CRCA and FRCA were adjusted to comply with the limits of ASTM C33 [63] (Figure 2b). The physical properties of CRCA and FRCA are listed in Table 1.

3.1.3. Recycled Concrete Powder

The RCP passed from sieve No. 100 during sieve analysis of BCWs is shown in Figure 1e. The RCP was analyzed by scanning electron microscope (SEM and EDS) to find its chemical components. The obtained chemical components from EDS analysis are listed in Table 2, while the photos of EDS and SEM are shown in Figure 3a,b, respectively.

Table 2. The chemical compositions of OPC and RCP.

Item	SiO_2	CaO	AL_2O_3	Fe_2O_3	MgO	SO_3	LOI
OPC	21.28	64.64	5.60	3.42	2.06	2.12	0.88
RCP	66.23	-	9.61	11.84	12.32	-	-

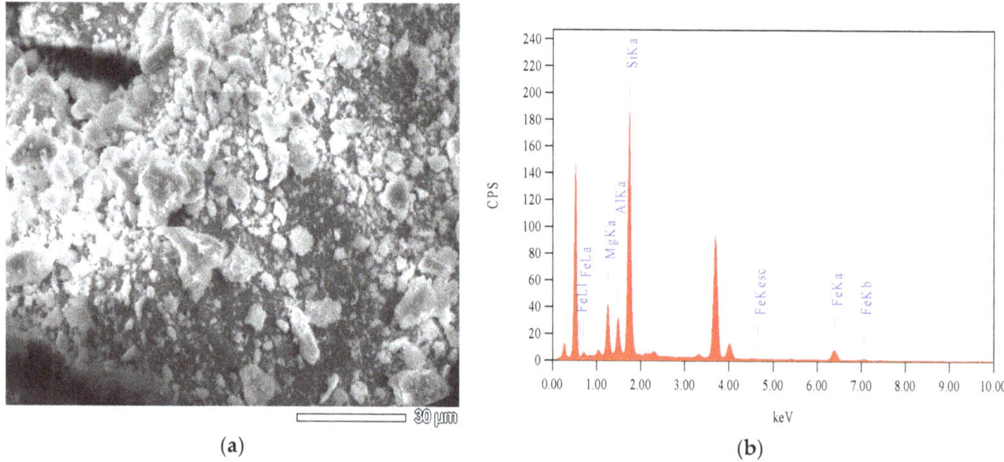

Figure 3. RCP components. (**a**) SEM; (**b**) EDS.

3.1.4. Cement

The OPC produced locally in one of the Cement Factories in KSA was used for all concrete mixes. The chemical composition of this cement as provided by the supplier is also given in Table 2.

3.2. Validation of the Material Chemical Composition as Natural Pozzolana

The chemical properties of the basaltic concrete powder are shown in Table 1. The chemical composition of RCP is underlain by silica (SiO_2). ASTM C618 recommends a sum of SiO_2, Fe_2O_3, and Al_2O_3 greater than 70% for the chemical composition requirements of natural pozzolana. According to Table 2, the sum of SiO_2, Fe_2O_3, and Al_2O_3 for RCP (87.68%) meets the ASTM C618 requirements for natural pozzolana. Therefore, the utilization of RCA as fine and coarse aggregate and RCP as a waste powder in the hydration process can help to produce sustainable concrete based on basalt waste.

3.3. Mix Proportions and Mixing Procedure

In this work, a control mix (M0) was designated according to the ACI specification [65] to obtain 35 MPa cylindrical concrete strength at 28 days. The M0 mix was designated to keep a slump of 50–80 mm and mechanical compaction was used. The CRCA and FRCA gradings were adopted to be like that for NA. The W/C ratio for M0 was adopted to be 0.5. For mixes with recycled aggregate, the recycled aggregate was used to partially replaced the natural aggregate by weight. All the aspects remained unchanged to study the effects of replacing this recycled aggregate with their different physical properties on the concrete mechanical properties and water absorption. Most previous studies neglected the humidity and water absorption when designing the concrete mixes with or without recycled materials [18,51–53]. The mix M0 was prepared, and its slump was obtained. The slump of M0 mix achieved the required slump with neglecting the humidity and water content of the aggregates. For mixes incorporating RCA, the additional water required for each mix to keep the same consistency of M0 mix was calculated and added during mixing (Figure 4a).

The mix (M0) consisted of natural sand, natural basalt and OPC with weight percentages of 643, 1016 and 400 kg/m^3 respectively (Table 3). The effect of the CRCA and FRCA on the W/C is also shown in Figure 4b. In order to produce sustainable concrete, eleven mixes (M1 to M11) were prepared in which the natural concrete constituents (aggregates and cement) were replaced by the recycled waste materials in different proportions. For the mixes M1-M4, the CRCA was used to replace the coarse NA with five percentages (0%, 20%, 40%, 60% and 80%, respectively). For mixtures M5-M8, natural sand was replaced with FRCA at five percentages (0%, 20%, 40%, 60%, and 80%, respectively). In addition, for mixes M9-M11, OPC was replaced with RCP at four percentages (0%, 5%, 10%, and 20%, respectively). Full details of the mix proportions for all mixes are summarized in Table 3. The aggregates and cement were dry mixed in the concrete mixer for 1 min. The RCP (if any) was then added and dry mixed for another 1 min. Then, the water was added gradually, and mixing was completed for another 2 min. The concrete mixtures were poured into 100 × 100 × 100 mm^3 cubes. The cubes were de-moulded after 24 h and cured in a water tank until the test age (7, 28, and 56 days).

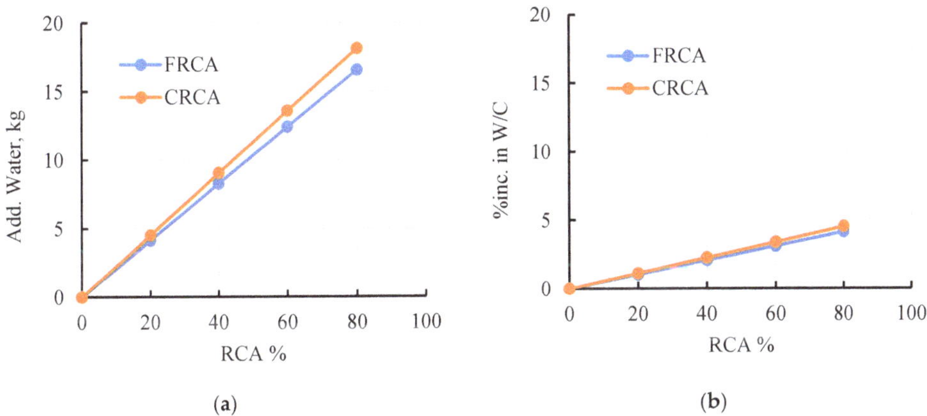

Figure 4. The relation between RCA% and (**a**) extra water; (**b**) % increase in W/C ratio.

Table 3. Concrete mixture proportion for 1 m^3.

Group	Mix ID	Recycled Aggregate (%)	RCP (%)	Aggregates (kg)				Water (kg)	Cement (kg)
				Natural Sand	Natural Basalt	FRCA	CRCA		
Control	M0	0	0	643	1016	0	0	200	400
CRCA	M1	20	0	643	812.8	0	203.2	200	400
	M2	40	0	643	609.6	0	406.4	200	400
	M3	60	0	643	406.4	0	609.6	200	400
	M4	80	0	643	203.2	0	812.8		
FRCA	M5	20	0	514.5	1016	128.6	0	200	400
	M6	40	0	385.9	1016	257.3	0	200	400
	M7	60	0	257.3	1016	385.9	0	200	400
	M8	80	0	128.6	1016	514.5	0	200	400
RCP	M9	0	5	643	1016	0	0	200	380
	M10	0	10	643	1016	0	0	200	360
	M11	0	20	643	1016	0	0	200	320

3.4. Test Method

To evaluate the compressive strengths of all mixes, the 100-mm cube specimens were compression tested at 7, 28, and 56 days according to BS EN 12390-3 [66]. At each age, the results of three tested cubes were averaged to obtain the compressive strength (f_{cu}). The concrete tensile strengths (f_{tu}) at 28 and 56 d were also obtained from the diagonal splitting test for the cubes with 100 mm edge length, as shown in Figure 5 and Equation (1) [67].

$$\sigma_{max} = \frac{2P}{\pi bd}\left[\left(1-\beta^2\right)^{\frac{5}{3}} - 0.0115\right] \quad (1)$$

where b and d are the width and diagonal length of the cube, and P is the maximum compressive load. In this study, the value of β is equal to 0.15. At each age, the results of three tested cubes were averaged to obtain the tensile strength (f_{tu}). The compression testing machine with a maximum capacity of 2000 tons was used to test the cubes in compression and in tension at a loading rate of 4 and 3 kN/s, respectively. Water absorption (*WA*) is one of the well-known concrete durability factors [47,68,69]. Moreover, *WA* of concrete was considered as a measure for resistance against carbonation and chloride migration [70]. *WA* was obtained according to ASTM C642 [71] to compare the durability of the mixes after 28 and 56 days. First, the concrete specimens (3 cubes at each age) were dried at a temperature (T) of 100–110 °C for more than 24 h to find the oven-dried mass (A) in grams. Subsequently, the samples were engrossed in water at T ≈ 21 °C for more than 48 h, then the surface-dried samples were weighed in grams (*B*), and then *WA* was calculated according to Equation (2).

$$WA\ \% = \left[\frac{(B-A)}{A}\right] \times 100 \quad (2)$$

(a)

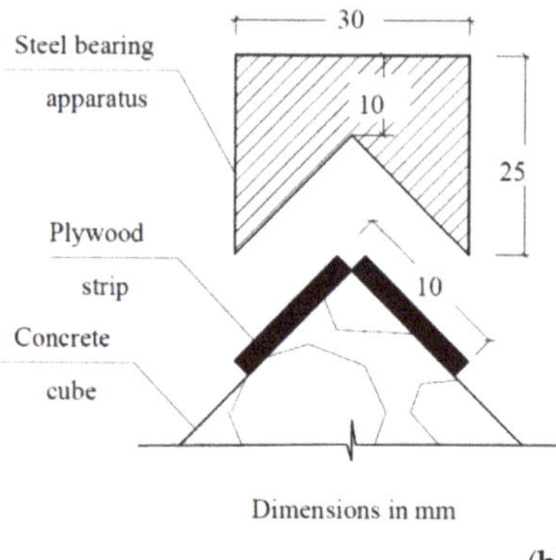

(b)

Figure 5. Details of the compression and diagonal cube splitting tests. (**a**) Compression test; (**b**) Details of diagonal cube splitting test.

4. Results and Discussions

4.1. Concrete Compressive Strength

The cubic concrete strengths (f_{cu}) at 7, 28, and 56 days of age ($f_{cu,7}$, $f_{cu,28}$, and $f_{cu,56}$, respectively) are shown in Table 4. The reduction values of concrete strength at 7, 28, and 56 days (μ_7, μ_{28}, and μ_{56}) due to partial replacement of natural basalt, sand and cement by the recycled waste are also presented in Table 4. The percentage increase in $f_{cu,28}$ compared to $f_{cu,7}$ ($\mu_{28/7}$) and $f_{cu,56}$ compared to $f_{cu,28}$ ($\mu_{56/28}$) for all the tested mixes are also presented in Table 4. The effect of recycled waste on compressive strengths is discussed in the following sections. The percentage compressive strength decreases at each curing time ($\mu_{cu,date}\%$) is calculated using Equation (3).

$$\mu_{cu,date}\% = \frac{f_{cu,date}(R) - f_{cu,date}(N)}{f_{cu,date}(N)} \times 100 \quad (3)$$

where $f_{cu,date}$ (R) and $f_{cu,date}$ (N) are the compressive strengths of the mixtures that integrate waste and the control mixture, respectively.

Table 4. The concrete compressive strength for all mixes at 7, 28 and 56 d.

Group	Mix ID	Specimen No.	$f_{cu,7}$ MPa	Mean (σ_μ) MPa	μ_7 %	$f_{cu,28}$ MPa	Mean (σ_μ) MPa	μ_{28} %	$f_{cu,56}$ MPa	Mean (σ_μ) MPa	μ_{56} %	$\mu_{28/7}$ %	$\mu_{56/28}$ %
Control		1	35.60			43.50			52.48				
	M0	2	33.10	34.07 (±1.34)	0.0	43.70	43.60 (±0.10)	0.0	50.48	50.02 (±2.72)	0.0	27.98	14.7
		3	33.50			43.60			47.10				
CRCA		1	28.10			38.50			44.93				
	M1	2	27.20	28.17 (±1.00)	17.3	39.40	39.33 (±0.80)	9.80	44.51	44.82 (±0.27)	10.4	39.64	13.9
		3	29.20			39.69			45.01				
		1	27.70			39.87			45.81				
	M2	2	30.10	29.33 (±1.20)	13.9	39.40	39.65 (±0.24)	9.10	45.78	45.33 (±0.80)	9.40	35.18	14.3
		3	28.80			39.69			44.40				
		1	28.70			41.00			46.60				
	M3	2	30.60	30.40 (±1.61)	10.8	39.57	40.30 (±0.72)	7.60	45.00	45.47 (±0.99)	9.10	32.58	12.8
		3	31.90			40.34			44.80				
		1	28.20			37.48			40.35				
	M4	2	29.70	28.37 (±1.26)	16.7	38.85	37.86 (±0.87)	13.2	40.98	40.86 (±0.66)	18.3	33.47	7.90
		3	27.20			37.25			40.25				
FRCA		1	30.60			40.54			42.83				
	M5	2	29.00	29.81 (±0.80)	12.5	40.50	39.81 (±1.22)	8.70	45.54	44.53 (±0.00)	11.0	33.57	11.8
		3	29.82			38.40			45.22				
		1	29.80			39.63			47.86				
	M6	2	28.20	29.01 (±0.80)	14.8	39.61	39.50 (±0.21)	9.40	54.71	49.78 (±0.00)	0.50	36.14	26.0
		3	29.04			39.26			46.49				
		1	32.50			40.64			46.49				
	M7	2	34.90	33.73 (±1.20)	1.0	44.40	41.60 (±2.46)	4.60	43.93	45.25 (±1.28)	9.50	23.33	8.80
		3	33.80			39.77			45.32				
		1	32.40			39.98			45.71				
	M8	2	32.30	32.30 (±0.10)	5.2	40.70	40.09 (±0.57)	8.10	43.52	45.04 (±1.32)	10.0	24.11	12.4
		3	32.20			39.58			45.89				
RCP		1	32.02			38.73			44.60				
	M9	2	31.95	32.01 (±0.06)	6.0	39.33	38.80 (±0.50)	11.0	45.44	44.81 (±0.55)	10.4	21.20	6.70
		3	32.07			38.34			44.40				
		1	22.34			43.39			46.90				
	M10	2	23.90	22.93 (±0.84)	32.7	43.57	43.30 (±0.32)	0.70	48.24	47.61 (±0.67)	4.80	88.81	3.70
		3	22.56			42.94			47.70				
		1	22.89			39.70			41.20				
	M11	2	20.42	21.57 (±1.24)	36.7	39.52	38.27 (±2.32)	12.2	43.00	42.43 (±1.07)	15.2	77.39	11.4
		3	21.41			35.59			43.10				

$\mu_{Tc}\% = \frac{f_{cu,Tc}(any\ mix)}{f_{cu,Tc}(Mix0)} \times 100$ where Tc = 7, 28, and 56 days.

4.1.1. Effect of the CRCA

For all mixes, the compressive strength increased with increasing curing age (Figure 6a). Comparing the strengths obtained at 7 and 56 days with those obtained at 28 days, they showed higher rates of strength increase when Tc was increased from 7 to 28 days than when Tc was increased from 28 to 56 days. This may be due to the higher cement hydration rates during the first 28 days compared to the late days.

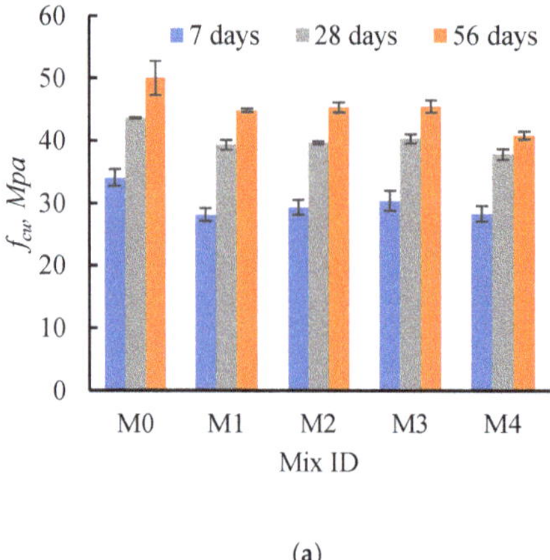

(a)

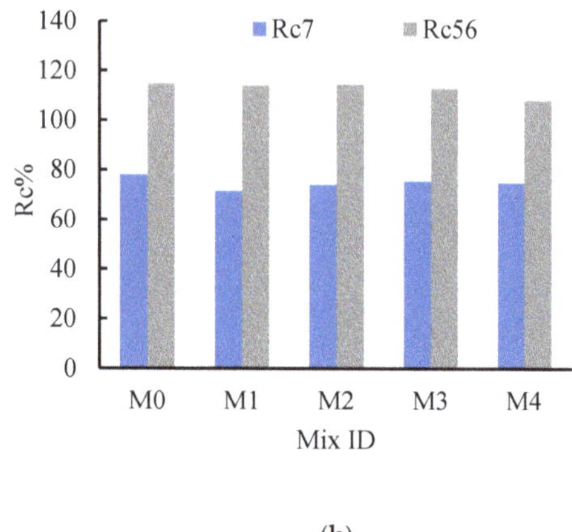

(b)

Figure 6. Effect of the CRCA%. (**a**) Concrete compressive strength; (**b**) The compressive strength ratio (Rc%).

The increase in strengths with increasing curing age was also related to CRCA%. When Tc increased from 7 to 28 days, the strength increased by 27.8%, 39.6%, 35.2, 32.5 and 33.4% for M0 to M4, respectively, illustrating the influence of CRCA% on early cement hydration. Consequently, the strength increased with increasing Tc from 28 to 56 d by 14.7%, 13.9%, 14.3%, 12.8% and 7.9% for M0 to M4, respectively. The low increase in strength at 56 days for the mixes integrating CRCA% than that of M0 ensured the effect of crushed concrete and its voids on the absorption of water required to complete cement hydration. The replacement of coarse NA with CRCA generally reduced the concrete strength. The CRCA are made of recycled basalt, and crushed concrete, which had a high void ratio compared to NA, and can affect concrete strength and cement hydration in addition to the powder. Strength reduction was strongly influenced by CRCA% and curing age. After 7 days, the strength reduction (μ_7%) is 17.3%, 13.9%, 10.8%, and 16.7% for mixtures with CRCA content of 20%, 40%, 60%, and 80%, respectively (mean (μ) = 14.8% and standard deviation (σ_μ) = ±2.6%). Subsequently, after 28 days, the strength reduction (μ28%) is 9.8%, 8.7%, 6.9% and 12.2% for mixes with CRCA content of 20%, 40%, 60% and 80%, respectively (μ = 9.9% and σ_μ = ±2.05%). Moreover, the strength reduction (μ56%) after 56 days is 10.4%, 9.4%, 9.1%, and 18.3% for mixes with CRCA content of 20%, 40%, 60% and 80%, respectively (μ = 11.8% and σ_μ = ±3.8%). These reductions were due to the CRCA defects that negatively affected their interfacial transition zones (ITZs)). These weak ITZs led to the propagation of the concrete cracks and reduced the strength values [72–77]. The lowest values of μ7, μ28 and μ56 are 10.8%, 6.9% and 9.1%, respectively (M3, CRCA = 60%). The mixtures with CRCA (M1-M4) experienced higher differences between $\mu_{28/7}$ and $\mu_{56/28}$ compared to M0 (Table 4). These observations summarize the weak ITZs between the cement paste and CRCA (C- CRCA) besides the CRCA voids. At Tc = 28 days, C-CRCA bond may be enhanced by the hydration of the RCP surrounding CRCA (RCP has a pozzolanic effect as mentioned earlier).

To assess the CRCA effects on the strength gained at 7 and 56 days compared to that at 28 days, the Rc7 (f_{cu7}/f_{cu28}) and Rc56 (f_{cu56}/f_{cu28}) were calculated (Figure 6b). From the figure, the Rc7 decreased for mixed integrated CRCA compared to M0. The values of Rc7% are 78.1%, 71.6%, 74%, 75.4%, and 74.9% for CRCA% of 0%, 20%, 40%, 60%, and 80%, respectively. Moreover, the values of R56% are 114.7%, 113.9%, 114.3%, 112.8% and 107.9% for the CRCA% of 0%, 20%, 40%, 60% and 80%, respectively. The comparison of the R7 and R56 values of M0 with those of M1–M5 highlights the small effects of CRCA% on the Rc7 and Rc56 values of concrete especially for M3. It is evident from [11,78] that the concrete strength decreased sharply when the coal waste and RCA aggregates replaced NA. Moreover, in [78] a strength reduction of about 26–32% was reported when the RCA aggregates replaced NA. Furthermore, compressive strength reduction of lower than 40% was observed when RAC was incorporated [79]. When the NA was replaced by RCA of 25–50%, the compressive strength reduction was 2–3% depending on the RCA source while when the RCA replacement was more than 50%, the reduction was 15–23% [22]. Based on the above results, basaltic CRCA could be considered as a good substitute for the coarse NA to produce sustainable concrete and reduces the environmental impact and cost of concrete production.

4.1.2. Effect of the FRCA

The effect of FRCA on concrete strength was also investigated (Figure 7). With increasing Tc, the strength increased for all mixes with FRCA (Figure 7a), although the trend was different from the CRCA mixes. The increase in strengths with increasing curing age was also related to the FRCA content. When Tc increased from 7 to 28 days, the strength increased by 33.57%, 36.14%, 23.33% and 24.11% for M5 to M8, respectively, highlighting the influence of CRCA% on early cement hydration. Consequently, the strength increased by 11.8%, 26%, 8.8% and 12.4% for M5 to M8, respectively, when Tc was increased from 28 to 56 days. Replacing sand with FRCA generally decreased the concrete strength. At 7 days, when FRCA replaced sand at 20%, 40%, 60% and 80%, the strength reduction was 12.5%, 14.8%, 1.0% and 5.2%, respectively (Mean (μ) = 8.38% and Standared deviasion (σ_μ) = ±5.5%). Conversely, the strength reduction for 28 d (μ28) was 8.7%, 9.4%, 4.6% and 8.1% when FRCA was added to 20%, 40%, 60% and 80%, respectively (μ = 7.68% and σ_μ = ±1.58%). Moreover, the strength reduction (μ56) after 56 days was 11.0%, 0.50%, 9.5% and 10.0% when FRCA was added at 20%, 40%, 60% and 80%, respectively (μ = 5.75% and σ_μ = ±4.22%). The mixes with integrated FRCA experience less reduction in concrete strength than those with integrated CRCA. This may be due to the pozzolanic effect of FRCA and its low voids. The lowest values of μ_7 and μ_{28} were 1.0 and 4.6% (M7, FRCA = 60%), respectively, while the lowest value of μ_{56} was 0.50% (M6, FRCA = 40%).

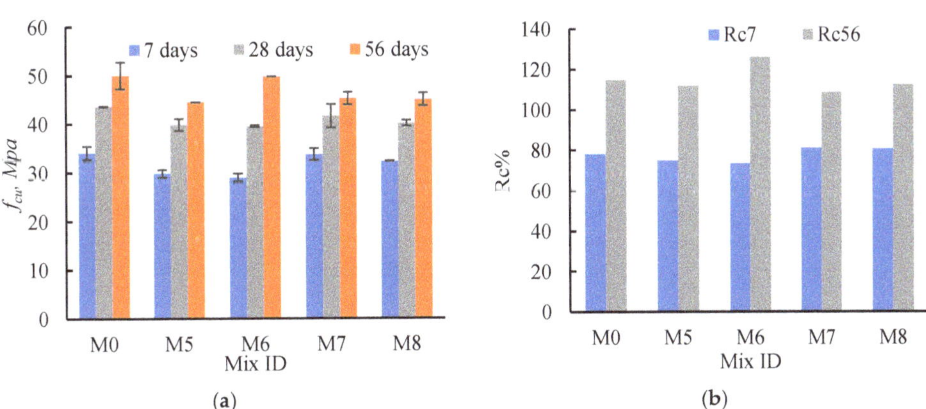

Figure 7. Effect of the FRCA%. (**a**) Concrete compressive strength; (**b**) The compressive strength ratio (Rc%).

The effect of FRCA on the strength gained at 7 and 56 days compared to that at 28 days (Rc7% and Rc56%) was calculated (Figure 7b). The figure shows that the Rc7 values were 78.1%, 74.9%, 73.5%, 81.1% and 80.6% when the FRCA was integrated by 0%, 20%, 40%, 60% and 80%, respectively (μ = 77.6% and σ_μ = ±3.03%). Moreover, the Rc56 values were 114.7%, 111.8%, 126.0%, 108.8% and 112.4% when the FRCA was added by 0%, 20%, 40%, 60% and 80%, respectively (μ = 114.7% and σ_μ = ±5.95%). M6 showed the highest Rc56 value, highlighting the pozzolanic effect of FRCA, which can maintain almost the same M0 strength (99.5%). The incorporation of FRCA into the concrete to replace the natural sand was more efficient than the RA and PA to restore the NAC strength [53]. The previous research studies [35,78] have shown that concrete with RCA has a higher pore structure than concrete with NA. When RCA (coarse and fine) replaced NA, a decrease in strength of about 26–32% was observed in [77] and the same finding was highlighted in [80]. The above observations underlined the suitability of FRCA to produce sustainable concrete with high FRCA replacement ratios.

4.1.3. Effect of the RCP

To produce green/sustainable concrete, the RCP was used to partially replace the cement component. The cement was partially substituted by 0%, 5%, 10% and 20% RCP. From Figure 8, the strength increased with increasing Tc value irrespective of the RCP content. When the Tc was increased from 7 to 28 days, the strength increased by 21.20%, 88.81% and 77.39% when the RCP was added with 5%, 10% and 20%, respectively. When the Tc was increased from 28 to 56 days, the strength increased by 6.7%, 3.7% and 11.4% when the RCP was added with 5%, 10% and 20%, respectively. The use of RCP as a replacement material for cement decreased the compressive strength by different percentages depending on the Tc and RCP percentage. After 7 d, the strength decreased by 6.0%, 32.7%, and 36.7% when 5%, 10%, and 20% RCP were added, respectively (μ = 25.12% and σ_μ = ±13.6%).

The use of RCP as a replacement material for cement decreased the compressive strength by different percentages depending on Tc and RCP%. After 7 days, the strength decreased by 6.0%, 32.7% and 36.7% (μ = 25.12% and σ_μ = ±13.6%) when 5%, 10% and 20% RCP were added, respectively. Conversely, the strength decreased by 11.0, 0.7 and 12.2% after 28 days when RCP was added by 5%, 10% and 20%, respectively (μ = 7.97% and σ_μ = ±5.17%). In addition, the attained strength decreased by 10.4%, 4.80% and 15.2% when RCP was integrated by 5%, 10% and 20%, respectively (μ = 10.13% and σ_μ = ±4.95%). The small reduction in strength for mixes with RCP (M9–M11) compared to M0, especially after 28 and 56 days, may be due to the late hydration of RCP. These results assured the pozzolanic effects of RCP, which covered about 99.3% and 95.2% of M0 strength after 28 and 56 days, respectively (RCP = 10%). Moreover, M9 achieved the lowest strength reduction after 7 d (6.0%) with an almost similar difference between μ28/7 and μ56/28 (Table 4). To compare the effect of RCP% on the obtained strength after 7 and 56 d compared to that after 28 d, the Rc7 and Rc56 were calculated (Figure 8b). From the figure, Rc7 was equal to 78.1%, 82.5%, 53% and 56.4% when the RCP was added by 0%, 5%, 10% and 20%, respectively (μ = 67.5% and σ_μ = ±12.97%). Moreover, the Rc56 was equal to 114.7%, 115.5%, 110.0% and 110.9% when the RCP was added by 0%, 5%, 10% and 20%, respectively (μ = 112.8% and σ_μ = ±2.38%). From all the above, basaltic RCP could be considered as a good substitute for cement to produce green/sustainable concrete, reduce environmental impact and produce concrete with low cost. As reported in [80–82], the replacement of cement with SF (5–25%) increases the concrete strength when the recommended superplasticizers are added. The reported and obtained results encourage the use of superplasticizers with the RCP that can regulate their hydration and increase the concrete strength as they have the SF effects on the concrete.

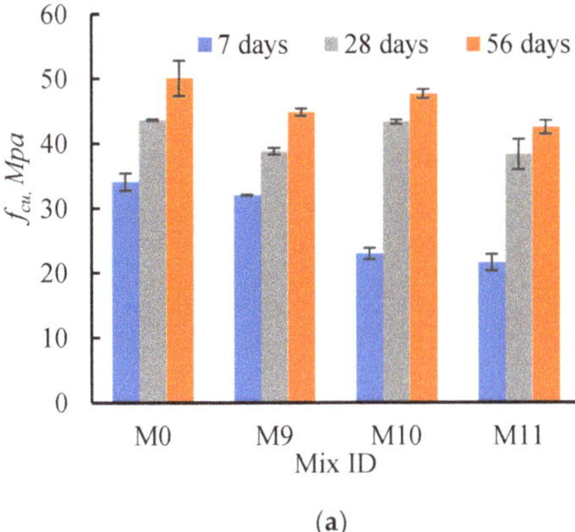

(a)

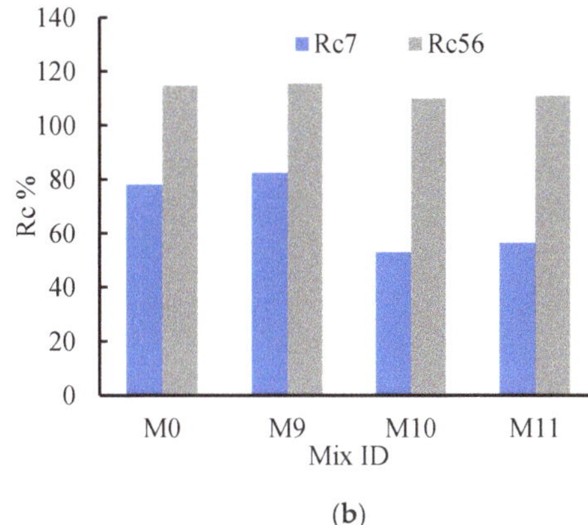

(b)

Figure 8. Effect of the fine RCP%. (**a**) Concrete compressive strength; (**b**) The compressive strength ratio (Rc%).

4.2. Concrete Tensile Strength

The cubic diagonal tensile strengths (f_{tu}) at ages 28 and 56 days ($f_{tu,28}$ and $f_{tu,56}$, respectively) are reported in Table 5 and Figures 9–11. Table 5 shows the increases and decreases in f_{tu} at 28 and 56 days ($\mu_{t,28}$ and $\mu_{t,56}$, respectively) as a result of the partial replacement of natural basalt, sand, and cement by CRCA, FRCA, and RCP. The percentage increase in $f_{tu,56}$ compared to $f_{tu,28}$ ($\mu_{t56/28}$) for all tested mixes was also calculated. In addition, the effect of waste concrete components on $f_{tu,28}/f_{cu,28}$ and $f_{tu,56}/f_{cu,56}$% was also adopted. The increased/decreased (I/d) percentage in f_{tu} ($\mu_{t,date}$%) at each curing time can be calculated using Equation (4).

$$\mu_{t,date}\% = \frac{f_{tu,date}(R) - f_{tu,date}(N)}{f_{tu,date}(N)} \times 100 \quad (4)$$

where $f_{tu,date}(R)$ and $f_{tu,date}(N)$ are the concrete tensile strengths for mixes with recycled components and that with NA, respectively.

Table 5. The concrete tensile strength for all mixes at 28 and 56 ages.

Group	Mix ID	Specimen No.	$f_{tu,28}$ MPa	Mean ($\sigma\mu$) MPa	μ_{t28}%	$f_{tu,56}$ MPa	Mean ($\sigma\mu$) MPa	μ_{t56} %	$\mu_{t56/28}$ %	$f_{tu,28}/f_{cu,28}$ %	$f_{tu,56}/f_{cu,56}$ %
Control	M0	1 2 3	3.48 3.58 3.51	3.52 (±0.05)	0.00	3.92 3.88 3.87	3.89 (±0.03)	0.00	10.45	8.08	7.78
CRCA	M1	1 2 3	3.45 3.37 3.39	3.40 (±0.04)	−3.42	3.86 3.53 3.69	3.69 (±0.16)	−5.02	8.61	8.65	8.24
	M2	1 2 3	3.51 3.31 3.47	3.43 (±0.10)	−2.63	3.56 3.57 3.65	3.59 (±0.05)	−7.60	4.81	8.65	7.93
	M3	1 2 3	3.54 3.39 3.49	3.47 (±0.08)	−1.40	3.82 4.32 3.77	3.97 (±0.31)	1.98	14.24	8.62	8.73
	M4	1 2 3	3.40 3.49 3.38	3.42 (±0.06)	−2.88	3.43 3.49 3.47	3.47 (±0.03)	−10.90	1.33	9.03	8.48

Table 5. *Cont.*

Group	Mix ID	Specimen No.	$f_{tu,28}$ MPa	Mean ($\sigma\mu$) MPa	μ_{t28}%	$f_{tu,56}$ MPa	Mean ($\sigma\mu$) MPa	μ_{t56} %	$\mu_{t56/28}$ %	$f_{tu,28}/f_{cu,28}$ %	$f_{tu,56}/f_{cu,56}$ %
FRCA	M5	1 2 3	3.66 3.29 3.38	3.44 (±0.20)	−2.24	4.00 3.78 3.78	3.85 (±0.13)	0.26	13.27	8.65	8.76
	M6	1 2 3	3.27 3.34 3.65	3.42 (±0.20)	−2.88	3.87 4.24 4.05	4.05 (±0.18)	5.40	19.78	8.67	8.24
	M7	1 2 3	3.67 3.61 3.70	3.66 (±0.04)	3.87	4.25 3.96 3.96	4.06 (±0.17)	5.40	12.08	8.79	9.06
	M8	1 2 3	3.21 3.95 3.73	3.63 (±0.36)	3.05	4.20 3.86 3.96	4.01 (±0.17)	3.09	10.49	9.05	8.90
RCP	M9	1 2 3	3.88 3.66 3.74	3.76 (±0.11)	6.81	4.02 3.99 4.03	4.01 (±0.03)	3.09	6.68	9.69	8.95
	M10	1 2 3	3.90 3.58 3.79	3.76 (±0.16)	6.81	3.88 3.97 3.83	3.90 (±0.07)	0.26	3.72	8.68	8.19
	M11	1 2 3	3.24 3.47 3.54	3.42 (±0.16)	−2.92	3.87 3.85 3.70	3.81 (±0.09)	−2.05	11.44	8.93	8.98

$\mu_{tTc}\% = \frac{f_{tu,Tc}(any\ mix)}{f_{tu,Tc}(Mix0)} \times 100$ where Tc = 7, 28 and 56 days.

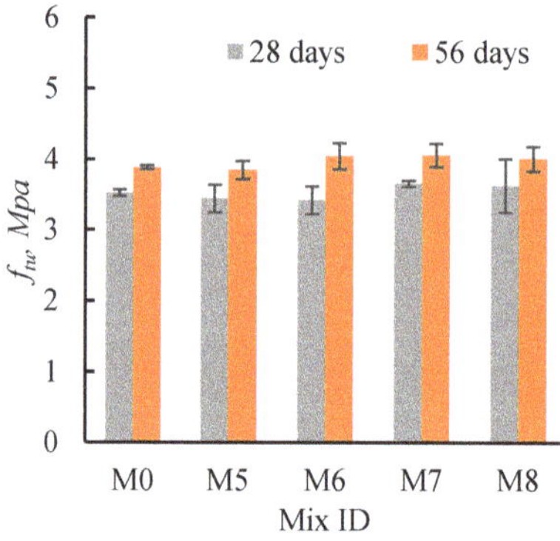

Figure 9. Effect of the CRCA% on the concrete tensile strength.

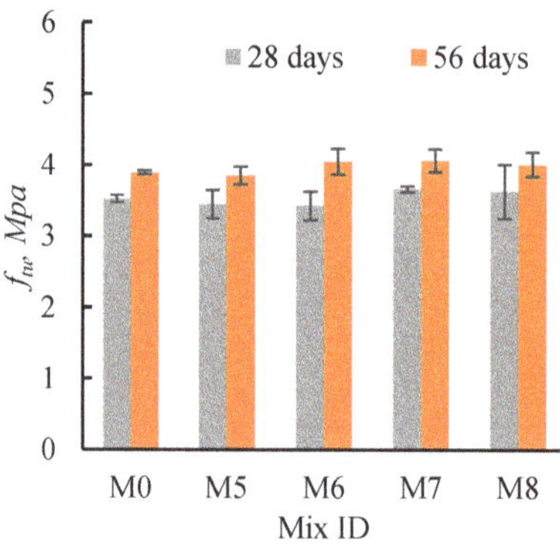

Figure 10. Effect of the FRCA on the concrete tensile strength.

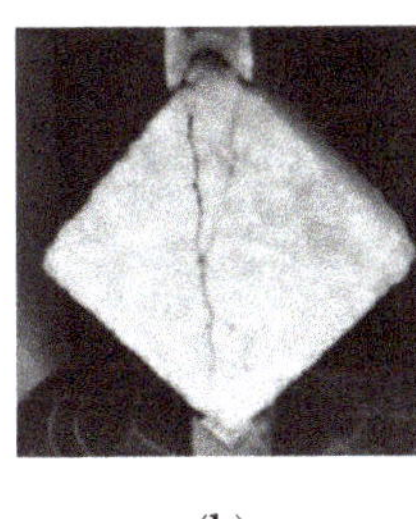

(a) (b) (c) (d)

Figure 11. Tensile diagonal failure for some cubes at 28 and 56 d. (**a**) Mix1 at 56 d; (**b**) Mix4 at 28 d; (**c**) Mix5 at 56 d; (**d**) Mix6 at 28 d.

4.2.1. Effect of CRCA

Concrete tensile strength was slightly affected by CRCA% and curing time. Increasing the Tc from 28 to 56 days increased the tensile strength at all CRCA% (Figure 9). The maximum value of increase in f_{tu} was for M3 (CRCA = 60%, $\mu_{t56/28}$ = 14.24%). In contrast, the tensile strength decreased with the addition of CRCA compared to mix M0, except for M3 at 56 d (μ_{t56} = 1.98%). The diagonal strength reduced by −3.42%, −2.63%, −1.4% and −2.88% after 28 days, while it reduced by −5.02%, −7.60%, 1.98% and −10.9% after 56 days when CRCA was added by 20%, 40%, 60% and 80%, respectively. M4 (CRCA = 80%) experienced the highest reduction in diagonal strength (μ_{t56} = −10.90%, Table 4). Although the voids in the CRCA negatively affected the tensile strength, the amount of RCP attached to the CRCA could help to reduce the effects of the voids as it produced more gel when hydrated. The comparison of $f_{tu,28}/f_{cu,28}$% and $f_{tu,56}/f_{cu,56}$% for the mixtures integrating CRCA with those of M0 ensured the lower effect of CRCA on tensile strength than on compressive strength. The pozzolanic effect of RCP can improve the bond between concrete particles when it contains much basalt powder. In contrast, the CRCA with little basalt powder can interfere with cement hydration by absorbing the hydration water and increasing the voids in the concrete. In addition, the mixes with integrated CRCA had more ITZs than those with NA, which contributed to the detachment of fracture surfaces under tensile loading. In addition, the higher porosity of CRCA concretes resulted in high lateral dilatation compared to concretes with NA, which led to a drastic reduction in compressive strength. The same poor effects of CRCA on compressive strength as on tensile strength were reported in [72,83,84]. Moreover, a tensile strength reduction of about 26–32% was found in [78] when RCA (coarse and fine) was added to replace NA. From the above, the low effect of CRCA makes it a good sustainable replacement material for NA to produce green concrete with low cost and environmental impact with lower tensile strength reduction (μ_{t28} = 3.42% (M1) and μ_{t28} = 10.9% (M4)).

4.2.2. Effect of FRCA

The FRCA was implemented to replace the natural sand in five percentages (0%, 20%, 40%, 60%, and 80%). The concrete diagonal strength was dependent on the FRCA percentage and curing time. Increasing Tc from 28 to 56 d increased the diagonal strength at all FRCA% (Figure 10) with a maximum enhancement value of 19.78% (M6, FRCA = 40%). Consequently, the diagonal strength of the mixtures with integrated FRCA was higher than that of M0 at 28 and 56 days, except for M5 and M6 at 28 days (Figure 9). After 56 days, diagonal strength increased by 0.26%, 5.40%, 5.40%, and 3.09% when FRCA was added by 20%, 40%, 60%, and 80%, respectively. Moreover, the highest reduction in diagonal strength compared to M0 was equal to −2.88% after 28 days (M6, Table 5). The diagonal strength may depend on the availability of RCP attached to the FRCA. In addition, the fine basalt aggregates may act as microfibers that help to suppress the internal macrocracks and increase the diagonal strength values.

Comparing the $f_{tu,28}/f_{cu,28}$% and $f_{tu,56}/f_{cu,56}$% for the mixes containing the FRCA with those of M0, the significant effect of the FRCA on the diagonal strength was greater than that on the compressive strength. The pozzolanic effect of RCP can improve the bond between concrete particles. The failure of the cubes followed the diagonal failure characteristics (Figure 11), as reported in [67]. Finally, the FRCA trivially diminished the f_{cu} values while amplified the f_{tu} values. The previous results and comparisons encouraged the use of recycled aggregates as a substitute material for the NA to produce green concrete.

In contrast to the effect of RA and PA, the incorporation of FRCA into concrete increased f_{tu} values compared to NAC [53].

4.2.3. Effect of the RCP

To emphasize the pozzolanic effect of RCP on concrete properties, the tensile strength of mixes integrating RCP was tested (Figure 12). From Figure 12, the diagonal strength for the mixes integrating RCP increased with increasing curing time regardless of the RCP content. When Tc was increased from 28 to 56 days, the diagonal strength increased by 6.68%, 3.72% and 11.44% when RCP was added by 5%, 10% and 20%, respectively. This could be due to the hydration of the remaining un-hydrated cement and the late hydration of the basalt fines and powder. Consequently, replacing the cement with 5% and 10% RCP increased the 28 days tensile strength by 6.81% and 6.81%, respectively, while the 56 days tensile strength increased by 3.09 and 0.26%, respectively, compared to M0. In contrast, for mixes integrating 20% RCP, the tensile strength at 28 and 56 days decreased by −2.92% and −2.05%, respectively, compared to that of M0. Conversely, the $f_{tu,28}/f_{cu,28}$% and $f_{tu,56}/f_{cu,56}$% increased with the incorporation of RCP compared to M0. The pozzolanic effect of RCP may enhance the bond between concrete particles. The results are in agreement with those in [35] which stated that the effect of RCA can be controlled by careful selection during concrete production. In fact, the loss or increase of tensile strength was dependent on the type, size, quantity, and quality of RCA.

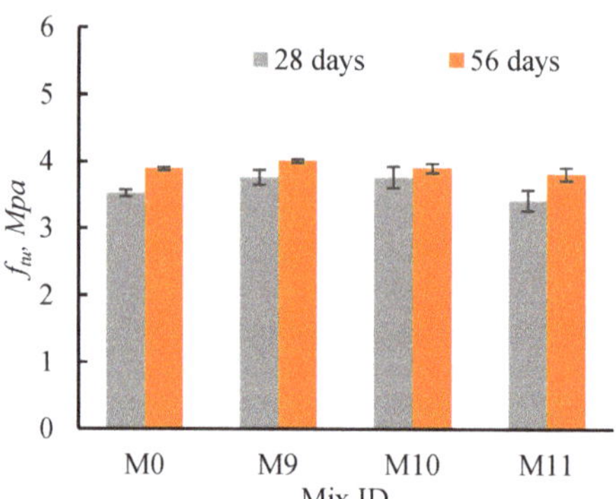

Figure 12. Effect of the RCP% on the concrete tensile strength.

4.3. Water Absorption

To examine the effect of CRCA, FRCA and RCP on concrete durability, the water absorption (WA) was calculated for all concrete mixes after 28 and 56 days (Figure 13). In general, all mixes had lower WA values after 56 days than their WA values after 28 days, except for mixes M2, M7, and M8 (Figure 13). This is because of the excessive hydration of the cement, which increased the C-S-H gel and decreased the voids. In addition, the WA values at the same Tc were dependent on the type and percentages of recycled materials. The WA of mixtures M1–M4 (with integrated CRCA) increased with increasing CRCA content at 28 d (Figure 13a). Similar effects of CRCA content on WA values were observed after 56 days for the same mixtures except for M2. The WA of mixtures M1-M4 was higher than that of mixture M0 (Figure 13a). This may be due to the high void content of

CRCA as it involves crushed mortar. The previous observations were previously reported in [85–87]. Similarly, at 28 and 56 d, mixtures M5–M8 had higher WA values than those of M0 (Figure 13b). Among the mixtures integrating FRCA, both mixtures M6 (FRCA = 40%) and M7 (FRCA = 60%) experienced the lowest WA values at 56 and 28 days, respectively. The WA of the mixes with FRCA may depend on the amount of crushed concrete and the hydration of the fines incorporated in the FRCA. Conversely, cement preplacement with 5% and 10% RCP had a small effect on WA values at 28 and 56 d, while increasing RCP to 20% increased WA (Figure 13c). When RCP replaced cement up to 10%, the effect of RCP on cement hydration was negligible and RCP could store almost the same C-S-H gel as cement, while increasing RCP to 20% may disturb cement hydration and affect the production of C-S-H gel in the mix.

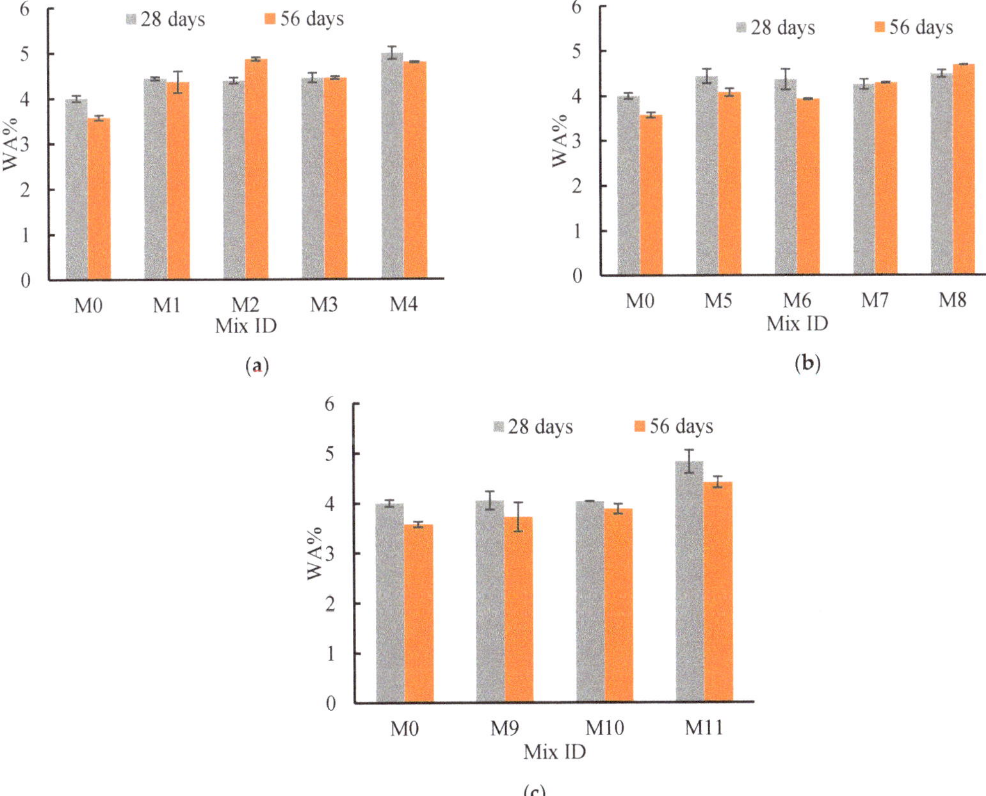

Figure 13. Effect of recycled concrete component on the WA %. (**a**) CRCA; (**b**) FRCA; (**c**) RCP.

5. Theoretical and Experimental Comparisons

The proposed model reported in [62] was used to predict the RAC compressive and tensile strength. The relationship between the theoretical findings and the obtained experimental results is shown in Figure 14. The theoretical compressive and tensile strength were evaluated using Equations (5) and (6), respectively as follows.

$$f'_{c,\ cube} = \frac{28.97 - 4.71 \times r^{1.69}}{\left(w_{eff}/c\right)^{0.63}} \tag{5}$$

$$f_{st} = \frac{2.12 - 0.31 \times r^{0.22}}{\left(w_{eff}/c\right)^{0.63}} \quad (6)$$

The model can predict with good accuracy the compressive and tensile strength of the RAC as shown in Figure 14.

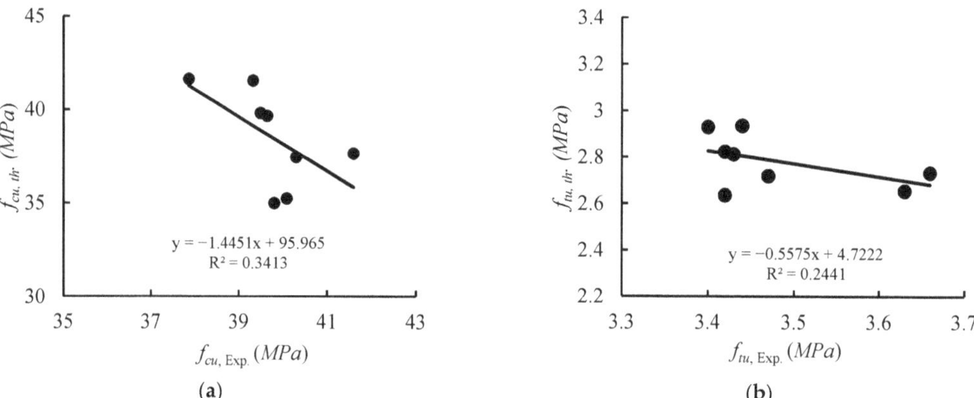

Figure 14. Experimental and theoretical comparison. (**a**) Compressive strength; (**b**) Tensile strength.

6. Conclusions

The strength and durability of green concrete produced with recycled basaltic concrete aggregates and powder were investigated. The effect of RCA, FRNA and RCP as substitutes for the concrete constituents (NA and cement) on the strength and durability of concrete was evaluated experimentally at different test ages. The conclusions obtained are given as follows:

- With increasing curing time, concrete compressive and tensile strengths increased regardless of the concrete components (with or without recycled materials). The concrete strength increased with increasing Tc from 7 to 28 days with a higher percentage than increasing Tc from 28 to 56 days. This may be because of the higher hydration rate of the cement during the first 28 days compared to the later days.
- Partial substitution of recycled aggregate for NA reduced concrete strength with varying percentages depending on the size and source of recycled aggregate, especially after 56 days. After 28 days, the maximum reduction in concrete strength when CRCA or FRCA was incorporated up to 80% was 7.6% and 4.6%, respectively, while the maximum reduction in strength was 13.2% and 9.4% for mixtures with CRCA and FRCA, respectively. After 56 days, the mixes with 40% FRCA achieved almost the same M0 compressive strength (99.5%).
- Increasing RCP from 5 to 10% enhanced the concrete strength at 28 and 56 days due to late hydration of RCP. After 28 days, the concrete strength was about 89% and 99.3% of the strength of M0, while the strength after 56 days was 89.6 and 95.2% of the strength of M0 when the RCP replaced the cement by 5% and 10%, respectively. The maximum strength reduction was 12.2% and 15.2% compared to M0 after 28 and 56 days, respectively, when the RCP = 20%.
- Compared to the M0 mixtures, the tensile strength increased or decreased depending on the proportion and type of recycled material and the curing time. After 56 days, the tensile strength augmented by 1.98%, 5.40% and 17.2% (maximum values) when the cement was substituted by CRCA, FRCA, and RCP, respectively.
- Generally, the values of WA decreased with increasing Tc from 28 to 56 days. The mixes in which coarse RCA was incorporated showed higher WA values than those of

M0. Consequently, WA decreased for mixes integrated FRCA and RCP compared to that of integrated CRCA.

- The obtained results and comparisons encourage the use of recycled aggregates and powders as a substitute material for NA and cement to produce sustainable/green concrete.
- The recycled aggregates could obtain nearly the same mechanical properties of mixes with natural aggregate. Conversely, the use of RCP to replace cement also reduces the cost and CO_2 emission from the cement production process. So, the use of recycled aggregates and powder to replace the natural aggregate and cement, respectively could be considered an efficient method to lower the structures' cost, their impact on the environment, and improves their sustainability.
- The transportation costs of natural aggregate and other construction materials have significant effects on the economic analysis. The use of recycled aggregate in re-building, demolition and retrofitting of structures could be very effective as the source of the old concrete is near to the construction site.

Author Contributions: Conceptualization, I.S. and M.A.; Data curation, I.S., U.I., A.A. and M.A.-H.; Formal analysis, I.S., U.I., M.A., A.A., A.S., M.A.-S., M.A.-H. and A.B.; Investigation, I.S., U.I., M.A., A.S., M.A.-S., M.A.-H. and A.B.; Methodology, I.S., U.I., M.A., A.A., A.S. and A.B.; Project administration, U.I. and A.B.; Resources, I.S., U.I., M.A., A.A., A.S., M.A.-S. and M.A.-H.; Visualization, A.A. and A.B.; Writing—review & editing, M.A., M.A.-S. and M.A.-H. All authors have read and agreed to the published version of the manuscript.

Funding: This study was funded by the Deanship of Scientific Research, Taif University, Saudi Arabia, (Research project number 1-441-108).

Institutional Review Board Statement: Not applicable.

Informed Consent Statement: Not applicable.

Data Availability Statement: Not applicable.

Conflicts of Interest: The authors declare no conflict of interest.

References

1. Xuan, D.; Poon, C.S.; Zheng, W. Management and sustainable utilization of processing wastes from ready-mixed concrete plants in construction: A review. *Resour. Conserv. Recycl.* **2018**, *136*, 238–247. [CrossRef]
2. Martínez-García, R.; de Rojas, M.I.S.; Del Pozo, J.M.M.; Fraile-Fernández, F.J.; Juan-Valdés, A. Evaluation of mechanical characteristics of cement mortar with fine recycled concrete aggregates (FRCA). *Sustainability* **2021**, *13*, 414. [CrossRef]
3. Pacheco-Torgal, F. High tech startup creation for energy efficient built environment. *Renew. Sustain. Energy Rev.* **2017**, *71*, 618–629. [CrossRef]
4. Tam, V.W.Y.; Soomro, M.; Evangelista, A.C.J. A review of recycled aggregate in concrete applications (2000–2017). *Constr. Build. Mater.* **2018**, *172*, 272–292. [CrossRef]
5. Ginga, C.P.; Ongpeng, J.M.C.; Daly, M.K.M. Circular economy on construction and demolition waste: A literature review on material recovery and production. *Materials* **2020**, *13*, 2970. [CrossRef] [PubMed]
6. Yu, B.; Wang, J.; Li, J.; Lu, W.; Li, C.Z.; Xu, X. Quantifying the potential of recycling demolition waste generated from urban renewal: A case study in Shenzhen, China. *J. Clean. Prod.* **2020**, *247*. [CrossRef]
7. Turk, J.; Cotič, Z.; Mladenovič, A.; Šajna, A. Environmental evaluation of green concretes versus conventional concrete by means of LCA. *Waste Manag.* **2015**. [CrossRef]
8. Liew, K.M.; Sojobi, A.O.; Zhang, L.W. Green concrete: Prospects and challenges. *Constr. Build. Mater.* **2017**. [CrossRef]
9. Zhao, Y.; Yu, M.; Xiang, Y.; Kong, F.; Li, L. A sustainability comparison between green concretes and traditional concrete using an emergy ternary diagram. *J. Clean. Prod.* **2020**, *256*, 120421. [CrossRef]
10. Dang, J.; Zhao, J. Influence of waste clay bricks as fine aggregate on the mechanical and microstructural properties of concrete. *Constr. Build. Mater.* **2019**, *228*, 116757. [CrossRef]
11. Karimaei, M.; Dabbaghi, F.; Sadeghi-Nik, A.; Dehestani, M. Mechanical performance of green concrete produced with untreated coal waste aggregates. *Constr. Build. Mater.* **2020**, *233*, 117264. [CrossRef]
12. Çakir, O. Experimental analysis of properties of recycled coarse aggregate (RCA) concrete with mineral additives. *Constr. Build. Mater.* **2014**. [CrossRef]
13. Majhi, R.K.; Nayak, A.N.; Mukharjee, B.B. Development of sustainable concrete using recycled coarse aggregate and ground granulated blast furnace slag. *Constr. Build. Mater.* **2018**. [CrossRef]

14. Majhi, R.K.; Nayak, A.N. Bond, durability and microstructural characteristics of ground granulated blast furnace slag based recycled aggregate concrete. *Constr. Build. Mater.* **2019**. [CrossRef]
15. De Juan, M.S.; Gutiérrez, P.A. Study on the influence of attached mortar content on the properties of recycled concrete aggregate. *Constr. Build. Mater.* **2009**. [CrossRef]
16. Casuccio, M.; Torrijos, M.C.; Giaccio, G.; Zerbino, R. Failure mechanism of recycled aggregate concrete. *Constr. Build. Mater.* **2008**. [CrossRef]
17. Corinaldesi, V. Mechanical and elastic behaviour of concretes made of recycled-concrete coarse aggregates. *Constr. Build. Mater.* **2010**. [CrossRef]
18. Pani, L.; Francesconi, L.; Rombi, J.; Mistretta, F.; Sassu, M.; Stochino, F. Effect of parent concrete on the performance of recycled aggregate concrete. *Sustainability* **2020**, *12*, 9399. [CrossRef]
19. Tabsh, S.W.; Abdelfatah, A.S. Influence of recycled concrete aggregates on strength properties of concrete. *Constr. Build. Mater.* **2009**. [CrossRef]
20. Malešev, M.; Radonjanin, V.; Marinković, S. Recycled concrete as aggregate for structural concrete production. *Sustainability* **2010**, *2*, 1204–1225. [CrossRef]
21. Rao, A.; Jha, K.N.; Misra, S. Use of aggregates from recycled construction and demolition waste in concrete. *Resour. Conserv. Recycl.* **2007**. [CrossRef]
22. Wagih, A.M.; El-Karmoty, H.Z.; Ebid, M.; Okba, S.H. Recycled construction and demolition concrete waste as aggregate for structural concrete. *HBRC J.* **2013**. [CrossRef]
23. Etxeberria, M.; Vázquez, E.; Marí, A.; Barra, M. Influence of amount of recycled coarse aggregates and production process on properties of recycled aggregate concrete. *Cem. Concr. Res.* **2007**. [CrossRef]
24. Zhang, W.; Ingham, J.M. Using Recycled Concrete Aggregates in New Zealand Ready-Mix Concrete Production. *J. Mater. Civ. Eng.* **2010**. [CrossRef]
25. Behera, M.; Bhattacharyya, S.K.; Minocha, A.K.; Deoliya, R.; Maiti, S. Recycled aggregate from C&D waste & its use in concrete—A breakthrough towards sustainability in construction sector: A review. *Constr. Build. Mater.* **2014**. [CrossRef]
26. Yang, H.; Qin, Y.; Liao, Y.; Chen, W. Shear behavior of recycled aggregate concrete after exposure to high temperatures. *Constr. Build. Mater.* **2016**. [CrossRef]
27. Otsuki, N.; Miyazato, S.; Yodsudjai, W. Influence of Recycled Aggregate on Interfacial Transition Zone, Strength, Chloride Penetration and Carbonation of Concrete. *J. Mater. Civ. Eng.* **2003**. [CrossRef]
28. Etxeberria, M.; Vázquez, E.; Marí, A. Microstructure analysis of hardened recycled aggregate concrete. *Mag. Concr. Res.* **2006**. [CrossRef]
29. Etxeberria, M.; Mari, A.R.; Vazquez, E. Recycled aggregate concrete as structural material. *Mater. Struct. Constr.* **2007**, *40*, 529–541. [CrossRef]
30. Li, W.; Long, C.; Tam, V.W.Y.; Poon, C.; Hui, W. Effects of nano-particles on failure process and microstructural properties of recycled aggregate concrete. *Constr. Build. Mater.* **2017**, *142*, 42–50. [CrossRef]
31. Katz, A. Properties of concrete made with recycled aggregate from partially hydrated old concrete. *Cem. Concr. Res.* **2003**, *33*, 703–711. [CrossRef]
32. Li, W.; Luo, Z.; Sun, Z.; Hu, Y.; Duan, W.H. Numerical modelling of plastic–damage response and crack propagation in RAC under uniaxial loading. *Mag. Concr. Res.* **2018**, *70*, 459–472. [CrossRef]
33. Silva, R.V.; de Brito, J.; Dhir, R.K. Establishing a relationship between modulus of elasticity and compressive strength of recycled aggregate concrete. *J. Clean. Prod.* **2016**, *112*, 2171–2186. [CrossRef]
34. Silva, R.V.; De Brito, J.; Dhir, R.K. The influence of the use of recycled aggregates on the compressive strength of concrete: A review. *Eur. J. Environ. Civ. Eng.* **2014**. [CrossRef]
35. Silva, R.V.; De Brito, J.; Dhir, R.K. Tensile strength behaviour of recycled aggregate concrete. *Constr. Build. Mater.* **2015**, *83*, 108–118. [CrossRef]
36. Majhi, R.K.; Nayak, A.N. Properties of Concrete Incorporating Coal Fly Ash and Coal Bottom Ash. *J. Inst. Eng. Ser. A* **2019**. [CrossRef]
37. Satpathy, H.P.; Patel, S.K.; Nayak, A.N. Development of sustainable lightweight concrete using fly ash cenosphere and sintered fly ash aggregate. *Constr. Build. Mater.* **2019**. [CrossRef]
38. Evangelista, L.; de Brito, J. Mechanical behaviour of concrete made with fine recycled concrete aggregates. *Cem. Concr. Compos.* **2007**. [CrossRef]
39. Flower, D.J.M.; Sanjayan, J.G. Green house gas emissions due to concrete manufacture. *Int. J. Life Cycle Assess.* **2007**. [CrossRef]
40. Estanqueiro, B.; Dinis Silvestre, J.; de Brito, J.; Duarte Pinheiro, M. Environmental life cycle assessment of coarse natural and recycled aggregates for concrete. *Eur. J. Environ. Civ. Eng.* **2018**. [CrossRef]
41. Mah, C.M.; Fujiwara, T.; Ho, C.S. Life cycle assessment and life cycle costing toward eco-efficiency concrete waste management in Malaysia. *J. Clean. Prod.* **2018**. [CrossRef]
42. Wijayasundara, M.; Mendis, P.; Crawford, R.H. Net incremental indirect external benefit of manufacturing recycled aggregate concrete. *Waste Manag.* **2018**. [CrossRef] [PubMed]
43. Marinković, S.; Radonjanin, V.; Malešev, M.; Ignjatović, I. Comparative environmental assessment of natural and recycled aggregate concrete. *Waste Manag.* **2010**. [CrossRef] [PubMed]

44. Braga, A. *Comparative Analysis of the Life Cycle Assessment of Conventional and Recycled Aggregate Concrete*; Instituto Superior Tecnico—University of Lisbon: Lisbon, Portugal, 2015. (In Portuguese)
45. Khatib, J.M. Properties of concrete incorporating fine recycled aggregate. *Cem. Concr. Res.* **2005**. [CrossRef]
46. Kou, S.C.; Poon, C.S. Properties of concrete prepared with crushed fine stone, furnace bottom ash and fine recycled aggregate as fine aggregates. *Constr. Build. Mater.* **2009**. [CrossRef]
47. Evangelista, L.; de Brito, J. Durability performance of concrete made with fine recycled concrete aggregates. *Cem. Concr. Compos.* **2010**. [CrossRef]
48. Fan, C.C.; Huang, R.; Hwang, H.; Chao, S.J. Properties of concrete incorporating fine recycled aggregates from crushed concrete wastes. *Constr. Build. Mater.* **2016**. [CrossRef]
49. Dapena, E.; Alaejos, P.; Lobet, A.; Pérez, D. Effect of Recycled Sand Content on Characteristics of Mortars and Concretes. *J. Mater. Civ. Eng.* **2011**. [CrossRef]
50. Zhao, Z.; Remond, S.; Damidot, D.; Xu, W. Influence of fine recycled concrete aggregates on the properties of mortars. *Constr. Build. Mater.* **2015**. [CrossRef]
51. Gholampour, A.; Zheng, J.; Ozbakkaloglu, T. Development of waste-based concretes containing foundry sand, recycled fine aggregate, ground granulated blast furnace slag and fly ash. *Constr. Build. Mater.* **2021**, *267*, 121004. [CrossRef]
52. Tamanna, N.; Tuladhar, R.; Sivakugan, N. Performance of recycled waste glass sand as partial replacement of sand in concrete. *Constr. Build. Mater.* **2020**, *239*, 117804. [CrossRef]
53. Steyn, Z.C.; Babafemi, A.J.; Fataar, H.; Combrinck, R. Concrete containing waste recycled glass, plastic and rubber as sand replacement. *Constr. Build. Mater.* **2021**, *269*, 121242. [CrossRef]
54. Uncik, S.; Kmecova, V. The effect of basalt powder on the properties of cement composites. *Procedia Eng.* **2013**, *65*, 51–56. [CrossRef]
55. Ren, P.; Li, B.; Yu, J.-G.; Ling, T.-C. Utilization of recycled concrete fines and powders to produce alkali-activated slag concrete blocks. *J. Clean. Prod.* **2020**. [CrossRef]
56. Shahrour, N.; Allouzi, R. Shear behavior of captive- and short- column effects using different basalt aggregate contents. *J. Build. Eng.* **2020**, *32*, 101508. [CrossRef]
57. Kurańska, M.; Barczewski, M.; Uram, K.; Lewandowski, K.; Prociak, A.; Michałowski, S. Basalt waste management in the production of highly effective porous polyurethane composites for thermal insulating applications. *Polym. Test.* **2019**, *76*, 90–100. [CrossRef]
58. Ashish, D.K.; Verma, S.K. Determination of optimum mixture design method for self-compacting concrete: Validation of method with experimental results. *Constr. Build. Mater.* **2019**. [CrossRef]
59. Jani, Y.; Hogland, W. Waste glass in the production of cement and concrete—A review. *J. Environ. Chem. Eng.* **2014**. [CrossRef]
60. Zaidi, K.A.; Ram, S.; Gautam, M.K. Utilisation of glass powder in high strength copper slag concrete. *Adv. Concr. Constr.* **2017**. [CrossRef]
61. Chen, G.; Lee, H.; Young, K.L.; Yue, P.L.; Wong, A.; Tao, T.; Choi, K.K. Glass recycling in cement production-an innovative approach. *Waste Manag.* **2002**. [CrossRef]
62. Xu, J.; Zhao, X.; Yu, Y.; Xie, T.; Yang, G.; Xue, J. Parametric sensitivity analysis and modelling of mechanical properties of normal- and high-strength recycled aggregate concrete using grey theory, multiple nonlinear regression and artificial neural networks. *Constr. Build. Mater.* **2019**, *211*, 479–491. [CrossRef]
63. ASTM International. *ASTM C33 Standard Specification for Concrete Aggregates*; ASTM International: West Conshohocken, PA, USA, 2010.
64. Ghorbani, S.; Sharifi, S.; Ghorbani, S.; Tam, V.W.; de Brito, J.; Kurda, R. Effect of crushed concrete waste's maximum size as partial replacement of natural coarse aggregate on the mechanical and durability properties of concrete. *Resour. Conserv. Recycl.* **2019**. [CrossRef]
65. American Concrete Institute. *ACI 211.1 Standard Practice for Selecting Proportions for Normal, Heavyweight, and Mass Concrete (ACI 211.1-91)*; American Concrete Institute: Farmington Hills, MI, USA, 2002.
66. British Standards Institution. *BS EN 12390-3:2009 Testing Hardened Concrete, Part 3: Compressive Strength of Test Specimens*; British Standards Institution: London, UK, 2009.
67. Ince, R. Determination of concrete fracture parameters based on peak-load method with diagonal split-tension cubes. *Eng. Fract. Mech.* **2012**. [CrossRef]
68. Zhang, S.P.; Zong, L. Evaluation of relationship between water absorption and durability of concrete materials. *Adv. Mater. Sci. Eng.* **2014**, *2014*. [CrossRef]
69. Gupta, N.; Gupta, A.; Saxena, K.K.; Shukla, A.; Goyal, S.K. Mechanical and durability properties of geopolymer concrete composite at varying superplasticizer dosage. *Mater. Today Proc.* **2020**. [CrossRef]
70. De Schutter, G.; Audenaert, K. Evaluation of water absorption of concrete as a measure for resistance against carbonation and chloride migration. *Mater. Struct. Constr.* **2004**, *37*, 591–596. [CrossRef]
71. ASTM International. *ASTM C642-13 Standard Test Method for Density, Absorption, and Voids in Hardened Concrete*; ASTM International: West Conshohocken, PA, USA, 2013.
72. Ali, B.; Qureshi, L.A. Durability of recycled aggregate concrete modified with sugarcane molasses. *Constr. Build. Mater.* **2019**. [CrossRef]

73. Ali, B.; Qureshi, L.A. Influence of glass fibers on mechanical and durability performance of concrete with recycled aggregates. *Constr. Build. Mater.* **2019**. [CrossRef]
74. Kou, S.C.; Poon, C.S.; Agrela, F. Comparisons of natural and recycled aggregate concretes prepared with the addition of different mineral admixtures. *Cem. Concr. Compos.* **2011**. [CrossRef]
75. Koushkbaghi, M.; Kazemi, M.J.; Mosavi, H.; Mohseni, E. Acid resistance and durability properties of steel fiber-reinforced concrete incorporating rice husk ash and recycled aggregate. *Constr. Build. Mater.* **2019**. [CrossRef]
76. Xiao, J.; Li, W.; Sun, Z.; Lange, D.A.; Shah, S.P. Properties of interfacial transition zones in recycled aggregate concrete tested by nanoindentation. *Cem. Concr. Compos.* **2013**. [CrossRef]
77. Wang, H.L.; Wang, J.J.; Sun, X.Y.; Jin, W.L. Improving performance of recycled aggregate concrete with superfine pozzolanic powders. *J. Cent. South Univ.* **2013**. [CrossRef]
78. Gorjinia, A.; Shafigh, P.; Moghimi, M.; Bin, H. The role of 0–2 mm fine recycled concrete aggregate on the compressive and splitting tensile strengths of recycled concrete aggregate concrete. *J. Mater.* **2014**, *64*, 345–354. [CrossRef]
79. Rahman, I.A. Assessment of Recycled Aggregate Concrete. *Mod. Appl. Sci.* **2009**, *3*, 47–54.
80. Pedro, D.; de Brito, J.; Evangelista, L. Evaluation of high-performance concrete with recycled aggregates: Use of densified silica fume as cement replacement. *Constr. Build. Mater.* **2017**. [CrossRef]
81. Wu, Z.; Shi, C.; Khayat, K.H. Influence of silica fume content on microstructure development and bond to steel fiber in ultra-high strength cement-based materials (UHSC). *Cem. Concr. Compos.* **2016**. [CrossRef]
82. Karthikeyan, B.; Dhinakaran, G. Influence of ultrafine TiO2 and silica fume on performance of unreinforced and fiber reinforced concrete. *Constr. Build. Mater.* **2018**. [CrossRef]
83. Hossain, F.M.Z.; Shahjalal, M.; Islam, K.; Tiznobaik, M.; Alam, M.S. Mechanical properties of recycled aggregate concrete containing crumb rubber and polypropylene fiber. *Constr. Build. Mater.* **2019**, *225*, 983–996. [CrossRef]
84. Ahmadi, M.; Farzin, S.; Hassani, A.; Motamedi, M. Mechanical properties of the concrete containing recycled fibers and aggregates. *Constr. Build. Mater.* **2017**, *144*, 392–398. [CrossRef]
85. Das, C.S.; Dey, T.; Dandapat, R.; Mukharjee, B.B.; Kumar, J. Performance evaluation of polypropylene fibre reinforced recycled aggregate concrete. *Constr. Build. Mater.* **2018**, *189*, 649–659. [CrossRef]
86. Lima, C.; Caggiano, A.; Faella, C.; Martinelli, E.; Pepe, M.; Realfonzo, R. Physical properties and mechanical behaviour of concrete made with recycled aggregates and fly ash. *Constr. Build. Mater.* **2013**, *47*, 547–559. [CrossRef]
87. Somna, R.; Jaturapitakkul, C.; Amde, A.M. Effect of ground fly ash and ground bagasse ash on the durability of recycled aggregate concrete. *Cem. Concr. Compos.* **2012**, *34*, 848–854. [CrossRef]

Article

Innovative Method for Seismic and Energy Retrofitting of Masonry Buildings

Luca Facconi [1], Sara S. Lucchini [1], Fausto Minelli [1,*], Benedetta Grassi [2], Mariagrazia Pilotelli [2] and Giovanni A. Plizzari [1]

[1] Department of Civil, Environmental, Architectural Engineering and Mathematics, University of Brescia, 25123 Brescia, Italy; luca.facconi@unibs.it (L.F.); sara.lucchini@unibs.it (S.S.L.); giovanni.plizzari@unibs.it (G.A.P.)
[2] Department of Mechanical and Industrial Engineering, University of Brescia, 25123 Brescia, Italy; benedetta.grassi@unibs.it (B.G.); mariagrazia.pilotelli@unibs.it (M.P.)
* Correspondence: fausto.minelli@unibs.it

Abstract: Masonry buildings built in Italy in the 60 s and 70 s of the last century frequently require energy and seismic renovation. To this end, the use of a retrofitting technique based on a multilayer coating may be applied on the building façades in order to improve its seismic and energy performances, leading to the partial or total fulfilment of structural and energy code provisions. The coating consists of a layer of Steel Fiber Reinforced Mortar combined with thermal insulation materials to get a composite package applied on the building façade. After a brief description of the proposed technique, the paper reports the results of seismic and thermal analyses carried out to prove the structural and energy performance of the retrofitting intervention.

Keywords: seismic retrofitting; multilayer coating; Steel Fiber Reinforced Mortar; energy performance of buildings; point thermal bridges; thermal behavior in summer; case study

Citation: Facconi, L.; Lucchini, S.S.; Minelli, F.; Grassi, B.; Pilotelli, M.; Plizzari, G.A. Innovative Method for Seismic and Energy Retrofitting of Masonry Buildings. *Sustainability* **2021**, *13*, 6350. https://doi.org/10.3390/su13116350

Academic Editor: Elena Mele

Received: 6 May 2021
Accepted: 27 May 2021
Published: 3 June 2021

1. Introduction

According to recent statistics [1], at least 30% of the current European building stock consists of buildings that are more than 50 years old. Residential buildings made with unreinforced masonry represent one of the most widespread construction typologies. Moreover, about 50% of the masonry structures built during the period 1961–1990 are located in the seismic prone areas of Southern Europe, which include, among the others, Italy, the Balkan Peninsula and Greece. Existing buildings have low energy performances that do not comply with some recent requirements, such as the Italian Ministerial Decree (DM) on 26 June 2015 [2]. The latter implements the European directive 2010/31/EU on the energy performance of buildings [3]. It is estimated that existing buildings are responsible for 36% of energy-related CO_2 emissions and that 40% of the primary energy consumption results from the building sector [4]. Besides the poor energy performance, existing buildings are frequently affected by a high seismic vulnerability due to the lack of seismic design codes at the time of their construction. Thus, upgrading existing masonry buildings has become an important target in order to reduce the potential economic losses and the human casualties resulting from seismic events.

When an earthquake hits a masonry structure, different failure mechanisms may occur, depending on the material and structural characteristics of the building. Out-of-plane collapses are the most frequent mechanisms in structures where there is not enough continuity between orthogonal bearing walls and floors are not adequately anchored to them. Therefore, any retrofitting or repairing intervention must firstly aim at eliminating all the vulnerabilities related to possible out-of-plane modes able to cause the failure of both façades and floor structures. In addition, buildings may experience in-plane failure modes due to the not sufficient shear resistance of masonry members. By focusing on in-plane mechanisms, different retrofitting strategies have been proposed over the last decades to improve the seismic behavior of masonry structures. Among other techniques such as FRP, textile, post-tensioning, etc., whose main features are well described and discussed elsewhere [5,6], the in-plane capacity and ductility of the walls can be improved by applying concrete or mortar coating on either one or both sides of masonry. Different research studies showed that a significant increase of the axial and lateral capacity could be obtained by spraying a cement-based mixture onto the surface of masonry in order to get an overlay with a thickness of about 40–50 mm. To control cracking, minimum reinforcement is provided to each shotcrete layer [7] by fixing a small diameter steel welded mesh on the wall surface. As an alternative to shotcreting, the masonry surface can be coated by a high-strength cement mortar layer reinforced with multiple layers of steel meshes. This technique, which is usually known as ferrocement, provides an improvement of the inelastic response of masonry that is strongly dependent on the mechanical and geometrical properties of the reinforcing mesh embedded in the mortar layer. An alternative coating-based technology developed in the last few years consists of applying a 20–50 mm thick layer of mortar reinforced with steel fibers randomly spread within the mortar matrix. The use of Steel Fiber Reinforced Mortar (SFRM) coating allows increasing the initial lateral stiffness and capacity of masonry walls [8]. Moreover, if relatively high (e.g., fiber volume fraction = 0.4%–0.6%) contents of steel fibers are used, the high tensile post-cracking strength and ductility of SFRM leads to a stable response of the retrofitted structure in the post-peak stage [8].

Energy improvement of masonry buildings can be partly attained by acting on the building envelope with the aim of increasing its thermal insulation. For this purpose, different kinds of insulation materials and technologies can be applied onto the masonry façades. In particular, the existing windows and doors can be replaced with more efficient ones. Thermal coatings are frequently used to enhance the energy performance of vertical walls. They consist of an insulating layer, which is preferably applied to the outer surface of the wall; this solution is more efficient than internal insulation because it leads to significantly lower thermal bridges and it preserves the existing thermal mass, thus optimizing the thermal inertia, which is a relevant issue in temperate climates.

In 2018, the European directive 2018/844/EU [9] was published, which amends the 2010/31/EU Directive on the energy performance of buildings and the 2012/27/EU Directive on energy efficiency [10]. Its aim is to speed up the deployment of energy-efficiency measures by integrating various aspects, focusing the attention on buildings with higher comfort levels and wellbeing for their occupants, according to the 2009 World Health Organization guidelines [11]. In Italy, it has been recently implemented with the Decree Presidency of the Italian Republic (DPR) n. 48 [12]. Among the innovations introduced by this regulation, only the request to check also the seismic risk when carrying out an energy requalification intervention is here highlighted.

Different research studies have recently shown that seismic and energy retrofitting can be successfully addressed by combining surface reinforcements, such as TRM, CFRP/GFRP or FRCM, with thermal insulation materials [13–16]. The present paper investigates a similar retrofitting strategy based on the combination of SFRM coating, which is able to provide the structure with improved seismic performance, with a thermal insulation layer applied only on its outer surface. The work has been subdivided into two main sections: the former focuses on the analysis of the behavior of the proposed technique based either on experimental tests or on numerical simulations. The latter part, which is more practically oriented, reports the analysis of an actual existing masonry building with the purpose of exploring the potential improvements resulting from retrofitting. To this end, pushover analyses have been performed to assess the seismic safety requirements recommended by the European and Italian structural codes. Finally, to estimate the sole impact of the modified external walls on the building thermal behavior, simulations with the EnergyPlus^{TM} [17] software before and after the application of the new materials have been also carried out.

2. Properties of the Multilayer Retrofitting System

The proposed retrofitting methodology combines two systems that are designed, respectively, to get the maximum performances from both a structural and thermal point of view. As shown in Figure 1, an SFRM coating is first applied on the outer surface of the building in order to improve its seismic behavior. A thermal insulating layer is then connected to the SFRM coating by nylon anchors. Finally, depending on the architectural requirements, a finishing plaster is laid on the insulating layer for decorative purposes as well as for protecting the substrate from moisture and water penetration. Note that the generic dimensions t_m, t_{coat} and t_{th} depicted in Figure 1 represent, respectively, the thickness of the masonry wall, the SFRM coating and the thermal insulating layer.

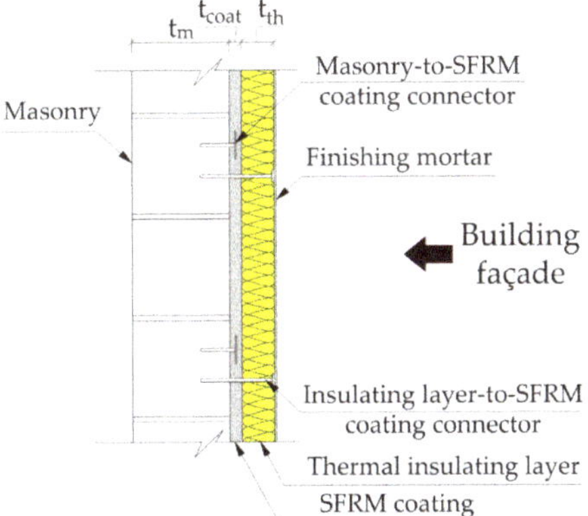

Figure 1. A schematic of a typical masonry wall cross-section retrofitted by the proposed multi-layer system.

2.1. SFRM Coating Application

The proposed technique is designed to be applied only to the outer surface of the building. Such a choice is made to minimize the occupant discomforts typical of the retrofitting interventions involving the bearing walls located inside the building. Note that a single layer of SFRM laid only on the façades may not be always enough to fulfill the high seismic demand that characterizes tall masonry buildings (i.e., buildings with more than 3 stories) or buildings located on high seismicity sites ($a_g > 0.25$ g). The latter cases may require additional strengthening interventions, including the inner surfaces of the perimeters walls as well as the internal bearing wall, when necessary.

As deeply discussed in previous works [18], the structural effectiveness of the SFRM coating is ensured provided that the following application procedure is fulfilled:

1. Existing plaster must be removed from the wall surface to improve the adhesion of the strengthening overlay to the masonry surface. If connection deficiencies between walls are observed, proper interventions (e.g., the installation of steel ties, local dismantling and reconstruction of masonry) must be undertaken to provide continuity at wall intersections.
2. As the wall surface is kept moist, a thin (about 5 mm thick) layer of mortar without steel fibers is applied on the wall to prepare a primer-bonding substratum. Masonry-to-coating connectors (i.e., about 6 connectors/m^2, according to a previous study [18]) are then installed with a center-to-center spacing of about 400–500 mm. The properties of connectors must be chosen depending on the masonry typology. If masonry is made with solid clay bricks of good quality, the connector may consist of a steel self-tapping screw inserted in a pilot hole drilled in the bricks. On the contrary, when hollow units form the masonry texture, the steel screw can be anchored to the units by a nylon plug able to expand within the block holes. For usual applications, steel

screws with a diameter of 6–8 mm can be generally adopted. The screw must be provided with a steel anchor plate that has the minimum dimensions of 50 × 50 × 1.8 (thickness) mm^3 that is placed in the middle of the coating thickness.

3. After installing the masonry-to-coating connectors, different layers of SFRM are applied to the wall until the required total thickness is obtained. Once the application procedure is completed, the wall surface is wet cured for about 3 days to mitigate shrinkage cracking.

2.2. Application of the Thermal Insulating Layer

The application of the thermal coating on top of the SFRM coating is performed by means of connecting elements. The thermal coating is fixed by nylon clips with a diameter of 10 mm applied approximately every 300 mm: the length through the wall must be sufficient to cross both the thermal coating and the SFRM coating. In this way, the thermal coating is fixed to the existing masonry wall as well. Finally, a layer of finishing plaster is applied. It is worth noting that the connecting elements act as point thermal bridges.

3. Seismic Performance of SFRM Coating and Validation of the Numerical Model

The assessment of the seismic behavior of buildings retrofitted with SFRM coating requires the adoption of numerical analysis tools able to take into account the fracture behavior of both SFRM and masonry. The present section reports the results of the nonlinear Finite Element (FE) analyses carried out to simulate the response of a test building before and after retrofitting. The aim of the simulations is to validate the numerical model used to perform the case study of Section 5 and highlight the structural improvements resulting from the retrofitting intervention. Note that the test building considered in the simulation is part of an experimental work described elsewhere [18,19]. Here, for the sake of brevity, a summary of the main properties of the specimen will be reported.

3.1. Specimen Description and Material Properties

Figure 2 reports the details of the full-scale building tested at the University of Brescia (Brescia, Italy) [18] and analyzed in this section. As one may note, the structure consists of four 200 mm thick Un-Reinforced Masonry (URM) walls supporting 2 composite wood–concrete floors. The URM is made with 250 (length) × 190 (height) × 200 (width) mm hollow clay blocks (void area: 60%) with vertical holes laid in a running bond pattern. Both the head and bed joints are filled with a 10 mm thick layer of a ready-mix cement-based mortar with a compressive strength of 8.9 MPa. The uniaxial compression tests performed on masonry wallets provided a mean compressive strength parallel and perpendicular to the holes equal to 2.93 MPa (CoV = 28.8%) and 0.59 MPa (CoV = 22.1%), respectively. A reinforced concrete (RC) chord running along the whole perimeter of the building connects the floors with the URM walls. Both the windows and the door are provided with RC-masonry lintels having the same thickness of the wall and a depth of 120 mm. A flexible wooden pitched roof is placed at the top of the building, over the second story floor. The 2 floors and the roof have, respectively, a self-weight of 1.82 kN/m^2 and 0.61 kN/m^2. A total overload of 2.6 kN/m^2 is applied only on the first floor.

The specimen was subjected to two horizontal cyclic forces acting at floor levels on both short sides of the building. The experimental test was subdivided into two different stages. During the first stage, the unstrengthened building was subjected to a series of loading cycles with increasing lateral drift up to the attainment of the maximum lateral capacity. Then, the damaged building was repaired by a 30 mm thick (t_{coat}) SFRM coating applied according to the procedure described in the previous section and detailed in Figure 2. Finally, the repaired building was re-tested under cyclic loading until the post-peak base shear was reduced to 70% of the maximum lateral resistance.

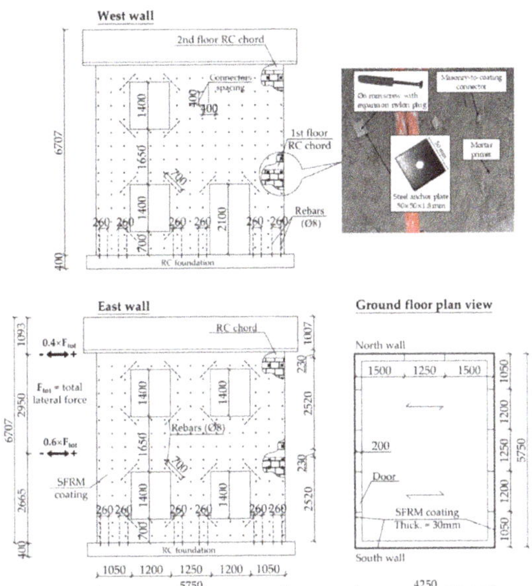

Figure 2. Details of the adopted seismic retrofitting technique applied to a full-scale test building [18,19] (dimensions in mm).

The SFRM was obtained by combining a commercial ready-mix cement-based mortar with 60 kg/m^3 (0.76% by volume) of double hooked-end steel fibers having a length of 32 mm, a diameter (Ø) of 0.4 mm and a tensile strength higher than 2100 MPa. Uniaxial compression tests on 40 mm cubes resulted in a mean compressive strength of SFRM equal to 35.1 MPa (CoV = 15.6%). To assess the fracture behavior of the SFRM, Three-Point Bending Tests (3PBTs) on notched beams having dimensions 150 (height) × 600 (length) × 150 (thickness) mm (notch depth = 25 mm) were carried out according to EN 14651 [20] (Figure 3a). The flexural stress (f)-CMOD (Crack Mouth Opening Displacement) curves obtained from the 3PBTs are shown in Figure 3b together with the strength at the limit of proportionality (f_L) and the mean residual strengths f_{R1}, f_{R2}, f_{R3} and f_{R4} corresponding, respectively, to CMOD values of 0.5 mm, 1.5 mm, 2.5 mm and 3.5 mm.

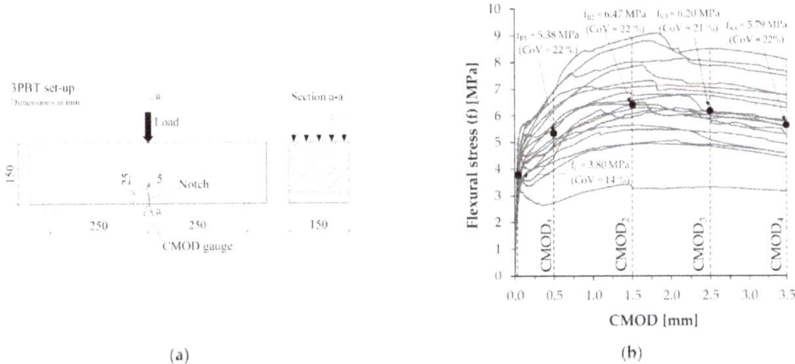

Figure 3. Three-Point Bending Tests on SFRM beams: (**a**) the test set-up; (**b**) experimental nominal flexural stress–CMOD curves.

3.2. Numerical Analysis and Comparison Against Experimental Results

The behavior of the full-scale building was simulated by 3D Non-Linear Finite Element Analyses (NLFEAs) implemented in the commercial program Midas FEA [21].

The 3D FE model of the building is depicted in Figure 4a. Both the masonry walls and the RC chords were modelled with 8-node hexahedral brick elements, whereas the SFRM coating was represented by 4-node quadrilateral plane stress elements. Despite the high computational cost, the solid elements allow the consideration of the whole 3D behavior of masonry by including the effect of the stress and displacement fields acting transversally to the wall plane. The latter may not be generally neglected, especially when considering multilayer members such as the walls retrofitted by the SFRM coating used in this study. As no de-bonding of coating from the masonry surface was generally observed in previous tests [8], the SFRM layer was considered as perfectly bonded to the masonry surface. The floors were simulated by 4-node quadrilateral plate elements. As the pitch roof used in the test had high in-plane flexibility, its interaction with the masonry walls resulted totally negligible.

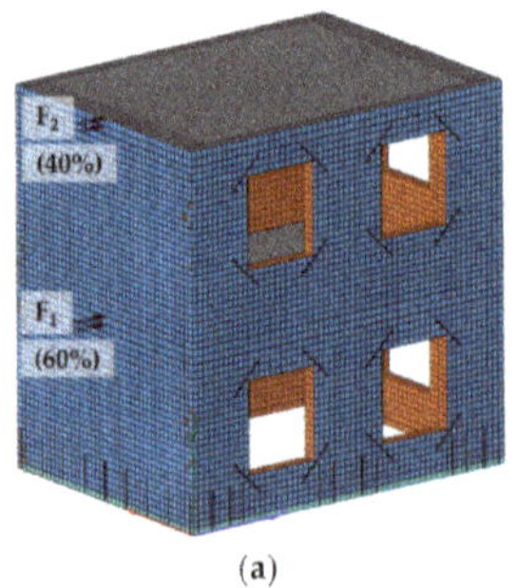

(a)

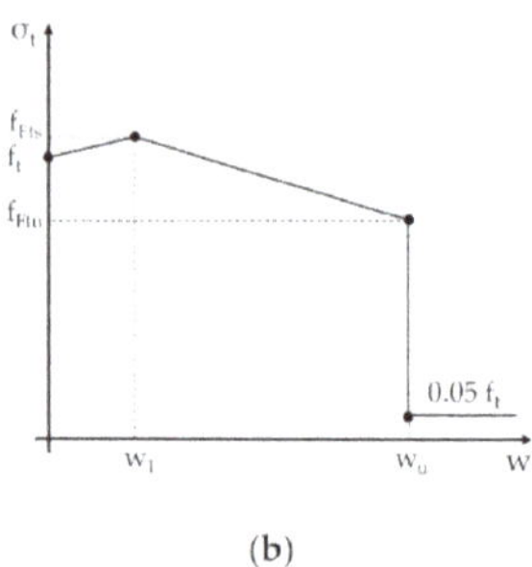

(b)

Figure 4. The numerical model of the test building: (**a**) mesh of repaired building; (**b**) post-cracking uniaxial tensile stress–crack width curve of SFRM.

A smeared crack model based on the "total strain rotating" approach [22] was used to simulate the fracture behavior of masonry and SFRM. About masonry, a uniaxial stress–strain parabolic law and Hordijk's model [23] were implemented to represent the compressive and the tensile behavior, respectively. The parabolic law depends on the compressive strength (f_c) and the compressive fracture energy (G_c), according to Feenstra's model [24]. Hordijk's stress-crack width law is fully defined by the tensile strength (f_t) and the mode-I fracture energy ($G^I{}_f$).

The compressive behavior of the SFRM was described by the Thorenfeldt model [25]. On the contrary, a linear curve with a constant slope equal to the elastic modulus (E) governed the behavior of SFRM in tension before cracking. After cracking, the uniaxial tensile stress (f_t)-crack width (w) law of Figure 4b was implemented. The latter was defined according to Model Code 2010 (MC2010) [26], which recommends the use of a linear law interpolating the following residual strengths:

$$f_{Fts} = 0.45 \cdot f_{R1}, \tag{1}$$

$$f_{Ftu} = f_{Fts} - w_u \cdot (f_{Fts} - 0.5 \cdot f_{R3} + 0.2 \cdot f_{R1}) / CMOD_3, \tag{2}$$

where f_{R1} and f_{R3} are the residual strengths provided by 3PBTs; w_1 = 0.5 mm; w_u = min ($l_{cs} \cdot 2\%$, 2.5 mm) is the ultimate crack width (in mm); l_{cs} is the characteristic length; $CMOD_3$ = 2.5 mm. As suggested by Rots [22], l_{cs} is equal to $V_{el}{}^{1/3}$ (V_{el} = element volume) for brick elements and $A_{el}{}^{1/2}$ (A_{el} = element area) for quadrilateral elements. The constitutive law reported in Figure 4b can be implemented in the "total strain rotating" model provided that the crack widths w_1 and w_u are divided by l_{cs} to get the corresponding uniaxial tensile strains. A summary of the mechanical parameters considered in the simulations is reported in Table 1.

Table 1. Material properties considered in the FE model of the test building.

Material	γ [kg/m^3]	E [MPa]	f_c [MPa]	G_c [N/mm]	f_t [MPa]	G^I_f [N/mm]	f_{Fts} [MPa]	f_{Ftu} [MPa]	w_1 [mm]	w_u [mm]
Masonry	745	2100	2.3	3	0.1	0.01	-	-	-	-
SFRM	2040	21,000	29	-	2.3	-	2.4	2.1	0.5	2.0

A Von Mises yield criterion including strain hardening was adopted for steel rebars anchoring the SFRM coating to the foundation of the building. The yield strength, the ultimate strength and the corresponding strain were assumed equal to 510 MPa, 610 MPa and 10%, respectively.

The analyses were performed by monotonically increasing the lateral load and by controlling the post-peak degrading response with the arc-length technique. Figure 5 reports the total base shear (V_b) against the lateral deflection (d_2) detected at the second-floor level. The grey curves represent the envelope of the experimental hysteretic responses exhibited by the masonry building before (curve MB-exp) and after repairing (curve MBR-exp). The black curves (MB-num, MBR-num) show the response predicted by the non-linear analyses. Table 2 summarizes the main analysis results related to both the positive (+) and the negative (−) loading direction, including the initial secant stiffness (K_s), the peak base shear (V_{peak}) and the corresponding relative increments ΔK and ΔV_{peak}, which referred to the results of the test on the unstrengthened building (MB-exp).

The experimental results show the significant increment of the lateral resistance and stiffness observed after repairing. As highlighted in Table 2, the capacity of the repaired structure was about 230% higher than that of the unstrengthened specimen. The initial lateral stiffness was also significantly improved as the relative increment ΔK ranged from 126% to 141%. Moreover, the final crack pattern observed on the East façade (Figure 6a) proves that the post-peak response of the repaired building was governed by the shear and flexural-shear mechanisms involving the first-story piers. The high number of small cracks (mean crack width <0.8 mm) spread over the surface of the coating is related to the internal stress redistribution promoted by the high tensile toughness of SFRM.

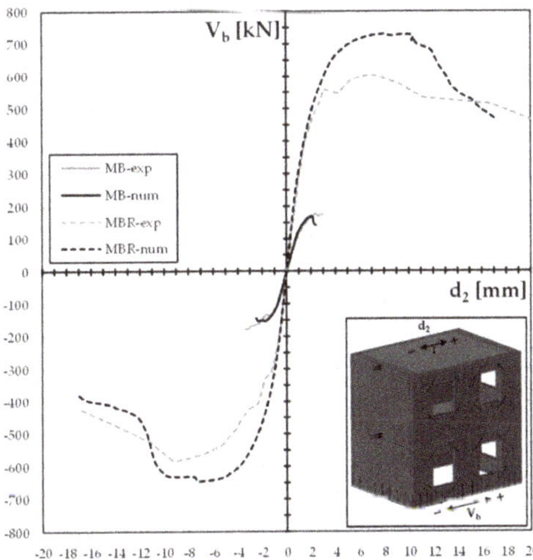

Figure 5. Base shear (V_b) vs. second-floor displacement (d_2) curves: numerical vs. experimental results.

Table 2. Main experimental and numerical results concerning the behavior of the test building.

	K^{+}_{S} [kN/mm]	ΔK^{+}_{S} [%]	K^{-}_{S} [kN/mm]	ΔK^{-}_{S} [%]	V^{+}_{peak} [kN]	ΔV^{+}_{peak} [%]	V^{-}_{peak} [kN]	ΔV^{-}_{peak} [%]
MB-exp	125	-	132	-	180	-	179	-
MB-num	132	+6	138	+5	169	−6	154	−14
MBR-exp	283	+126	318	+141	605	+236	584	+226
MBR-num	334	+167	337	+155	730	+306	645	+260

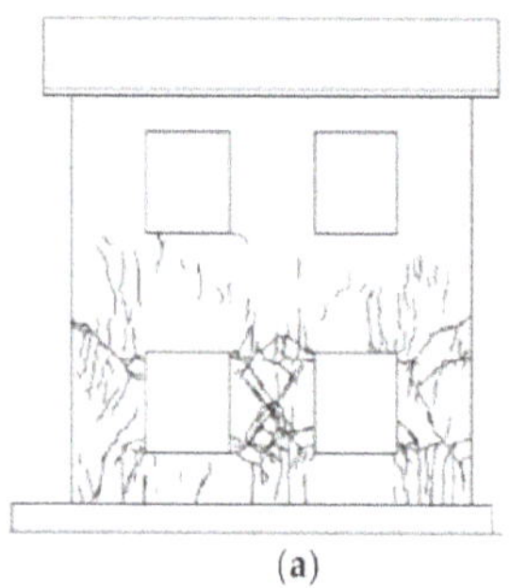

(a)

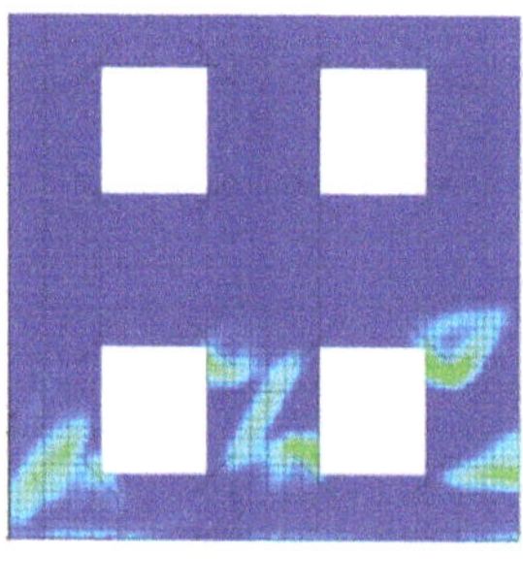

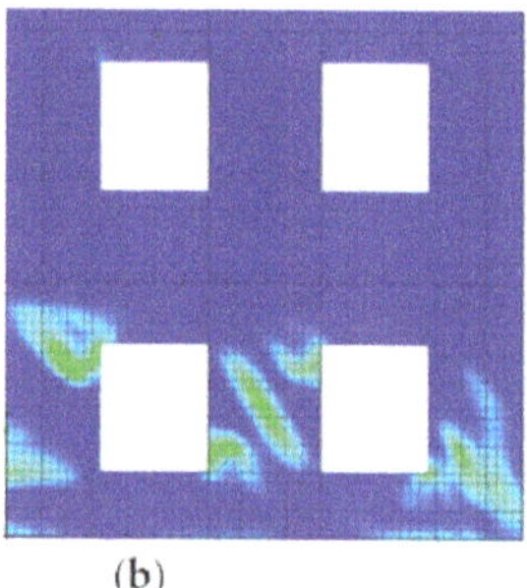

(b)

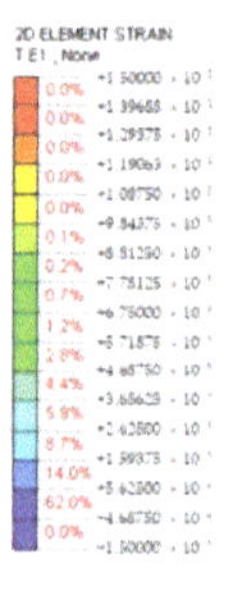

Figure 6. (**a**) Final crack pattern observed on the East façade of the repaired building. (**b**) Numerical contours of principal tensile post-cracking strains acting on the East façade of the repaired building at d_2 = 9.4 mm.

The NLFEAs were able to capture all the main failure mechanisms exhibited by the test building before and after retrofitting. The simulations accurately predicted both the lateral capacity and the initial stiffness of the unstrengthened specimen (see Table 2). On the contrary, both the initial stiffness and the capacity were over-predicted of a quantity ranging from 6% to 21% as compared to those observed in the experimental test MBR-exp. Such an overestimation is due to the partial inability of the MC2010′s constitutive law (Figure 4b) to accurately simulate the post-cracking response of SFRM. As discussed in previous work [19], a better prediction of the repaired building response can be obtained by implementing the tensile constitutive law resulting from the inverse analysis of the 3PBTs (Section 3.1; Figure 3b) results. The contours of the principal tensile strains, plotted in Figure 6b, well agree with the final crack pattern depicted in Figure 6a. It appears that the central pier exhibited a diagonal shear failure, whereas the two slender piers located on both sides of the wall presented flexural mechanisms combined with diagonal cracks starting from the corners of the openings. In conclusion, the adopted finite element model was effective in predicting the response of the test building in spite of the slight overestimation of the lateral capacity.

4. Thermal Characterization of Coating

The typical characteristics of buildings built in Italy in the 60 s and 70 s are collected in the UNI/TR 11552 technical report [27]. In a previous research activity on thermal and acoustic properties of insulating materials [28], three packages were proposed as thermal coatings, which are here analyzed as possible thermal complements to the SFRM layer. The packages had originally been chosen to try to use natural and sustainable materials as much as possible while meeting the criteria for the energy retrofit of the opaque walls in relation to the different Italian climate zones (from A to F with increasing degree-days). As stated by the DM 26 June 2015 [2], when performing energy upgrading interventions, the resulting thermal transmittance U of the opaque vertical envelope components must fulfill the following limit values: 0.40 W/(m^2 K) for climate zones A and B, 0.36 W/(m^2 K) for climate zone C, 0.32 W/(m^2 K) for climate zone D, 0.28 W/(m^2 K) for climate zone E and 0.26 W/(m^2 K) for climate zone F. Two of the proposed packages are suitable for climatic zones A to E, allowing them to meet the limits of thermal transmittance for all construction types of the period of interest. Tailored solutions may be necessary in some cases only for zone F. The first package adopted an innovative solution based on the use of a 50 mm thick (t_{th}) layer of a commercial panel made of needled fiberglass and silica aerogel. The second package (t_{th} = 120 mm) was composed of two superimposed layers made with "light" and "heavy" wood fiber, respectively. The "light" wood fiber layer, having a thickness of 80 mm, formed the inner part of the package laid on the SFRM coating surface. To improve the summer performance, the 40 mm thick "heavy" wood fiber layer was applied on the outer side of the "light" layer. The third package consisted of an 80 mm thick (t_{th}) wood fiber layer that met the thermal transmittance requirements for all construction types of the period of interest only for climatic zones A–C. The thermal properties of the insulating materials are collected in Table 3.

Table 3. Thermal properties of the insulating materials.

	γ [kg/m^3]	c [J/ (kg K)]	k [W/ (m K)]
Aerogel	180	1030	0.016
"Light" wood fiber	110	2100	0.038
"Heavy" wood fiber	265	2100	0.048
Wood fiber	140	2100	0.040

The three packages were designed so that they can also improve summer comfort, reducing and postponing the thermal peak of the central hours of the day. In [28], both transmittance, according to ISO 6946 standard [29], and periodic thermal transmittance, according to ISO 13786 [30], were evaluated on a bare wall made of 200 mm thick hollow clay blocks. As discussed in Section 2.2, the thermal coating was connected to the wall by nylon clips. These connectors act as point thermal bridges. While linear thermal bridges must be included in the calculations (their influence is shown, for example, in [31,32]), point thermal bridges are usually neglected because their influence is minimal, as shown for example in [33]. However, since, in this case, the connections are numerous, their influence was investigated for the three packages installed on the 200 mm thick masonry wall retrofitted by a 30 mm thick SFRM layer. The three analyzed multilayer packages are therefore:

- Bare wall + SFRM + N.1: hollow blocks (t_m = 200 mm), SFRM coating (t_{coat} = 30 mm), aerogel layer (t_{th} = 50 mm).
- Bare wall + SFRM + N.2: hollow blocks (t_m = 200 mm), SFRM coating (t_{coat} = 30 mm), thermal layer (t_{th} = 120 mm) formed by "light" wood fiber (80 mm) combined with "heavy" wood fiber (40 mm).
- Bare wall + SFRM + N.3: hollow blocks (t_m = 200 mm), SFRM coating (t_{coat} = 30 mm), wood fiber layer (t_{th} = 80 mm).

By means of the software COMSOL Multiphysics [34], conduction was simulated through three solid calculation domains that represented the behaviour of a wall panel having total dimensions of 1080 × 1200 mm^2. Because two planes of symmetry characterized the whole panel, appropriate boundary conditions were adopted so that only one-quarter of the whole panel (i.e., 540 mm width and 600 mm height) required discretization. The masonry-to-SFRM coating connectors (see Section 3) were placed with a spacing of 360 mm in the horizontal direction and 400 mm in the vertical direction. The nylon connectors for the thermal coating were positioned with a horizontal and vertical spacing of 270 mm and 300 mm, respectively. A typical analysis domain is shown in the schematic of Figure 7.

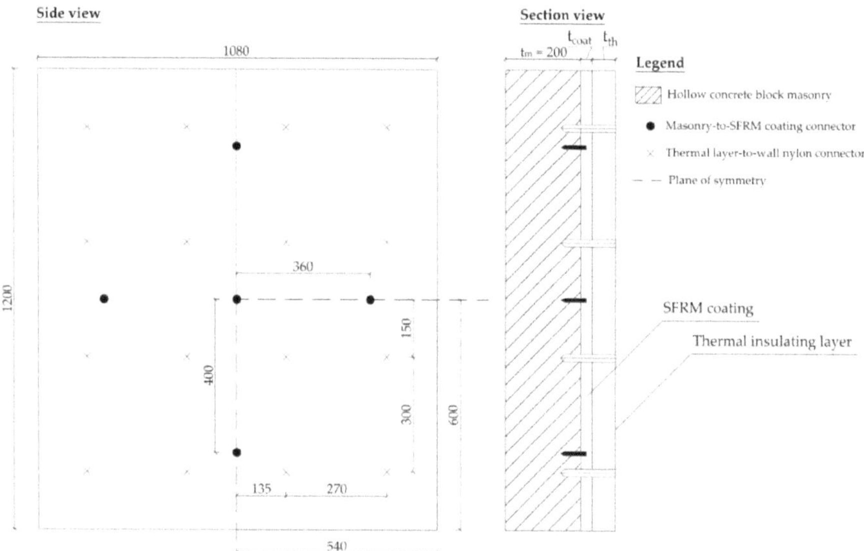

Figure 7. Schematic of the calculation domain for the multilayer wall (dimensions in mm). The dashed lines represent the quarter model implemented in the simulations.

Convective boundary conditions were imposed on the internal and external surfaces of the multilayer wall, while the lateral surfaces were considered adiabatic. The internal and external heat transfer coefficients were chosen in accordance with the ISO 6946 [29] standard. The grid was built with elements with a maximum size of 20 mm and a minimum size of 1.5 mm. Where possible, the grid was obtained by extrusion in the direction normal to the wall, taking care to place at least three rows of cells in each layer.

For each multilayer package:

- a steady simulation imposing a difference of temperature ($T_i - T_e$) of 20 K was carried out to determine the total resistance R_{tot} and the transmittance U of the wall;
- an unsteady simulation was carried out to determine the periodic thermal transmittance Y_{ie}, imposing:

$$T_e = T_m + T_a \sin(2\pi t/P)$$
$$T_i = T_m,$$

where T_m is 300 K; T_a is 100 K; P is 86,400 s.

In order to allow a stabilized periodic regime, ten consecutive days were simulated. Instead of the thermal conductance L_{ie} introduced by the standard ISO 13786 [30] for inhomogeneous walls, the periodic thermal transmittances Y_{ie} were determined by using the value of the average specific heat flow on the internal wall; therefore, the result can be seen as mean equivalent value.

In order to make sure that the differences between analytical and numerical values are not due to the numerical resolution, preliminary simulations were carried out for the case of walls without connection elements, showing a perfect match between simulated and analytical values. By comparing the homogeneous (analytically determined) values with numerical simulations of the wall with connectors, summarized in Table 4, it can be seen that point thermal bridges have a negligible influence, despite their large number. This result justifies the assumption to neglect point thermal bridges in the thermal calculations.

Table 4. Thermal transmittance and periodic thermal transmittance of the analyzed multilayer partitions; analytical values (an): homogeneous layers; numerical values (num): walls with connectors.

Package	t_{tot} * [mm]	U (an) [W/m^{2}K]	Y_{ie} (an) [W/m^{2}K]	U (num) [W/m^{2}K]	Y_{ie} (num) [W/m^{2}K]
Bare wall	200	1.038	0.643	—	—
Bare wall + SFRM + N.1	280	0.242	0.038	0.246	0.030
Bare wall + SFRM + N.2	350	0.254	0.029	0.256	0.024
Bare wall + SFRM + N.3	310	0.335	0.042	0.336	0.042

* $t_{tot} = t_m + t_{coat} + t_{th}$ is the total thickness.

Regarding the periodic thermal transmittance Y_{ie}, it can be noted that package N.2 represents the best solution, as was expected given the presence of "heavy" wood fiber on the outer side of the wall. However, all three packages provide good summer performance. The values of Y_{ie} can, in fact, be compared with the limit value of 0.10 W/(m^2 K) required by the DM 26 June 2015 [2] in case of major renovations for sites where the value of the average monthly irradiance on the horizontal plane during the month of highest summer insolation is greater than or equal to 290 W/m^2: In this case, it is required that the value of the mass per unit area M_s is greater than 230 kg/m^2, or that the value of the periodic thermal transmittance Y_{ie} is lower than 0.10 W/(m^2 K) for all the opaque vertical walls excluding those included in the northwest/north/northeast quadrant.

5. Case Study

The case study reported in the present section refers to the retrofitting intervention performed on an existing residential masonry building constructed in the middle of the 1960s. As shown in Figure 8, the building is a two-story house having plan dimensions of 11.4 × 10.0 m^2 and a total height of about 8.2 m. The bearing structure is made of 330 mm thick concrete hollow unit masonry walls constructed to support the ribbed one-way concrete slabs forming the two floors and the pitched roof. The building is located in L'Aquila, a city situated in central Italy.

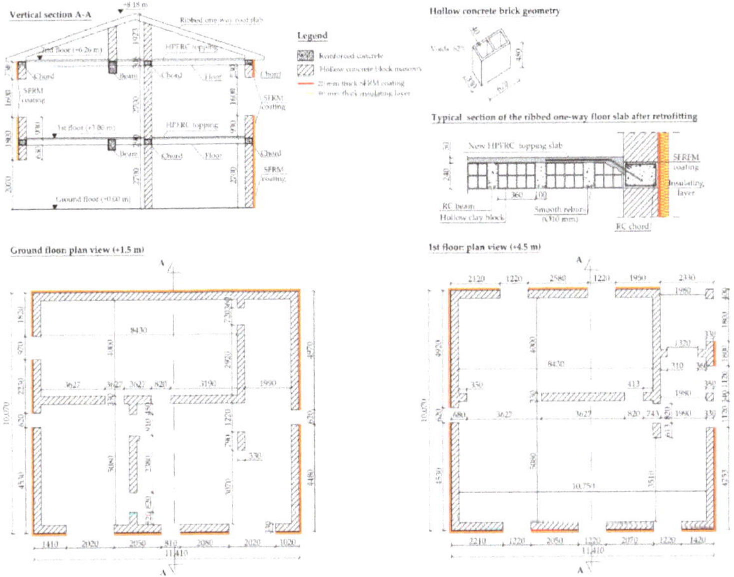

Figure 8. Properties of the existing masonry building considered in the case study (dimensions in mm).

The following sections report and discuss the main features related to the procedure adopted to design the retrofitting intervention.

5.1. Seismic Verification of the Structure

5.1.1. Numerical Modelling

A series of 3D non-linear pushover analyses were carried out to assess the seismic performance of the structure. To compare the improvement resulting from retrofitting, two different models, named "CS" and after "CS-R(20)," were implemented in the finite element code Midas FEA [21] to simulate the building before and after retrofitting. The masonry and SFRM coating were modeled by 8-nodes hexahedral solid elements and by 4-node quadrilateral plane stress elements, respectively, adopting the same constitutive models for the test building analyzed in Section 3. Since data from field investigations were not available, the mean mechanical properties of masonry were determined according to the provisions reported in the Guidelines of MIT 2009 (Table C8A.2.1) [35]. As recommended by EN1998-3 [36] for the Limited Knowledge Level (KL1), the mean properties of masonry were divided by the Confidence Factor CF = 1.35 to get the design properties considered in the analyses (Table 5). About SFRM, the characteristic values of the properties reported in Section 3.1 and Section 3.2 were used in the simulation of the retrofitted building (Table 5). The same constitutive laws described in Section 3.2 were used to represent the inelastic behavior of materials.

Table 5. Design properties of materials considered in the pushover analyses.

Material	γ [kg/m^3]	E [MPa]	f_c [MPa]	G_c [N/mm]	f_t [MPa]	G^I_f [N/mm]	f_{Fts} [MPa]	f_{Ftu} [MPa]	w_u [mm]
Masonry	1200	600	1.1	0.12	0.05	0.005	-	-	-
SFRM	2040	21,000	22	-	1.7	-	1.5	1.4	2.0

The finite element model of the building is shown in Figure 9. The floors were assumed as infinitely stiff and rigidly connected to the masonry walls. In addition to the self-weight of the walls, total vertical loads of 4.7 kN/m^2 and 5.2 kN/m^2, which included both the permanent and the variable actions, were applied to the first and second floor, respectively.

Figure 9. The pushover analysis of the concrete unit masonry building: finite element mesh and lateral load distribution.

When performing a pushover analysis, one should consider two systems of horizontal forces applied at the story levels and acting in the two orthogonal directions coinciding with the principal axes of the building:

- a system of forces proportional to masses (i.e., typical of soft ground story response);
- a system of forces proportional to the first mode shape of the building (i.e., it is able to represent the structural dynamic amplification).

The results reported herein refer to the analyses concerning the most critical condition, which is the one including a lateral load distribution proportional to masses and oriented in the +X direction shown in Figure 9. As suggested by the Italian structural code NTC2018 [37] as well as by EN1998-1 [38], possible uncertainties in the distribution of masses were taken into account by an additional mass eccentricity (i.e., accidental eccentricity) corresponding to 5% of the width of the floor perpendicular to the direction of the acting seismic action.

5.1.2. Discussion of Analysis Results

A preliminary linear modal analysis was performed to determine the mass participation and the basic parameters required by the pushover analysis. Table 6 reports some of the obtained results related to the first mode, namely the fundamental period (T_1), the modal participation mass in X-direction (M_x) as well as the modal participation factor (Γ) [39]. Note that the first mode involved an amount of mass higher than 55% both for the unstrengthened and the retrofitted building. The fundamental period resulting from the dynamic analysis was compared with those provided by two empirical equations reported by structural codes [37,40]. The first-mode period $T'_{1\text{-NTC}}$ was calculated as

$$T'_{1\text{-NTC}} = 2 \cdot d^{1/2}, \quad (3)$$

where d [m] is the lateral elastic displacement of the top of the building obtained from the application of gravity loads in horizontal direction. On the contrary, the period $T''_{1\text{-NTC}}$ resulted from the following equation:

$$T''_{1\text{-NTC}} = 0.05 \cdot H^{3/4}, \quad (4)$$

where H is the total height of the structure calculated from either a rigid basement or the foundation. As highlighted by Table 6, both equations overestimated by 60% ($T'_{1\text{-NTC}}$ of CS) to 240% ($T''_{1\text{-NTC}}$ of CS-R(20)) the fundamental period resulting from the numerical models. However, one should consider that the empirical equations above are very rough and should be used only where the initial sizing of structural elements cannot easily be made [41]. As an alternative to the empirical equations reported above, one may perform an in situ dynamic characterization of the building able to provide a more accurate estimation of the fundamental period.

Table 6. Main results of eigenvalue analyses related to first mode and parameters of the SDOF idealized bi-linear curve.

	T_1 [s]	$T'_{1\text{-NTC}}$ [s]	$T''_{1\text{-NTC}}$ [s]	M_x [%]	M_x [kg]	Γ [-]	T* [s]	M* [kg]	K* [kN/mm]	F^*_y [kN]	D^*_u [mm]
CS	0.073	0.117	0.196	59	130371	1.20	0.173	171280	227	760	9.8
CS-R(20)	0.058	0.119	0.196	66	151754	1.20	0.116	176190	517	1386	15.7

The capacity curves resulting from the non-linear analyses of the building (i.e., the MDOF structure) are reported in Figure 10 (see the dotted curves). As compared to the reference structure (CS_MDOF), the repaired building (CS-R(20)_MDOF) exhibited both an initial stiffness and maximum capacity improvement equal to 128% and 92%, respectively. Moreover, the conventional ultimate displacement capacity d_u (see Table 7 and the cross marks depicted in Figure 10), which corresponded to the 20% reduction of the maximum capacity on the post-peak response according to [37], was 1.6 times higher than that exhibited by the unstrengthened building.

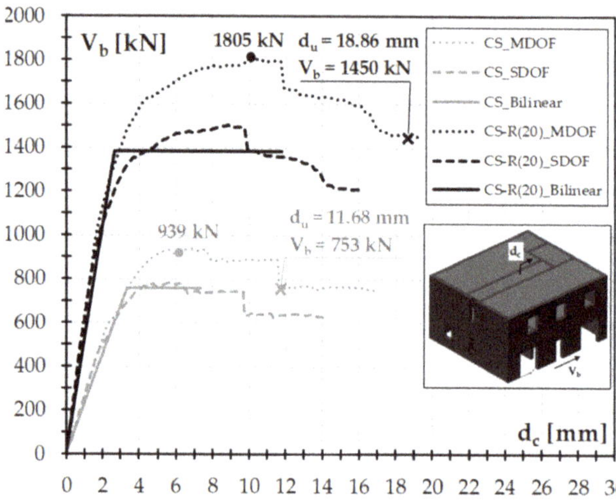

Figure 10. Base shear (V_b) vs. second-floor displacement (d_c) curves and the bi-linear idealized response.

Table 7. Safety verification and q-factor.

	d_u [mm]	$d_{C\text{-}SLV}$ [mm]	S_e (T*) [g]	$d_{e,max}$ [mm]	q_u [-]	$d_{D\text{-}SLV}$ [mm]	α_u/α_1 [-]	μ [-]	q_{NTC} [-]	q_{DUC} [-]
CS	11.68	7.31	0.755	5.62	1.70	13.00	1.39	2.93	2.43	2.02
CS-R(20)	18.86	11.78	0.633	2.12	1.00	2.12	1.43	5.86	2.49	2.52

The results of the pushover analyses were used to perform the seismic safety check according to the method proposed by NTC 2018 [37]. First, the force (V^*_b) and the displacement (d^*_c) of the SDOF system were calculated as:

$$V^*_b = V_b/\Gamma, \quad (5)$$

$$d^*_c = d_c/\Gamma, \quad (6)$$

where V_b and d_c are, respectively, the total base shear and the displacement of the control node of the MDOF system. The idealized bi-linear SDOF capacity curve was obtained by the stiffness of the elastic branch (k*) resulting from the following equation (see NTC2018; clause 7.8.1.6):

$$k^* = 0.7 \cdot V^*_{b,max}/d_{(0.7 \cdot V^*bu)}, \quad (7)$$

in which $V^*_{b,max}$ is the maximum base shear of the SDOF system, and $d_{(0.7 \cdot V^*b,max)}$ is the displacement corresponding to 70% of $V^*_{b,max}$. The yield resistance (F^*_y) of the idealized SDOF capacity curve was evaluated by the equal energy assumption, assuming $d^*_u = d_u/\Gamma$. Table 6 reports the parameters obtained from the equations above.

As suggested by NTC2018, safety assessment considered the Life Safety at Ultimate Limit State (ULS)". Figure 11a reports the elastic spectrum for the city of L'Aquila for a return period T_R = 475 years, whereas Figure 11b shows the SLV performance points on the corresponding elastic demand spectrum for both the unstrengthened and the retrofitted structure. For the sake of clarity, the demand spectrum of Figure 11b was limited to a maximum displacement of 35 mm. The displacement demand $d_{D\text{-}SLV}$ represented by the performance points (see Figure 11b and Table 7) was calculated by the following relation [40]:

$$d_{D\text{-}SLV} = d_{e,max} \cdot [1 + (q_u - 1)\, T_c/T^*]/q_u \geq d_{e,max} \text{ with } T^* < T_c, \quad (8)$$

which depends on the parameters described below (see also Table 6): $d_{e,max} = S_e(T^*) \cdot (T^*/2\pi)^2$ is the target displacement of an SDOF system with unlimited elastic behavior; T^* = fundamental period of the idealized system; m^* = equivalent mass of the structure; $q_u = S_e(T^*) \cdot m^*/F^*_y \geq 1$ is the ratio between the elastic response force and the yield force of the equivalent SDOF system; $S_e(T^*)$ = the elastic spectrum of L'Aquila corresponding to the period T^*; T_c = 0.74 s is the corner period between the short and medium period range of the elastic spectrum.

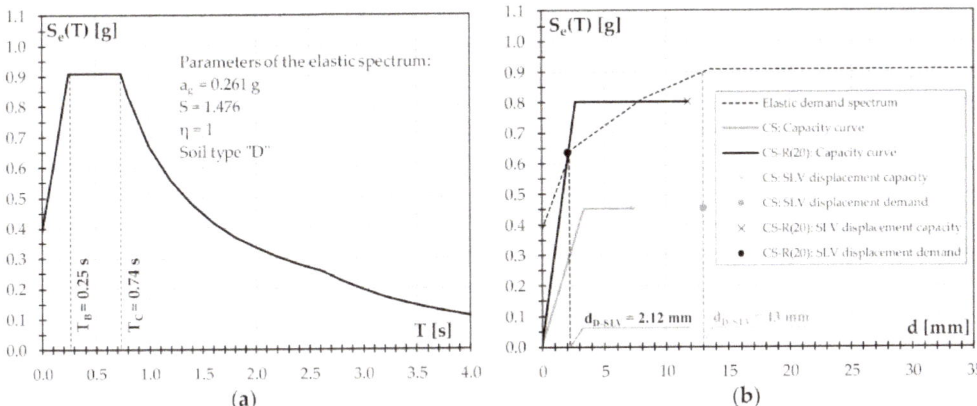

Figure 11. Seismic assessment of the structure before and after retrofitting: (**a**) elastic spectrum of acceleration of L'Aquila; (**b**) a comparison between displacement capacity and demand. The performance points are represented by dot marks.

Each performance point was compared with the corresponding SLV capacity displacement $d_{C\text{-}SLV}$ (Table 7) calculated as:

$$d_{C\text{-}SLV} = 3/4 \cdot d^*_u, \quad (9)$$

The curves of Figure 11b highlight that the unstrengthened building (CS curve) did not fulfill the safety requirement as the capacity displacement was lower than the displacement demand. After retrofitting, the SLV performance point was satisfied as the capacity displacement of the retrofitted building (CS-R(20) curve) was more than five times higher than the displacement demand.

It is worth remarking that the increase in stiffness and capacity due to the SFRM coating was so high that the performance point fell within the elastic branch of the bi-linear capacity curve.

Figure 12 shows the contours of principal tensile post-cracking strains of the building (i.e., the MDOF system) at a lateral deflection d_2 equal to $\Gamma \cdot d_{D\text{-}SLV}$, where $d_{D\text{-}SLV}$ corresponds to the displacement demand for the SDOF system. As expected, the crack pattern of the unstrengthened building was much more severe than that exhibited by the retrofitted building. In fact, as proved by Figure 12a, most of the piers located on the ground floor of the CS building presented a damage pattern typical of active diagonal shear mechanisms. On the contrary, the lower displacement demand, together with the higher initial stiffness, allowed the CS-R(20) building to remain mostly undamaged (Figure 12b). Besides small flexural cracks starting from the corners of the openings, only the internal pier showed incipient flexural mechanisms. This confirms the benefit provided by the proposed technique in terms of both better control of the cracking process and higher structure durability.

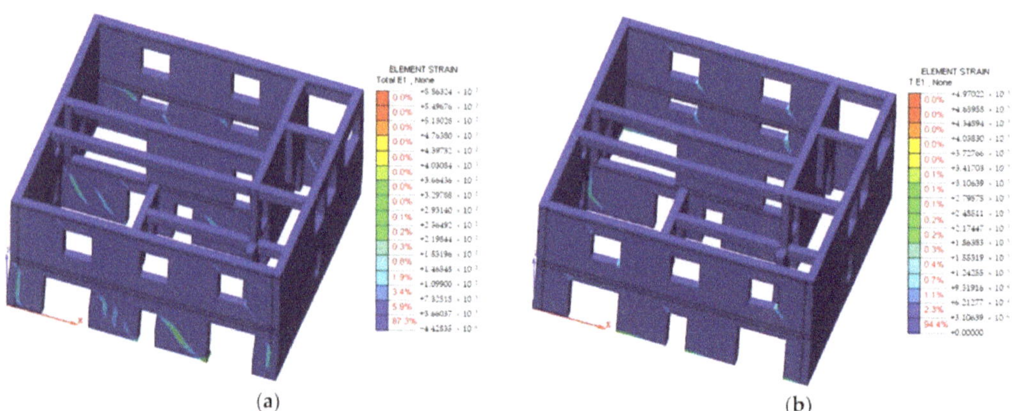

Figure 12. Contours of the principal tensile post-cracking strains corresponding to a second-floor lateral displacement $d_c = \Gamma \cdot d_{D\text{-}SLV}$: (**a**) the unstrengthened building; (**b**) the strengthened building.

The pushover analyses were finally used to estimate the behavior factor q of the structure before and after the retrofitting. The q-factor typically used in linear elastic analysis can be estimated as follows:

$$q = q_0 \cdot \alpha_u / \alpha_1, \quad (10)$$

where q_0 is the basic value of the behavior factor, whereas α_u/α_1 (Table 7) is an over-strength factor defined as follows: α_1 is the multiplier of the horizontal seismic load for which the first pier reaches its ultimate shear or flexural resistance; α_u is the 90% of the multiplier of the horizontal seismic load for which the structure reaches its maximum resistance.

According to the Guidelines of MIT 2019 [40], q_0 can be assumed equal to 1.75 for URM structures. As an alternative, q_0 can be computed by the following equation [37,38]:

$$q_0 = 1 + (\mu - 1)\, T^*/T_C \text{ with } T^* < T_c \quad (11)$$

depending on the ductility $\mu = d^*_u \cdot k^*/F^*_y$ (Table 7) of the bi-linear SDOF system. Note that both the ratio α_u/α_1 and the ductility μ were obtained from the pushover analyses described above. The factors q_{NTC} and q_{DUC} reported in Table 7 were calculated by Equation (10): the former resulted from the implementation of the factor q_0 proposed by the NTC2018, and the latter was obtained, including the factor q_0 calculated by means of the Equation (11). As the q-factor predicted by the NTC2018 depends mainly on the over-strength factor, the factor q_{NTC} of the unstrengthened structure was very close (i.e., 2% lower) to that of the strengthened building. It is worth noting that both values of q_{NTC} are slightly lower than the maximum value (i.e., q = 2.6) recommended by the Guidelines of MIT 2019 [40] for buildings made with hollow block masonry. By including the structure ductility in the calculation of the q-factor, the factor q_{DUC} of the retrofitted structure result was 25% higher than that of the unreinforced building. Moreover, the value of q_{DUC} exhibited by the retrofitted structure was about 2.5. Based on the previous results, one may conclude that the methods adopted to determine the q-factor provided consistent results. In addition, the present case study suggests that the upper limit of q = 2.6 recommended by the Italian code for existing URM buildings may be suitable also for structures retrofitted by SFRM coating.

5.2. Energy Analysis

L'Aquila, with 2514 degree-days, is located in the climate zone E. Thus the thermal transmittance *U* must fulfil the limit value of 0.28 W/(m^2 K) for any intervention on opaque vertical walls. To give an idea of the figures, if an ETICS solution with $\lambda = 0.035$ W/(m K) expanded polystyrene is considered, an 80 mm thick insulation layer would be sufficient to fulfil the limits. The only case in which this limit is not expressly required is that of a major first-level renovation that also involves the heating system. In that case, the overall heat transfer transmission coefficient value H'_T must fulfil a limit value depending on the surface to volume ratio S/V and on the climate zone. Moreover, for sites characterized by the average monthly irradiance on the horizontal plane during the month of highest summer insolation higher than or equal to 290 W/m^2, it is required that either the mass per unit area (M_s) is greater than 230 kg/m^2 or the periodic thermal transmittance (Y_{ie}) is lower than 0.10 W/(m^2 K). The latter limitations apply to all opaque vertical walls, excluding those included in the northwest/north/northeast quadrant. In fact, even in the case of major second-level renovations, which may not involve the heating system, the thermal transmittance limit must be fulfilled, and a limit (i.e., 0.65 W/(m^2 K) for the climatic zone E) on the global transmission coefficient is added.

The vertical walls of the building consisted of a 330 mm thick masonry wall provided with a 15 mm thick internal layer of plaster. Masonry and plaster had a thermal conductivity equal to 0.27 W/(m K) and 0.8 W/(m K), respectively. Package N.1, presented in Section 4, provided a U-value of 0.22 W/(m^2 K) and package N.2 a U-value of 0.23 W/(m^2 K), whereas package N.3 provided a thermal transmittance of 0.29 W/(m^2 K). Therefore, to respect the law limits, package N.1 or package N.2 can be chosen. As reported in the recent works by Dickson and Pavía [42] and Kumar [43], the main benefit of aerogel-based insulation lies in the low thickness needed to fulfil thermal requirements, which makes it especially interesting in high-rent areas. The considerable initial cost of these materials is justified in cold climates, where building walls with a higher thermal resistance are more cost-effective. In the case at hand, the climate is temperate, and the wall thickness is not an issue; thus, the main parameters that can drive the choice are sustainability and cost. A comparison between embodied energy and embodied carbon of the two materials [43] reveals sensibly higher values for aerogel than for wood fiber, which turns out to be more sustainable in terms of global warming, ozone depletion and acidification potential, albeit not reaching the levels of other materials such as cellulose [42]. As concerns costs, the total market price of the two 80 mm and 40 mm layers of "light" and "heavy" wood fibers is 20 €/m^2, against over 300 €/m^2 for the considered aerogel-based material. In the light of these considerations, package N.2 is chosen over N.1.

Regarding ceilings, roofs and floors, typical details of the building's construction period were adopted. Doors were assumed to be made of wood and 70 mm thick and windows made of a single 3 mm thick glass layer, having thermal transmittance of 5.89 W/(m^2 K). As a further improvement, the replacement of single-glazed windows with double-glazed windows with argon cavity (thickness 3 mm – 9 mm – 3 mm, thermal transmittance 2.59 W/(m^2 K)) was also simulated.

In this case, the global transmission coefficient of the vertical walls was estimated before and after the renovation, finding a reduction from 1.53 W/(m^2 K) to 0.54 W/(m^2 K), which allowed to meet the limit value of 0.65 W/(m^2 K). To take into account linear thermal bridges, an increase factor for thermal transmittance equal to 0.10 for opaque walls before the renovation and equal to 0.05 after the renovation was used.

With EnergyPlus^{TM} [17] software, the annual dynamic behavior of the building was simulated before the renovation, after the application of the multilayer package consisting of the SFRM coating and the chosen thermal coating, and finally after the further replacement of the windows. An ideal loads air system with a heating thermostat set-point was used to calculate the annual heating energy needs and to evaluate the behavior of the building in summer.

5.2.1. Numerical Modelling

The building was schematized as in Figure 13. Three zones were created: only the central body of the building was heated, while the roof space and the basement were considered unheated zones.

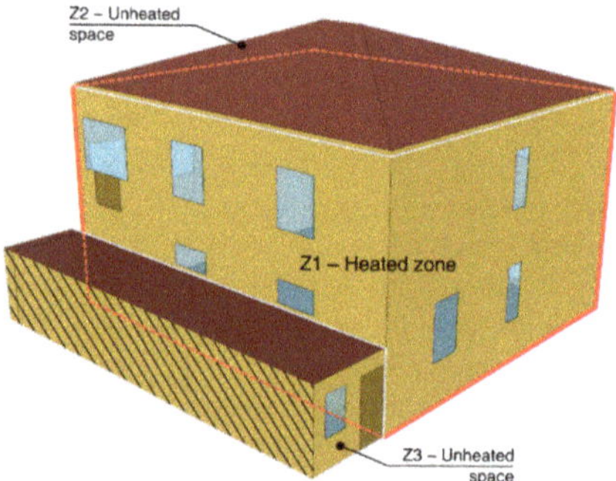

Figure 13. Scheme of the building: Z1 heated zone, Z2 and Z3 unheated zones.

The settings used for simulations in EnergyPlus™ [17] were:

- typical meteorological year 2005–2014 for L'Aquila extracted from EU Photovoltaic Geographical Information System [44];
- constant temperature of the surfaces in contact with the ground, equal to 18 °C;
- internal gains equal to 450 W in the zone Z1, as stated from UNI/TS 11300-1:2014 [45];
- air changes per hour equal to 0.5 h^{-1} in the zone Z1, as stated form UNI/TS 11300-1:2014 [45];
- temperature between 20 °C and 40 °C (thermostat constant heating and cooling set-points) to model an ideal heating system and no cooling system;
- calculation timesteps of 15 min.

5.2.2. Results and Discussion of the Thermal Analysis

The energy needs of this building during the year are shown in Figure 14, and the total values are summarized in Table 8. It can be observed that the seismic and energy retrofitting of the vertical walls reduces the energy needs by 17.1%. If the replacement of the windows is added, the savings rise to 29.3%.

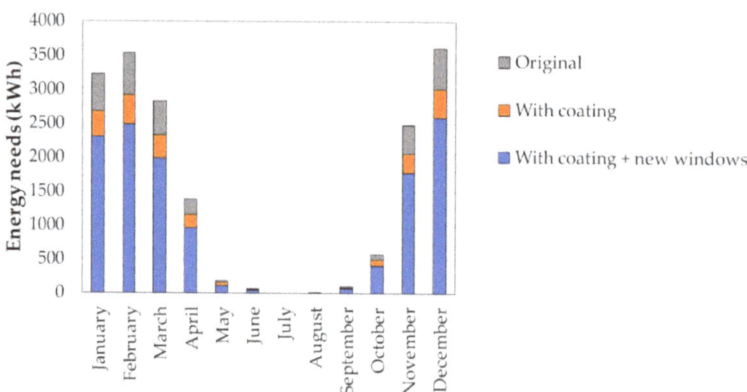

Figure 14. The energy needs of the building during the year.

Table 8. The energy needs and energy savings of the building with respect to the original configuration.

	Original	With Coating	With Coating + New Windows
Energy needs, kWh/(m^2 y)	87.23	72.28	61.69
Savings, %	—	17.1%	29.3%

Regarding the summer performances, Figure 15 shows that the retrofitting intervention modifies the operative temperature profile, which appears to be lower after the addition of the coating. A further benefit in terms of maximum operative temperature reduction, particularly in daytime, can be obtained by replacing single glazing with better-performing windows.

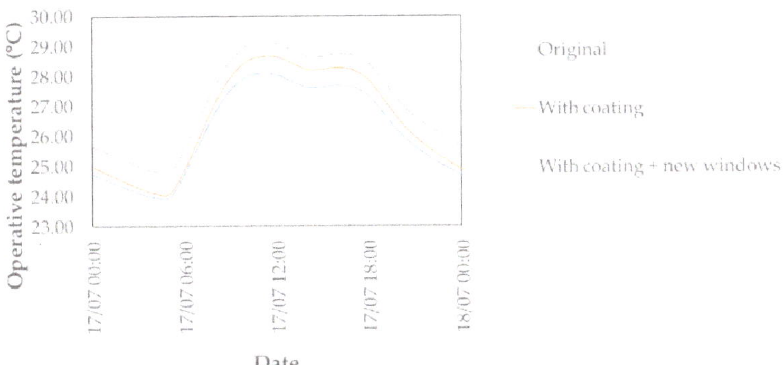

Figure 15. The operative temperature of the building during the hottest day of the year.

6. Conclusions

An integrated synergistic structural and energy upgrading system, especially designed for masonry buildings, was presented and discussed in this paper. The proposed technique combines a rather light and effective layer of SFRC mortar and an insulation layer; both applied only on the external masonry surfaces of buildings.

Some experimental evidence on the efficiency under a structural level was described first. The energy analyses of different insulation layers were therefore characterized in combination with the structural coating. An actual case study regarding a small building was finally proposed, and the benefits, both in terms of structural safety and energy savings, were depicted.

It can be certainly concluded that the idea of a combined retrofitting of masonry buildings, by implementing a structural and an insulation layer, applied only on the outer vertical surfaces to limit the occupant discomforts, is a promising and ready-to-use technique. The seismic performance is extremely improved, and the corresponding collapse mechanisms are postponed.

After evaluating the low effect of point thermal bridges due to layer connectors, it is estimated that the addition of the coating reduces the energy needs in winter and the peak operative temperature in summer. The replacement of poor-performing windows is strongly recommended in this case to fully exploit the potential of the retrofitting measures.

As proved by others [15,46], the use of an integrated retrofitting solution, such as that proposed herein, is generally able to minimize the economic losses since the initial investment is paid back faster. Future work will be devoted to the investigation of the economic efficiency of the proposed retrofitting method applied to buildings placed in different geographic areas.

Author Contributions: Data curation, M.P.; Formal analysis, L.F.; Funding acquisition, G.A.P.; Investigation, S.S.L. and B.G.; Methodology, F.M. and M.P.; Software, S.S.L. and B.G.; Supervision, L.F.; Writing—original draft, L.F., S.S.L. and B.G.; Writing—review and editing, F.M., M.P. and G.A.P. All authors have read and agreed to the published version of the manuscript.

Funding: The authors gratefully thank the University of Brescia and Tecnologia e Ricerca Italiana S.r.l. (TRI S.r.l) for the financial support to the "Health & Wealth" project "Sismacomf", whose research activities led to the results described in this paper. The financial contribution provided by the RELUIS-DPC project 2014–2018 is gratefully acknowledged.

Institutional Review Board Statement: Not applicable.

Informed Consent Statement: Not applicable.

Data Availability Statement: All data, models, and code generated or used during the study appear in the published article.

Conflicts of Interest: The authors declare no conflict of interest.

References

1. Economidou, M.; Atanasiu, B.; Despret, C.; Maio, J.; Nolte, I.; Rapf, O.; Laustsen, J.; Ruyssevelt, P.; Staniaszek, D.; Strong, D.; et al. *Europe's Buildings under the Microscope. A Country-by-Country Review of the Energy Performance of Buildings*; Buildings Performance Institute Europe BPIE: Brussels, Belgium, 2011.
2. Italian Ministry of Economic Development. *DM 26 giugno 2015. Applicazione Delle Metodologie di Calcolo Delle Prestazioni Energetiche e Definizione Delle Prescrizioni e Dei Requisiti Minimi Degli Edifici*; Supplement to the Official Gazette of the Italian Republic; Italian Ministry of Economic Development: Roma, Italy, 2015; n. 162 of 15 July 2015, Ordinary Supplement n. 39. (In Italian)
3. European Parliament. *2010/31/EU Directive of the European Parliament and of the Council on the Energy Performance of Buildings (EPBD Recast)*; European Parliament: Brussels, Belgium, 2010.
4. European Commission. Factsheet: Energy Performance in Buildings Directive. Available online: https://ec.europa.eu/energy/sites/default/files/documents/buildings_performance_factsheet.pdf (accessed on 4 May 2021).
5. Oliveira, D.V.; Basilio, I.; Lourenço, P.B. Experimental bond behavior of FRP sheets glued on brick masonry. *J. Compos. Constr.* **2010**, *15*, 32–41. [CrossRef]
6. Giaretton, M.; Dizhur, D.; Garbin, E.; Ingham, J.M.; da Porto, F. In-Plane Strengthening of Clay Brick and Block Masonry Walls Using Textile-Reinforced Mortar. *J. Compos. Constr.* **2018**, *22*, 04018028. [CrossRef]
7. ElGawady, M.A.; Lestuzzi, P.; Badoux, M. Retrofitting of masonry walls using shotcrete. In Proceedings of the NZSEE Conference, Napier, New Zealand, 10–12 March 2006; Volume 45.
8. Facconi, L.; Minelli, F.; Lucchini, S.; Plizzari, G. Experimental study of solid and hollow clay brick masonry walls retrofitted by steel fiber-reinforced mortar coating. *J. Earthq. Eng.* **2018**, *24*, 381–402. [CrossRef]
9. European Parliament. *2018/844/EU Directive of the European Parliament and of the Council of 30 May 2018 amending Directive 2010/31/EU on the Energy Performance of Buildings and Directive 2012/27/EU on Energy Efficiency*; European Parliament: Brussels, Belgium, 2018.
10. European Parliament. *2012/27/EU Directive of the European Parliament and of the Council on Energy Efficiency, Amending Directives 2009/125/EC and 2010/30/EU and Repealing Directives 2004/8/EC and 2006/32/EC.*; European Parliament: Brussels, Belgium, 2012.
11. World Health Organization. *WHO Guidelines for Indoor Air Quality: Dampness and Mould*; World Health Organization: Geneva, Switzerland, 2009.
12. Presidency of the Italian Republic. DPR 10 giugno 2020 n. 48. Attuazione della direttiva (UE) 2018/844 del Parlamento europeo e del Consiglio, del 30 maggio 2018, che modifica la direttiva 2010/31/UE sulla prestazione energetica nell'edilizia e la direttiva 2012/27/UE sull'efficienza energetica. In *Official Gazette of the Italian Republic*; Presidency of the Italian Republic: Rome, Italy, 2020; n. 146 of 10 June 2020. (In Italian)
13. Bournas, D.A. Concurrent seismic and energy retrofitting of RC and masonry building envelopes using inorganic textile-based composites combined with insulation materials: A new concept. *Compos. Part B Eng.* **2018**, *148*, 166–179. [CrossRef]
14. Karlos, K.; Tsantilis, A.; Triantafillou, T. Integrated Seismic and Energy Retrofitting System for Masonry Walls Using Textile-Reinforced Mortars Combined with Thermal Insulation: Experimental, Analytical, and Numerical Study. *J. Compos. Sci.* **2020**, *4*, 189. [CrossRef]
15. Mistretta, F.; Stochino, F.; Sassu, M. Structural and thermal retrofitting of masonry walls: An integrated cost-analysis approach for the Italian context. *Build. Environ.* **2019**, *155*, 127–136. [CrossRef]
16. Longo, F.; Cascardi, A.; Lassandro, P.; Aiello, M.A. Thermal and Seismic Capacity Improvements for Masonry Building Heritage: A Unified Retrofitting System. *Sustainability* **2021**, *13*, 1111. [CrossRef]
17. *EnergyPlus™, Version 9.4.0*; U.S. Department of Energy's (DOE) Building Technologies Office (BTO) and Managed by the National Renewable Energy Laboratory (NREL): Washington, DC, USA.

18. Lucchini, S.S.; Facconi, L.; Minelli, F.; Plizzari, G.A. Cyclic Test on a Full-Scale Unreinforced Masonry Building Repaired with Steel Fiber-Reinforced Mortar Coating. *J. Struct. Eng.* **2021**, *147*, 04021059. [CrossRef]
19. Lucchini, S.S.; Facconi, L.; Minelli, F.; Plizzari, G.A. Retrofitting a full-scale two-story hollow clay block masonry building by steel fiber reinforced mortar coating. In Proceedings of the 3rd FRC International Workshop Fibre Reinforced Concrete: From Design to Structural Applications, Lake Garda, Italy, 28–30 June 2018; Extended Abstract on pp. 114–115. ISBN 978-88-89252-44-4.
20. UNI EN 14651:2007. *Test Method for Metallic Fibre Concrete—Measuring the Flexural Tensile Strength (Limit of Proportionality (Lop), Residual)*; UNI EN: Milan, Italy, 2007.
21. Midas FEA v1.1. Advanced Nonlinear and Detail FE Analysis System. *Midas Information Technology Co. Ltd.* Available online: https://www.cspfea.net/ (accessed on 15 July 2020).
22. Rots, J.G. Computational Modeling of Concrete Fracture. Ph.D. Thesis, University of Technology of Delft, Delft, The Netherlands, 1988.
23. Hordijk, D.A. Local Approach to Fatigue of Concrete. Ph.D. Thesis, University of Technology of Delft, Delft, The Netherlands, 1991.
24. Feenstra, P.H. Computational Aspects of Biaxial Stress in Plain and Reinforced Concrete. Ph.D. Thesis, University of Technology of Delft, Delft, The Netherlands, 1993.
25. Thorenfeldt, E.; Tomaszewicz, A.; Jensen, J.J. Mechanical properties of high-strength concrete and applications in design. In Proceedings of the Symposium on Utilization of High-Strength Concrete, Trondheim, Norway; 1987.
26. International Federation for Structural Concrete. *Fib Bulletin 65. Model Code 2010*; Final Complete Draft; International Federation for Structural Concrete (fib): Lausanne, Switzerland, 2012.
27. UNI/TR 11552:2014. *Opaque Envelope Components of Buildings—Thermo-Physical Parameters*; UNI EN: Milan, Italy, 2014.
28. Neri, M.; Pilotelli, M.; Piana, E.A.; Lezzi, A.M. Prediction of the Acoustic and Thermal Performance of a Multilayer Partition. In Proceedings of the 4th IBPSA-Italy Conference on Building Simulation Applications, BSA 2019, Bolzano, Italy, 19–21 June 2020; pp. 275–282.
29. ISO (International Organization for Standardization). *ISO 6946 Building Components and Building Elements—Thermal Resistance and Thermal Transmittance—Calculation Methods*; ISO: Geneva, Switzerland, 2017.
30. ISO (International Organization for Standardization). *ISO 13786 Thermal Performance of Building Components—Dynamic Thermal Characteristics—Calculation Methods*; ISO: Geneva, Switzerland, 2017.
31. Tenpierik, M.; van Der Spoel, W.; Cauberg, H. An Analytical Model for Calculating Thermal Bridge Effects in High Performance Building Enclosure. *J. Build. Phys.* **2008**, *31*, 361–387. [CrossRef]
32. Luscietti, D.; Gervasio, P.; Lezzi, A.M. Computation of linear transmittance of thermal bridges in precast concrete sandwich panels. *J. Phys. Conf. Ser.* **2014**, *547*, 012014. [CrossRef]
33. Benedetti, M.; Gervasio, P.; Luscietti, D.; Pilotelli, M.; Lezzi, A.M. Point Thermal Transmittance of Rib Intersections in Concrete Sandwich Wall Panels. *Heat Transfer Eng.* **2019**, *40*, 1075–1084. [CrossRef]
34. COMSOL. *Reference Manual, Version 5.6*; COMSOL Corporation: Stockholm, Sweden, 2020.
35. *Circolare n. 617 del 2 febbraio 2009 del Ministero delle Infrastrutture e dei Trasporti. Istruzioni per l'applicazione delle Nuove Norme Tecniche delle Costruzioni di cui al DM 14 gennaio 2008*; Ministero delle Infrastrutture e dei Trasporti (MIT): Rome, Italy, 2009. (In Italian)
36. UNI EN 1998-3:2005. *Eurocode 8: Design of Structures for Earthquake Resistance—Part 3: Assessment and Retrofitting of Buildings*; UNI EN: Milan, Italy, 2005.
37. NTC 2018: Decreto del Ministero delle Infrastrutture e dei Trasporti del 17 gennaio 2018. *Aggiornamento delle Norme Tecniche per le Costruzioni*; Ministero delle Infrastrutture e dei Trasporti (MIT): Rome, Italy, 2018. (In Italian)
38. UNI EN 1998-1:2013. *Eurocode 8: Design of Structures for Earthquake Resistance—Part 1: General Rules, Seismic Actions and Rules for Buildings*; UNI EN: Milan, Italy, 2013.
39. Fajfar, P.; Gašperšič, P. The N2 method for the seismic damage analysis of RC buildings. *Earthq. Eng. Struct. Dyn.* **1996**, *25*, 31–46. [CrossRef]
40. *Circolare, n. 7 del 21 gennaio 2019 del Ministero delle Infrastrutture e dei Trasporti. Istruzioni per l'applicazione dell'«Aggiornamento Delle "Norme Tecniche per le Costruzioni"» di Cui al DM 17 Gennaio 2018*; Ministero delle Infrastrutture e dei Trasporti (MIT): Rome, Italy, 2019. (In Italian)
41. Paulay, T.; Priestley, M.J.N. *Seismic Design of Reinforced Concrete and Masonry Buildings*; Wiley-Interscience: New York, NY, USA, 1992; 768p.
42. Dickson, T.; Pavía, S. Energy Performance, Environmental Impact and Cost of a Range of Insulation Materials. *Renew. Sustain. Energy Rev.* **2021**, *140*, 110752. [CrossRef]
43. Kumar, D.; Alam, M.; Zou, P.X.W.; Sanjayan, J.G.; Memon, R.A. Comparative Analysis of Building Insulation Material Properties and Performance. *Renew. Sustain. Energy Rev.* **2020**, *131*, 110038. [CrossRef]
44. Photovoltaic Geographical Information System (PVGIS). Available online: https://ec.europa.eu/jrc/en/pvgis (accessed on 10 November 2020).
45. UNI/TS 11300-1:2014. Prestazioni Energetiche Degli Edifici—Parte 1: Determinazione Del Fabbisogno Di Energia Termica Dell'edificio per La Climatizzazione Estiva Ed Invernale. Ente Nazionale Italiano di Unificazione (UNI): Milan, Italy; Rome, Italy. (In Italian)
46. Gkournelos, P.D.; Bournas, D.A.; Triantafillou, T.C. Combined seismic and energy upgrading of existing reinforced concrete buildings using TRM jacketing and thermal insulation. *Earthq. Struct.* **2019**, *16*, 625–639.

Article

Performance Evaluation of a Prestressed Belitic Calcium Sulfoaluminate Cement (BCSA) Concrete Bridge Girder

Nick Markosian [1], Raed Tawadrous [2], Mohammad Mastali [3], Robert J. Thomas [4] and Marc Maguire [3,*]

1 Calder Richards Structural Consulting Engineers, Salt Lake City, UT 84101, USA; nick@crceng.com
2 eConstruct, Orlando, FL 32817, USA; raed.tawadrous@gmail.com
3 Durham School of Architectural Engineering and Construction, University of Nebraska-Lincoln, Omaha, NE 68182, USA; mmastali2@unl.edu
4 Department of Civil Engineering, Clarkson University, Potsdam, NY 13699, USA; rthomas@clarkson.edu
* Correspondence: marc.maguire@unl.edu

Abstract: Belitic calcium sulfoaluminate (BCSA) cement is a sustainable alternative to Portland cement that offers rapid setting characteristics that could accelerate throughput in precast concrete operations. BCSA cements have lower carbon footprint, embodied energy, and natural resource consumption than Portland cement. However, these benefits are not often utilized in structural members due to lack of specifications and perceived logistical challenges. This paper evaluates the performance of a full-scale precast, prestressed voided deck slab bridge girder made with BCSA cement concrete. The rapid-set properties of BCSA cement allowed the initial concrete compressive strength to reach the required 4300 psi release strength at 6.5 h after casting. Prestress losses were monitored long-term using vibrating wire strain gages cast into the concrete at the level of the prestressing strands and the data were compared to the American Association of State Highway and Transportation Officials Load and Resistance Factor Design (AASHTO LRFD) predicted prestress losses. AASHTO methods for prestress loss calculation were overestimated compared to the vibrating wire strain gage data. Material testing was performed to quantify material properties including compressive strength, tensile strength, static and dynamic elastic modulus, creep, and drying and autogenous shrinkage. The material testing results were compared to AASHTO predictions for creep and shrinkage losses. The bridge girder was tested at mid-span and at a distance of 1.25 times the depth of the beam (1.25*d*) from the face of the support until failure. Mid-span testing consisted of a crack reopening test to solve for the effective prestress in the girder and a flexural test until failure. The crack reopen effective prestress was compared to the AASHTO prediction and AASHTO appeared to be effective in predicting losses based on the crack reopen data. The mid-span failure was a shear failure, well predicted by AASHTO LRFD. The 1.25*d* test resulted in a bond failure, but nearly developed based on a moment curvature estimate indicating the AASHTO bond model was conservative.

Keywords: prestressed concrete; prestress losses; bridges; flexural strength; shear strength; drying and autogenous shrinkage; creep; sustainability

Citation: Markosian, N.; Tawadrous, R.; Mastali, M.; Thomas, R.J.; Maguire, M. Performance Evaluation of a Prestressed Belitic Calcium Sulfoaluminate Cement (BCSA) Concrete Bridge Girder. *Sustainability* **2021**, *13*, 7875. https://doi.org/10.3390/su13147875

Academic Editor: Jorge de Brito

Received: 21 May 2021
Accepted: 11 July 2021
Published: 14 July 2021

1. Introduction

Calcium sulfoaluminate (CSA) cements are a class of hydraulic cements based on Ye'elimite ($C_4A_3\bar{S}$) which were first patented by Klein in 1964 and 1966 [1,2]. Depending on the specific chemistry and blend, CSA cements can be shrinkage-compensating, rapid-setting, or both. CSA cements include accelerating additives, shrinkage-compensating additives, and standalone cements. Type K shrinkage-compensating cements, for example, rely on up to 5% Ye'elimite for their expansive characteristics. Standalone CSA cements, which are used as single cements rather than blended with Portland cement, have been developed in both the United States and China. In China, "high-CSA" cements that contain Ye'elimite as the predominate phase have been used for many years.

In the United States, Ost et al. [3] introduced standalone belitic calcium sulfoaluminate (BCSA) cements in the 1970s. BCSA cements are rapid setting cements comprising 20–30% Ye'elemite and 30–60% belite interground with 5–25% calcium sulfate, by mass. Rapid early strength development is the hallmark property of BCSA cement, but it is also increasingly viewed as a potential sustainable alternative to Portland cement [4]. The embodied energy and carbon dioxide emissions are significantly lower than Portland cement, owing to several factors including:

1. Lower clinkering temperature, which reduces the energy and emissions associated with firing the kiln [5];
2. Less calcium oxide in the clinker, which reduces the release of CO_2 associated with calcination;
3. Increased friability, which reduces the energy consumption of grinding [6]; and
4. Higher proportion of interground components (e.g., calcium sulfate) that do not require firing, further reducing the embodied energy and emissions associated with clinkering.

Further, a variety of industrial and municipal wastes can be used as precursors, reducing the environmental impact of mining and quarrying raw materials for cement production [7,8].

Figure 1 shows a comparison of the embodied CO_2 emissions for the component phases of Portland and BCSA cements, based on an analysis by Gartner [9]. Accordingly, alite and belite are the main responsible phases for releasing CO_2 into the atmosphere. Generally, Portland cement comprises different cement phases, including 63% alite, 15% belite, 8% tricalcium aluminate, 9% ferrite, 5% others; in comparison, BCSA cement ingredients by 22–71% belite (depends on the sulfate content), 3–7% ferrite, 65–15% calcium sulfoaluminate cement (Ye'elimite). Considering CO_2 emission and the content of each phase of Portland and BCSA cement, Gartner and Sui concluded that BCSA cement could have up to 30% lower CO_2 emission than Portland cement, depending on the phase composition [10].

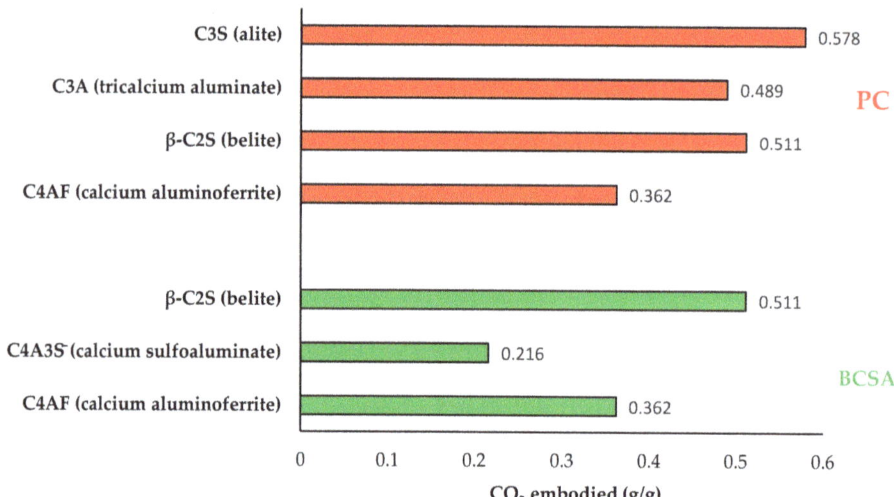

Figure 1. Embodied CO_2 in Portland and BCSA cement, by phase.

BCSA has been used for decades in the United States for accelerated airfield reconstruction [11], pavement repair [12], and structural rehabilitation [13]. These applications rely on the high early strength development of BCSA cement to accelerate construction and limit its impact on the end user. More recently, researchers have pondered the use of BCSA for structural applications. Specifically, the rapid-setting and sustainable characteristics of BCSA cement make it attractive for precast concrete construction. BCSA's rapid-setting behavior accelerates release times and reduces reliance on energy-intensive steam-curing protocols. Furthermore, the reduced carbon footprint and embodied energy of BCSA cement would help the precast concrete industry achieve its sustainability goals and reduce its reliance on Portland cement.

While there is a large body of research on the material-scale performance and durability of BCSA cement concrete, there is very limited research on its performance in precast structural members. Research on the performance of precast BCSA concrete elements has been led by Ramseyer and Floyd at the University of Oklahoma, and later by Murray et al. at the University of Arkansas [14,15]. Murray at al. demonstrated construction of precast, prestressed BCSA concrete beams [15]. Successful construction of beams in as little as two hours from water addition to mold removal validated the feasibility of using BCSA cement to accelerate precast operations. Results showed that transfer and development lengths were in good agreement with AASHTO LRFD predictions, and load tests showed good agreement with predicted strengths, but prestress losses were far smaller than predicted.

Bowser studied the bond behavior of 0.6 in prestressing strands in eight 18-ft prestressed BCSA cement concrete beams and four prestressed Portland cement concrete (PCC) beams [16]. The researchers concluded that transfer lengths were approximately the same between BCSA and PCC beams; that ACI [17] and AASHTO [18] codes predicted transfer lengths in BCSA cement concrete reasonably well; that strands developed adequate bond relative to ACI 318 [17] requirements; and that prestress losses were significantly over-predicted for BCSA concrete beams.

The studies mentioned above are the only known articles that discuss the performance of prestressed BCSA concrete members. These papers have been of significant utility in demonstrating the feasibility of using BCSA cement concrete in a precast setting, and in validating (or invalidating) codified predictions of beam performance. However, the results of these studies must be validated for structural elements with widely varying composition and geometry before they can be generalized. Put simply: Much more data on the performance of full-scale BCSA cement concrete structural elements are needed before these materials can be adopted within concrete practice. To that end, this paper:

1. Demonstrates accelerated construction of a precast, prestressed voided deck slab bridge girder made with BCSA cement concrete;
2. Quantifies the performance of said girder in terms of material properties, transfer length, prestress losses, crack initiation and reopening, flexural strength, and shear strength; and
3. Compares measured performance with that predicted by with relevant ACI [17] and AASHTO [18] building codes.

2. Experimental Investigation

2.1. Materials

Concrete

Several trial mixes were conducted prior to reaching the final mix design with adequate setting time, workability, and compressive strength. Initially, the recommended ratios for cement, water, aggregates, and admixtures by the BCSA cement manufacturer were followed. The mix was modified by adjusting the ratios of the different materials to allow for the concrete to be mixed at the batch plant, delivered to the site, and placed in the forms without setting. Table 1 shows the final mix used to cast the full-scale specimens (details on full-scale specimens are given below). The final mix design reached the target compressive strength of 5000 psi in 4 h and had a final setting time of 1 h. Laboratory mixtures gave good workability, but the site-mixed concrete was too stiff to cast. A total of 200 lb (40 lb/yd^3) of water was added to the mixture to bring the mixture to a 6-in slump. The added water increased the water-to-cement ratio to 0.457, a 15.7% increase in water.

Table 1. BCSA concrete mix design.

BCSA Mixture Proportions		
BCSA	700	lb/cy
Water	280	lb/cy
Coarse aggregate	1650	lb/cy
Fine aggregate	1150	lb/cy
Air	6	%
Superplasticizer (GCP ADVA Cast 555)	98	fl oz/cy
Air entrainer (GCP Daravair 1000)	52.5	fl oz/cy
Retarder (GCP Recover)	182	fl oz/cy

Concrete was mixed in a twin-shaft high-speed mixer and charged into a drum truck for delivery to the prestressing bed. Full-scale specimens were cast, consolidated, and finished by plant personnel. Concrete cylinders (4 × 8 in), prisms (4 × 4 × 14 in), and other specimens were cast for measurements of setting time, mechanical properties, and shrinkage.

Early age measurements of setting time, autogenous shrinkage, and compressive strength were made at the precast plant, beginning immediately after casting. Setting time was measured by the mortar penetrometer method in accordance with ASTM C403 [19]. Compressive strength was measured in accordance with ASTM C39 [20] every 30 min following the measured final setting time. Autogenous shrinkage was measured by the corrugated tube method in accordance with ASTM C1698 [21].

Later-age measurements of mechanical properties and drying shrinkage were made after transportation of the specimens back to the laboratory. Compressive strength was measured in accordance with ASTM C39 [20]. Splitting tensile strength was measured in accordance with ASTM C496 [22]. Static modulus of elasticity was measured in accordance with ASTM C469 [23]. Dynamic modulus of elasticity, which is a nondestructive estimate of the modulus of elasticity of concrete based on the theory of wave propagation through solids, was measured in accordance with ASTM C215 [24]. Each of the preceding tests was performed on the date of full scale testing (concrete age of 110 days). Drying shrinkage was measured in accordance with the specifications of ASTM C157 [25] beginning at the final setting time and continuing until the specimens reached apparent equilibrium.

Table 2 lists the material properties of the concrete at the time of releasing the prestressing strands and on the day of full-scale testing. The results represent the average measured values of three specimens. The split tension tests showed high variability during the early age (first day); refer to Markosian for detailed material testing results [26].

Table 2. Concrete material properties.

	f'_c (psi)	f_{sp} (psi)	E	E_d (ksi)
Prestress release	4300	347	2200	3800
Full-scale testing	9760	-	2500	4000

2.2. *Voided Deck Slab Bridge Girder*

2.2.1. Casting

A precast, prestressed voided deck slab bridge girder was cast at a PCI-certified precast concrete plant. Figure 2 shows the geometry and detailed reinforcement of the beam. The beam was 22 ft long, 4 ft wide, and 21 in thick with three 10 in diameter polystyrene tubes used to create the voids as shown in Figure 2. The specimen was prestressed (pretensioned) using 22–0.6 in diameter Gr. 270 ksi low relaxation strands. The prestressing strands were detensioned by torching the strands (sudden release) once the required concrete release compressive strength was reached.

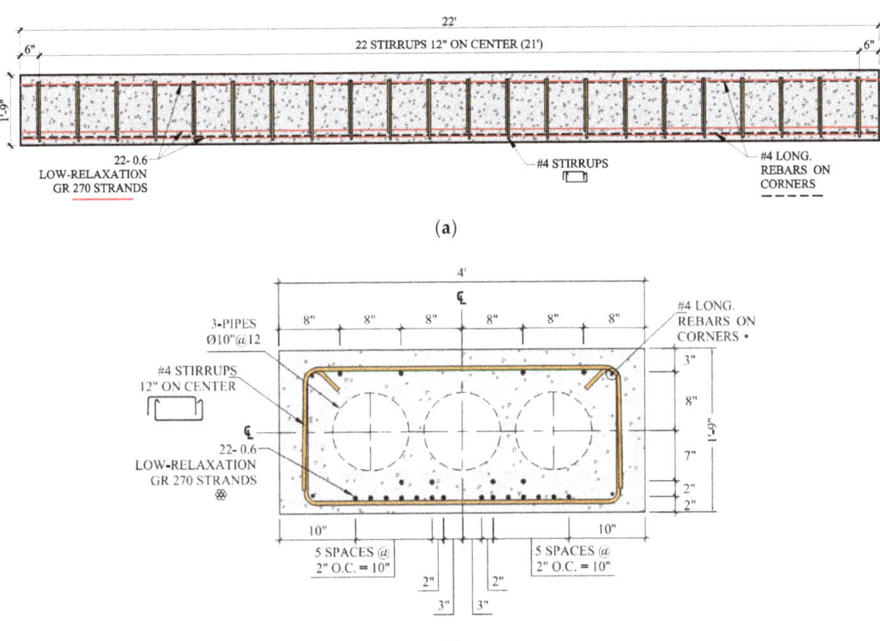

Figure 2. Full-scale specimen: (**a**) elevation and (**b**) cross section.

Detachable mechanical strain gages (DEMECs) were used at both ends of the specimen to measure the concrete surface strains at the level of prestressing strands to measure the transfer length. The DEMECs were placed at 2 in from the edge of the specimen and spaced at 4 in on center for 60 in. Prestressing losses were measured using vibrating wire strain gages (VWSGs) cast into the beam. A series of 8 total VWSGs were cast into the specimen and attached to the strands at both the top and bottom strands using zip ties, as shown in Figure 3. Four VWSGs were placed at mid-span: two on the top strands and two on the bottom. The remaining four gages were attached to top and bottom strands at quarter and three-quarter points along the length of the member. The VWSGs had both strain and temperature sensors, and readings were recorded once every 30 s for the lifetime of the beam before the final destructive tests. Camber was measured using an engineer's level and steel ruler accurate to one-tenth of a millimeter to calculate the change in elevation along the length of the member. Camber measurements were compared to the expected camber given in the AASHTO and PCI Bridge Design provisions. Both initial and long-term camber were measured [18,27].

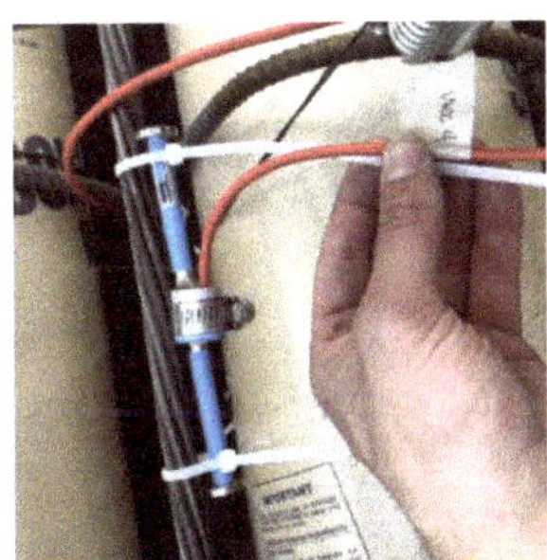

Figure 3. Vibrating wire strain gage placement.

2.2.2. Testing

Figure 4 shows the midspan test setup along with the instrumentation plan used to test the full-scale prestressed voided bridge deck panel in flexure. The specimen was simply supported on pin and roller supports with a span length of 21-ft. A 400-kip load cell was concentrically placed under the 400-kip hydraulic ram, and two potentiometers were placed on each side of the deck at mid-span to measure the deflection during the test. Strand end slip was measured at both ends of the specimen using linear variable differential transformers (LVDTs) mounted to the strands, as shown in Figure 5. The LVDTs have a +/−2 in range and were calibrated to an accuracy of 0.0005 in. The crack initiation test consisted of a concentrated load centered over the bridge deck panel to induce cracking near the midspan of the specimen. The crack initiation test procedure consisted of incrementally increasing load and measuring the change in strain for each loading increment. Once a crack was initiated at midspan, the load was released and four strain transducers with 3 in gage length and accuracy of 1 $\mu\varepsilon$ were mounted across the initial crack for the crack reopening test using the same test setup used for the initial crack test. The crack reopening test was performed to evaluate the effective prestressing force in the specimen. The load was increased constantly until the crack reopened. After the crack reopening test was completed, the load was increased to failure to test the specimen for prestressing strand development length.

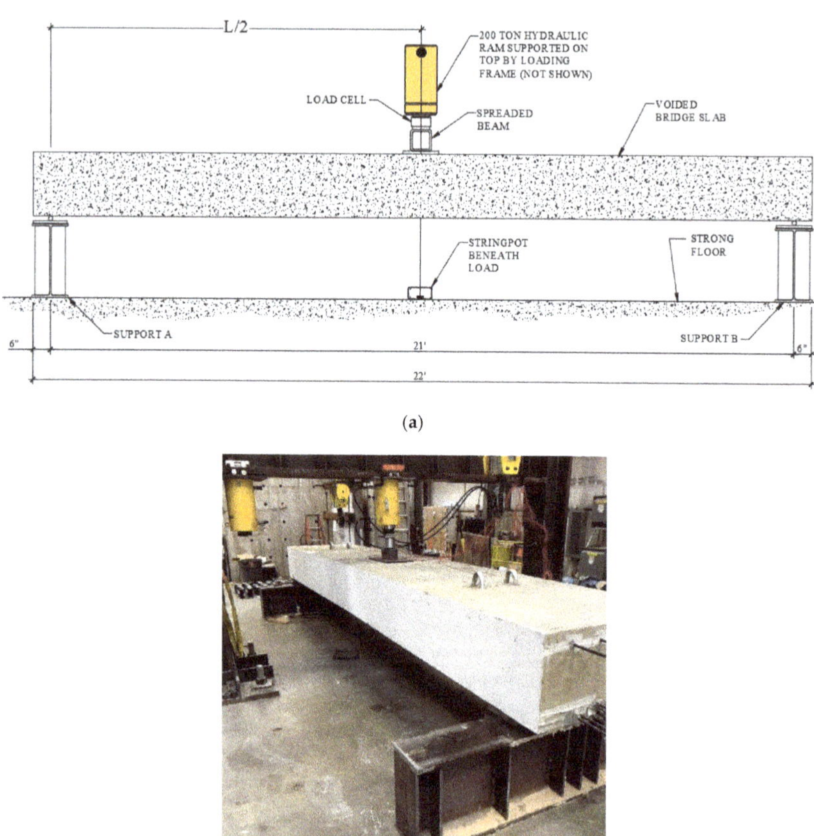

(**a**)

(**b**)

Figure 4. Midspan test setup and instrumentation: (**a**) test setup sketch and (**b**) test setup photo.

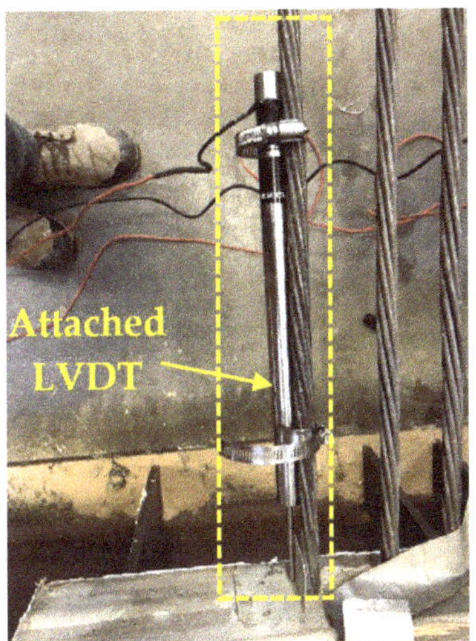

Figure 5. Strand end slip measurement using LVDT.

A test with the load placed at 1.25*d* from the face of the support, where *d* was the depth of the total section, was conducted after saw cutting the damaged half of the specimen during the midspan flexural test. Figure 6 shows the 1.25*d* test setup and instrumentation plan. The load was applied using two 400-kip hydraulic rams across the full width of the bridge deck using a spreader beam. Four LVDTs were mounted on the prestressing strands to measure the strand end slip. Wire potentiometers were placed on either side of the beam to measure the deflection at the applied load location.

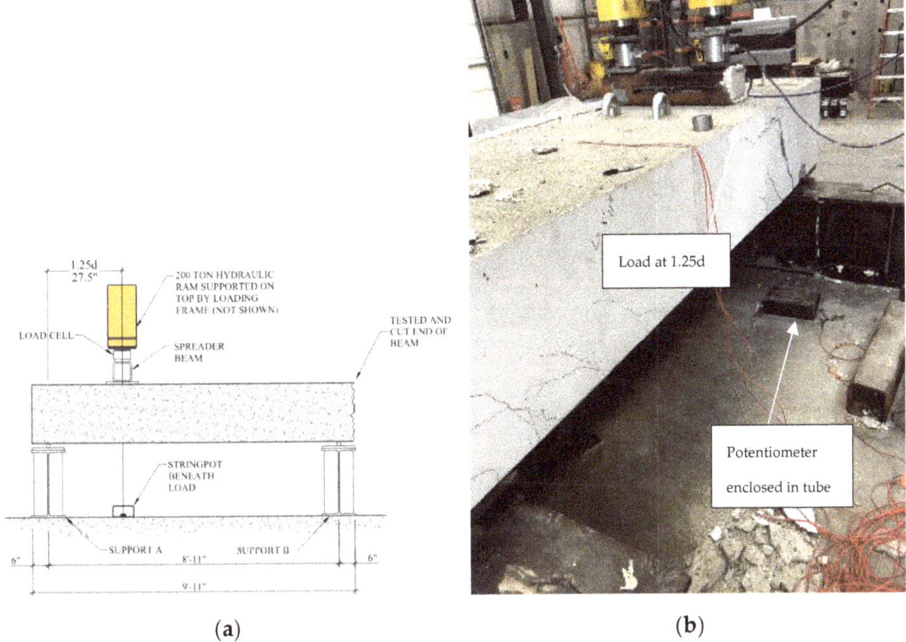

Figure 6. 1.25*d* test setup and instrumentation: (**a**) test setup sketch, and (**b**) test setup photo.

3. Experimental Results

3.1. Transfer Length

The transfer length of the prestressing strands was measured at the live and dead ends of the specimen using DEMEC strain gauges and strand end slippage immediately after the prestressing release. Table 3 lists the measured and predicted transfer length values using AASHTO LRFD and ACI 318-14 [AASHTO, PCI, Markosian, ACI]. The measured values using DEMEC strain gauges used the least squares regression procedure along with the 95% average maximum strain (95% AMS) method to calculate the strain development profile. The transfer length was also calculated using strand end slippage using Guyon's formula (Equation (1)) [28]:

$$\ell_t = \alpha \frac{E_{ps}}{f_{ps}} \Delta_{es} \quad (1)$$

Table 3. Measured and predicted transfer length.

Beam End	Measured Transferred Length (in.)		Predicted Transfer Length (in.)	
	DEMEC	Strand end slip	AASHTO LRFD (60 L_d)	ACI 318-14 (50 L_d)
Live end	15.9	23	36	30
Dead end	15.9	22.9	36	30

The average measured transfer lengths using DEMEC and end slippage procedures were 15.90 and 22.95 in, respectively. The measured transfer lengths were significantly less than the predicted values of 36 and 30 in using AASHTO LRFD and ACI provisions, respectively. Measured transfer lengths at both ends using both methods yielded the same values. Calculated transfer length using the strand end slip method (Equation (1)) provided 44.6% longer transfer length than that of using DEMEC method. A similar trend was reported by Bowser [16]. However, measured transfer lengths using DEMEC and strand end slip methods are well below the predicted values by ACI 318-14 [17] and AASHTO LRFD [18].

3.2. Prestress Losses

3.2.1. Creep

The BCSA cement concrete creep was measured in accordance with ASTM C512 and compared to AASHTO predicted creep values, as shown in Figure 7a. AASHTO LRFD [18] predicted creep values were calculated at a constant time step to form the creep strain–time relationship using Equation (2). Measured humidity of 50% was used to calculate the predicted values.

$$\varepsilon_{CR} = \Delta f_{pCR} / E_{ps} \quad (2)$$

3.2.2. Shrinkage

Figure 7b shows measured and predicted shrinkage strain using AASHTO LRFD code provisions. Total shrinkage strain was calculated by combining autogenous and drying shrinkage. AASHTO LRFD provisions counts only for drying shrinkage using Eq. (3). The measured drying and autogenous shrinkage were conducted in accordance with ASTM C157 and ASTM C1698 standards, respectively.

$$\varepsilon_{SR} = \Delta f_{pSR} / E_{ps} \quad (3)$$

3.2.3. Total Prestressing Losses

Figure 8 shows the measured and predicted losses using AASHTO LRFD code provisions [18]. Predicted losses were computed using a relative humidity of 20% (measured by electronic psychrometer during casting). Measured losses were calculated using VWSG readings at midspan and quarter points along the length of the specimen. VWSG readings were corrected for temperature using the manufacturer recommended procedures. Effective prestressing stress was computed using Equation (4):

$$f_{pe} = (\varepsilon_{pi} - \varepsilon) \times E_{ps} \tag{4}$$

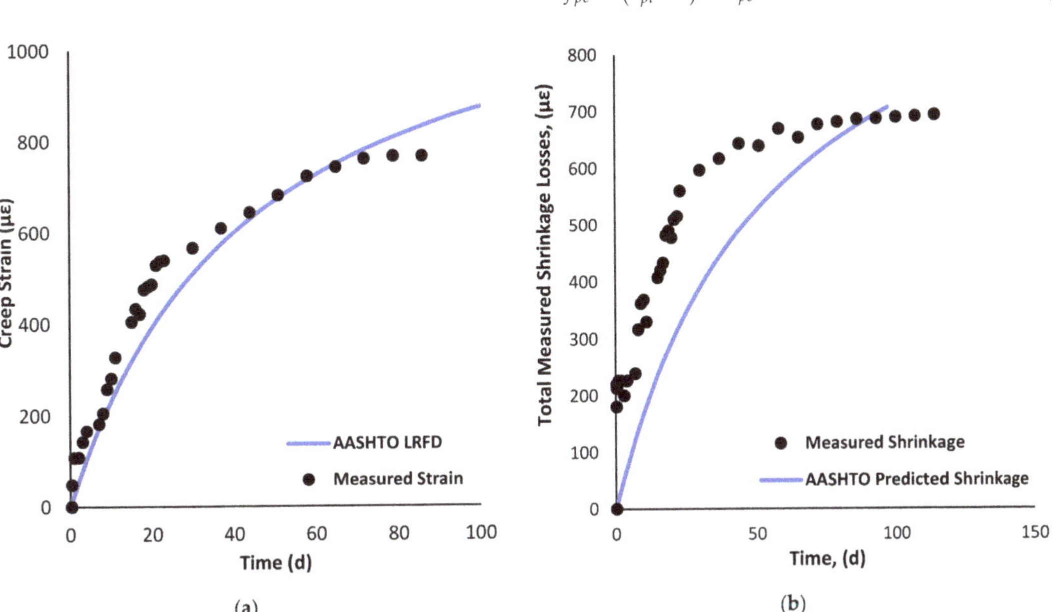

Figure 7. Measured vs. predicted: (**a**) creep strain; (**b**) shrinkage strain.

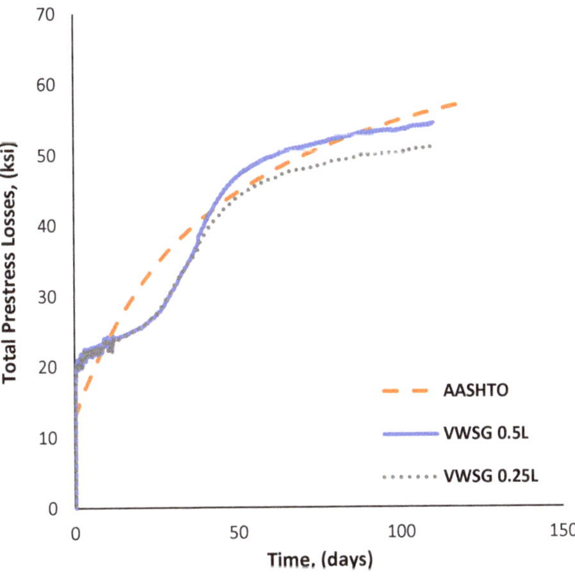

Figure 8. Average measured VWSG vs. predicted prestress losses.

3.2.4. Camber

Camber was measured immediately after transfer and over time until the time of testing. Measured camber after release was 0.28 in and at the time of testing was 1.26 in, which was 110 d after casting. The predicted initial and long-term camber using long-term PCI multipliers were 0.4 in and 0.972 in, respectively. The initial camber prediction was overestimated by 0.12 in and the long-term camber was underestimated by 0.288 in. This variation is not uncommon as the predicted values are approximate and the normal variations of the parameters used can cause ±20% deviation compared to predictions from Section 13.2.5 of the PCI Design Manual [27].

3.3. Beam Testing

3.3.1. Crack Initiation and Reopening

The load was applied monotonically at a constantly increasing rate until an initial crack was observed at midspan. Figure 9 shows the applied load–deflection relationship for the crack initiation test. The recorded load at the initiation of the first crack was 159.8 kips with a corresponding deflection of 0.008 in. Prestressing strand end slip was monitored during the crack initiation test and no end slippage was observed. After the observations of the crack initiation test were made, the load was released and the specimen was prepared for the crack reopening test. The applied load was increased until the load–deflection relationship experienced a nonlinear behavior and the initial crack was visible. Figure 10 shows the crack reopening load verses concrete strain across the crack. A least square regression analysis was performed on the load–strain relationship to identify the cracking load point at the intersection of the linear and nonlinear load–strain relationship. The measured crack reopening load was 83.12 kips with a strain reading of 190 $\mu\varepsilon$. The crack reopening load was used to estimate the effective prestressing stress in the beam using Equation (5):

$$f_{pe} = \frac{\frac{M_{tot} y_b}{I_n}}{\frac{A_p}{A_n} + \frac{A_p e_n y_b}{I_n}} \quad (5)$$

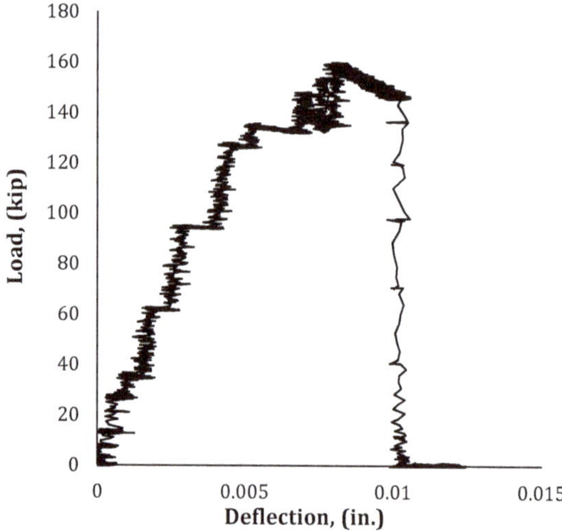

Figure 9. Load–deflection relationship for crack initiation test.

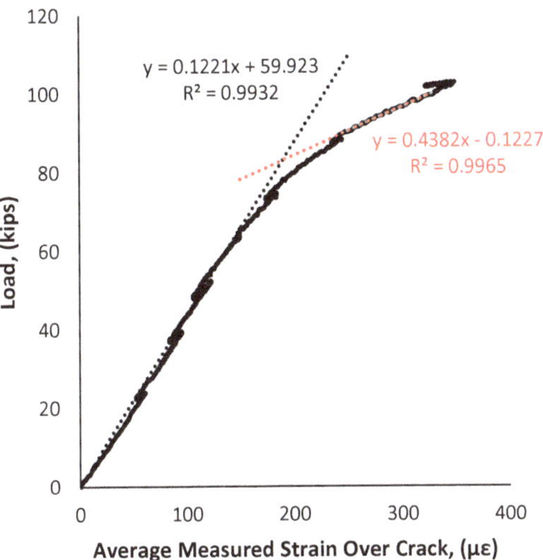

Figure 10. Crack reopening load–strain relationship.

The estimated effective prestressing stress was 138.2 ksi.

3.3.2. Load at Midspan

The crack reopening test was followed by an increased load until failure. Figure 11a shows the load–deflection relationship. Two distinct behaviors were observed, linear increase in deflection as the load resistance increased up to 160 kips, then non-linear deflection increase with lower rate of load resistance increase up to failure. Failure occurred at a maximum load of 258.6 kips, with a corresponding deflection of 2.1 in. The specimen failed suddenly in shear at mid-span following a diagonal crack at mid-span, as shown in Figure 11b. The failure was brittle as the load resistance suddenly dropped after the formation of the diagonal shear crack. Like the crack initiation test, the strand end slip was monitored throughout the test until failure and no strand end slippage was observed.

(a) (b)

Figure 11. (**a**) Load–deflection relationship and (**b**) failure mode when loaded at midspan.

3.3.3. Load at 1.25*d*

Figure 12a,b show the load-deflection and load-strand end slip relationships, respectively, for the 1.25*d* test. The specimen sustained a maximum applied load of 484.9 kips at a corresponding deflection of 0.194 in before significant strand slip. Flexural cracks directly under the load and shear-flexural cracks on the shear span were observed prior to failure, as shown in Figure 13. As the maximum load approached, the load resistance dropped as the strand slipped and flexural shear crack started to propagate parallel to the longitudinal axis of the member. The strand end slip was noticeable, as seen in Figure 12b, and audible during the test. The slipping of the strands corresponded to a loud metallic popping sound, followed by a noticeable jump in the slip at the time of the sound. Strand end slip corresponding to the maximum load of 484.9 kips was 0.0924 in. Final strand end slip was measured as 0.143 in after the load was released.

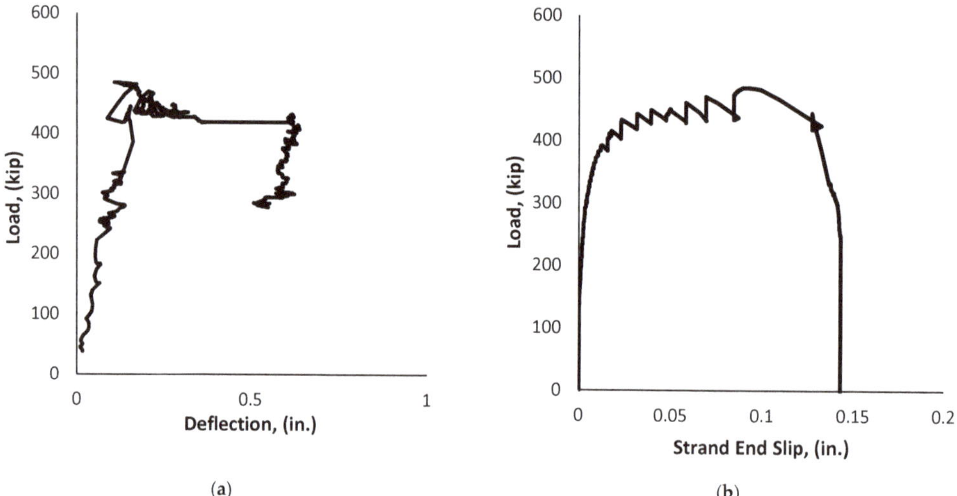

Figure 12. (**a**) Load–deflection relationship and (**b**) load-strand end slip when loaded at 1.25*d*.

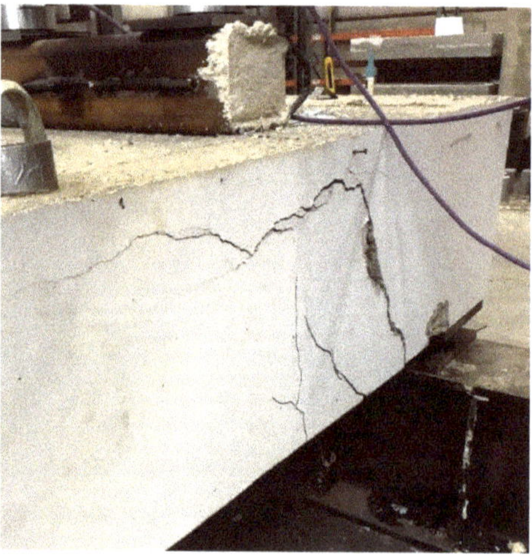

Figure 13. Failure mode when loaded at 1.25*d*.

4. Discussion

Measured creep strain for the BCSA cement was well predicted by AASHTO LRFD provisions, as shown in Figure 7. At an early loading age (first 60 days) predicted creep was nearly similar to the measured strains, however, after the first 60 days, the predicted creep strains values were overestimated.

The measured total shrinkage, which accounts for drying and autogenous shrinkage, was compared to predicted shrinkage values using AASHTO LRFD provisions. The predicted shrinkage strain values underestimated the measured values at an early age because AASHTO LRFD provisions considers drying shrinkage only and does not include autogenous shrinkage, which is significant in BCSA cement concrete. However, the long-term shrinkage strain was overestimated. The major reason for the underestimation in the early age of the concrete is that the autogenous shrinkage is measured without coarse aggregates. The coarse aggregates in concrete are not subject to shrinkage during curing as the paste in the concrete and it provides shrinkage restraint. Therefore, autogenous shrinkage strain, as measured using the standard test, is not necessarily accurate to include in the calculation of shrinkage losses. A more appropriate measure for this purpose would be to modify the autogenous shrinkage specimen such that it can accommodate large aggregate.

Prestress losses were measured using VWSGs cast into the concrete, material testing, and crack reopening test and the measured effective prestressing stress were 145.7, 149.5, and 138.2 ksi, respectively. Effective prestress for BCSA cement concrete was higher than the AASHTO predicted value (139.1 ksi) for all measured values of effective prestress except that of the crack reopening test. The average measured effective prestress for BCSA cement concrete was 144.5 ksi +/− 5.8 ksi. If the average measured effective prestress is taken as the actual effective prestress, then AASHTO predicts effective prestress for BCSA cement concrete with reasonable accuracy for the period investigated. Based on the material testing, it is likely that the AASHTO equations to estimate prestress losses will overestimate long-term losses. Further investigation is warranted.

The moment capacity of the specimen was calculated using moment curvature analysis, using Response 2000 software, and strain compatibility procedures [29]. The measured moment strength due to the applied point load at mid-span was 1290 kip-ft, while the predicted capacity using moment curvature analysis and strain compatibility were 1466.2 and 1458.4 kip-ft, respectively. Figure 14 shows the moment–curvature relationship calculated at mid-span. The specimen did not achieve the predicted nominal flexural strength due to shear failure at mid-span. The predicted shear strength using the modified compression field theory (MCFT) per AASHTO LRFD shear design provisions was 116.8 kips, while the applied shear at failure was 129.3 kips. Full bridge analysis showed that the ultimate moment at mid-span was associated with a corresponding ultimate shear of 25 kips, which is much lower than the applied shear force (129.3 kips), due to the concentrated point load at mid-span. Therefore, the observed shear failure mode will not occur in real bridge loading configuration.

Moment capacity at 1.25d was calculated using AASHTO LRFD provisions for flexural strength and using the bond-slip model. The bond-slip model developed in Figure 15 compares the AASHTO development length model to the measured development and transfer length. The AASHTO model shows a rapid stress increase in the prestressing strands from the end of the beam to the transfer length, with a slower gain in stress along the development length. The measured stress in the strands was found by performing a strain compatibility analysis using the power stress–strain formula [30] for calculating stress in the strands (f_{ps}). The value of f_{ps} given in the bond model represents the maximum stress in the strands at any location along the beam. A bond-slip failure occurs when the stress in the strands exceeds the maximum stress as given in the bond model. The measured stress in the strands due to the maximum applied load of 484.9 kips at the applied load location was calculated as 257 ksi. However, the AASHTO model predicted the maximum stress in the strands as 104.35 ksi. The extrapolated development length was calculated by extending the measured stress–strain curve at the same slope until reaching a maximum stress of 270 ksi. Additionally, the measured shear strength at 1.25*d* was compared to the predicted value using AASHTO LRFD MCFT procedures. The measured shear strength was 367 kips, while the predicted nominal shear strength was 227 kips. The demand for a bridge using the full-length girder only requires an ultimate shear at the critical section of 133 kips, which is significantly lower than the applied shear of 367 kips. Measured shear resistance was significantly higher than the predicted and required shear strength, which is not unusual for deck panel bridge type when testing close to the disturbed region [31].

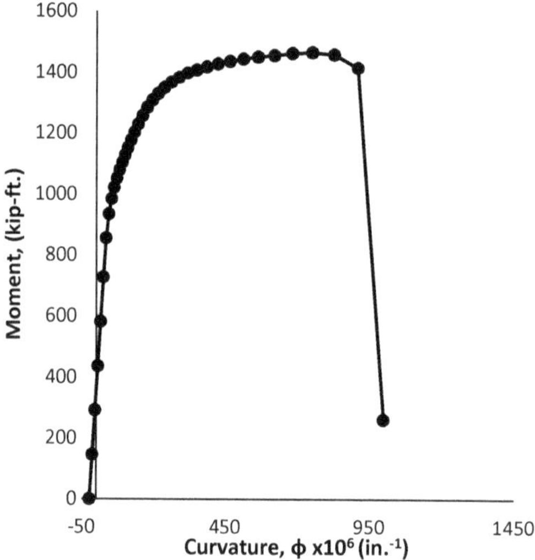

Figure 14. Moment–curvature relationship obtained from Response 2000.

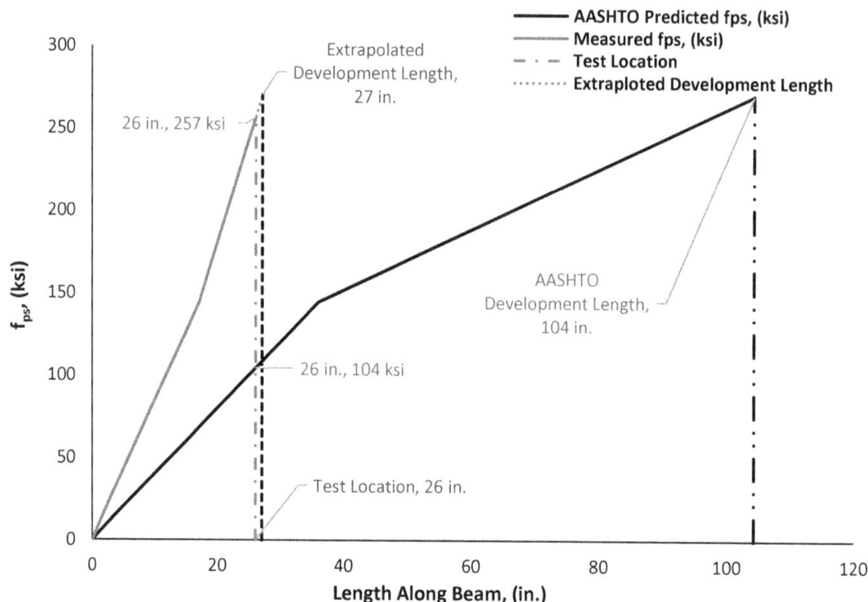

Figure 15. Prestressing strand bond model.

5. Summary and Conclusions

A precast, prestressed voided deck slab bridge girder mimicking a full-scale in-service bridge in Utah was cast using BCSA cement concrete. The full-scale specimen was tested in a variety of configurations, and the resulting performance was compared with the predictions from relevant design codes. The results of the study demonstrate the feasibility of using BCSA cement concrete—a sustainable alternative to traditional Portland cement concrete—in precast, prestressed concrete construction. However, the results also suggest that existing building codes do not always accurately predict the beam performance. Specifically, the results of this study suggest:

- Creep and shrinkage strains were generally well predicted using AASHTO for BCSA cement concrete during the period of monitoring, but if trends continued, likely would overestimate both shrinkage and creep. Based on these results AASHTO predictions for losses are likely accurate enough for design and will be conservative for long-term predictions.
- Autogenous shrinkage for BCSA cement is significantly larger than in Portland cement, therefore it is important to include this behavior when estimating losses, though this is not currently explicitly covered in AASHTO LRFD provisions.
- The average measured effective prestress for BCSA cement concrete was 144.5 ksi +/− 5.8 ksi and showed good agreement with the AASHTO LRFD-predicted effective prestress for the time period predicted indicating safe use of AASHTO LRFD effective prestress provisions for BCSA concrete.
- The maximum applied shear for the mid-span test was 129.0 kips, while the AASHTO predicted nominal shear strength was 116.8 kips which is commensurate with the observed shear failure. Bridge loading analysis indicated shear envelope design is only 25 kips, which is much lower than the applied shear force (129.3 kips), due to the concentrated point load. Therefore, the observed failure mode will not occur in real bridge loading configuration and was only observed because of the testing conditions. These results indicate prediction of shear capacity for BCSA concrete members is accurate and conservative for the situation tested.
- No strand-end slip was observed during the mid-span testing indicating strands were fully developed for the applied moment of 1290 kip-ft with a development length of 11 ft.
- During the 1.25*d* test, the specimen failed at 484.9 kips after considerable strand slip, indicating bond failure. The calculated strand stress (f_{ps}) at failure was 256.9 ksi, which significantly exceeded the predicted value of 104.35 ksi using AASHTO LRFD transfer and development lengths criteria.
- Using the measured transfer length and the strand stress results from the 1.25*d* bond failure test, the expected embedment length to achieve 270 ksi in the strand is 27 in, much shorter than the 104 in predicted by the AASHTO LRFD bond length model. This indicates the use of AASHTO LRFD can be conservative in predicting the transfer and development length for BCSA concrete for the strength and strands used in this study.

Author Contributions: Conceptualization, M.M. (Marc Maguire) and R.J.T.; methodology, M.M. (Marc Maguire) and R.J.T.; formal analysis, N.M., R.T.; investigation, N.M. and M.M. (Marc Maguire); writing—original draft preparation, N.M. and R.T.; writing—review and editing, all; visualization, N.M., R.T., M.M. (Mohammad Mastali); supervision, M.M. (Marc Maguire) and R.J.T.; project administration, M.M. (Marc Maguire); funding acquisition, M.M. (Marc Maguire) and R.J.T. All authors have read and agreed to the published version of the manuscript.

Funding: This paper is based on research supported by the Mountain Plains Consortium Project Number 560. Any observations, findings, recommendations, or conclusions presented in this publication are those of the authors and do not necessarily reflect the views of the Mountain Plains Consortium. The APC was funded by the University of Nebraska-Lincoln.

Institutional Review Board Statement: Not applicable.

Informed Consent Statement: Not applicable.

Data Availability Statement: The data presented in this study are available in the article.

Acknowledgments: The authors are grateful to Olympus Precast for their donation of the full-scale bridge member and use of their facilities.

Conflicts of Interest: The authors declare no conflict of interest. The funders had no role in the design of the study; in the collection, analyses, or interpretation of data; in the writing of the manuscript, or in the decision to publish the results.

Nomenclature

A_n	transformed area of cross section
A_p	area of prestressing strands
d	total depth of concrete section
E_c	static modulus of elasticity of concrete
E_d	dynamic modulus of elasticity of concrete
e_n	eccentricity of prestressing strands about center of gravity of cross section
E_{ps}	static modulus of elasticity of prestressing strand
f_{pe}	effective prestressing stress
f_{ps}	stress in prestressing strand just prior to transfer
f_{sp}	splitting tensile strength of concrete
I_n	transformed moment of inertia of cross section
ℓ_t	transfer length
M_{tot}	total applied moment including self-weight
y_b	distance from bottom concrete tension fiber to center of gravity of cross section
α	shape factor of the bond stress distribution constant ($\alpha = 2$ *or* 3 for constant or linear stress distribution, respectively). Constant stress distribution was assumed in this study.
Δ_{ps}	prestressing strand end slip
ε	measured strain in prestressing strand
ε_{pi}	initial strain in prestressing strand
ε_{CR}	predicted creep strain
Δf_{pCR}	prestress loss due to creep calculated using AASHTO LRFD Eq. 5.9.5.4.2b-1
Δf_{pSR}	prestress loss due to shrinkage calculated using AASHTO LRFD Eq. 5.9.5.4.2a-1

References

1. Klein, A. Calcium Aluminosulfate and Expansive Cements Containing Same. U.S. Patent 3,155,526, 3 November 1964.
2. Mehta, P.K.; Klein, A. Investigations on the Hydration Products in System 4CaO-3Al2O3-SO3-CaSO4-CaO-H2O. In *Highway Research Board Special Report*; Highway Research Board: Washington, DC, USA, 1966; pp. 328–352.
3. Ost, B.W.A.; Schiefelbein, B.; Summerfield, J.M. Very High Early Strength Cement. U.S. Patent 3,860,433, 14 January 1975.
4. Thomas, R.; Maguire, M.; Sorensen, A.; Quezada, I. Calcium Sulfoaluminate Cement. *Concr. Int.* **2018**, *40*, 65–69.
5. Habert, G. Assessing the environmental impact of conventional and 'green' cement production. In *Eco-efficient Construction and Building Materials*; Pacheco-Torgal, F., Cabeza, L.F., Labrincha, J., de Magalhães, A., Eds.; Woodhead Publishing: Sawston, UK, 2014; pp. 199–238.
6. Glasser, F.P.; Zhang, L. High-performance cement matrices based on calcium sulfoaluminate-belite compositions. *Cem. Concr. Res.* **2001**, *21*, 1881–1886. [CrossRef]
7. Ambroise, J.; Péra, J. Immobilization of calcium sulfate contained in demolition waste. *J. Hazard. Mater.* **2008**, *151*, 840–846. [CrossRef] [PubMed]
8. Jewell, R.B. Influence of Calcium Sulfoaluminate Cement on the Pullout Performance of Reinforcing Fibers: An Evaluation of the Micro-Mechanical Behavior. Ph.D. Thesis, Civil Engineering, University of Kentucky, Lexington, KY, USA, 2015.
9. Gartner, E. Industrially interesting approaches to "low-CO2" cements. *Cem. Concr. Res.* **2004**, *34*, 1489–1498. [CrossRef]
10. Gartner, E.; Sui, T. Alternative cement clinkers. *Cem. Concr. Res.* **2018**, *114*, 27–39. [CrossRef]
11. Bescher, E.; Stremfel, J.; Ramseyer, C. The Role of Calcium Sulfoaluminate in Concrete Sustainability. In Proceedings of the Twelfth International Conference on Recent Advances in Concrete Technology and Sustainability Issues, Prague, Czech Republic, 1 October 2012.
12. Bescher, E.; Kim, J. Belitic Calcium Sulfoaluminate Cement: History, Chemistry, Performance, and Use in the United States. In Proceedings of the 1st International Conference on Innovation in Low Carbon Cement and Concrete Technology, London, UK, 25 June 2019.

13. Maggenti, R.; Gomez, S.; Luena, R. Bridge Hinge Reconstruction. *Struct. Mag.* **2015**, 30–32. Available online: https://www.structuremag.org/wp-content/uploads/2014/12/F-BridgeHinge-Maggenti-Jan151.pdf (accessed on 14 July 2021).
14. Floyd, R.W.; Ramseyer, C. Behavior of Precast, Prestressed Calcium Sulfoaluminate Cement Concrete Beams. In Proceedings of the PCI Convention and National Bridge Conference, Nashville, TN, USA, 5 March 2016.
15. Murray, C.D.; Floyd, R.W.; Ramseyer, C.C.E. Using belitic calcium sulfoaluminate cement for precast, prestressed concrete beams. *PCI J.* **2019**, *64*, 55–67. [CrossRef]
16. Bowser, T. Development Length of 0.6 in. In *Prestressing Strands in Precast, Prestressed Calcium Sulfoaluminate Cement Concrete, M.S. in Civil Engineering*; University of Oklahoma: Norman, OK, USA, 16 December 2016.
17. ACI Committee 318. *Building Code Requirements for Structural Concrete (ACI 318-14) and Commentary (ACI 318R-14). Farmington Hills*; American Concrete Institute: Michigan, MI, USA, 2014.
18. American Association of State Highway and Transportation Officials (AASHTO). In *LRFD Bridge Design Specifications*, 7th ed.; AASHTO: Washington, DC, USA, 2014.
19. ASTM C403/C403M-16. *Standard Test Method for Time of Setting of Concrete Mixtures by Penetration Resistance*; ASTM International: West Conshohocken, PA, USA, 2016.
20. ASTM C39/C39M-18. *Standard Test Method for Compressive Strength of Cylindrical Concrete Specimens*; ASTM International: West Conshohocken, PA, USA, 2018.
21. ASTM C1698-14. *Standard Test Method for Autogenous Strain of Cement Paste and Mortar*; ASTM International: West Conshohocken, PA, USA, 2014.
22. ASTM C496/496M-17. *Standard Test Method for Splitting Tensile Strength of Cylindrical Concrete Specimens*; ASTM International: West Conshohocken, PA, USA, 2017.
23. ASTM C469/C469M-14e1. *Standard Test Method for Static Modulus of Elasticity and Poisson's Ratio of Concrete in Compression*; ASTM International: West Conshohocken, PA, USA, 2014.
24. ASTM C215-19. *Standard Test Method for Fundamental Transverse, Longitudinal, and Torsional Resonant Frequencies of Concrete Specimens*; ASTM International: West Conshohocken, PA, USA, 2019.
25. ASTM C157/C157M-17. *Standard Test Method for Length Change of Hardened Hydraulic-Cement Mortar and Concrete*; ASTM International: West Conshohocken, PA, USA, 2017.
26. Markosian, N. Calcium Sulfoaluminate Cement Concrete for Prestressed Bridge Girders: Prestressing Losses, Bond, and Strength Behavior. Master's Thesis, Utah State University, Logan, UT, USA, 2019.
27. *PCI Bridge Design Manual*, 3rd ed.; Precast/Prestressed Concrete Institute: Chicago, IL, USA, 2014.
28. Guyon, Y. *Prestressed Concrete*; John Wiley & Sons: New York, NY, USA, 1960.
29. Bentz, E.C. Sectional Analysis of Reinforced Concrete Members. Ph.D. Thesis, University of Toronto, Toronto, ON, Canada, 2000.
30. Tawadrous, R.; Morcous, G. Shear Strength of Deep Hollow-Core Slabs. *ACI Struct. J.* **2018**, *115*, 699–709. [CrossRef]
31. Devalapura, R.K.; Tadros, M.K. Stress-Strain Modeling of 270 ksi Low-Relaxation Prestressing Strands. *PCI J.* **1992**, *32*, 100–106. [CrossRef]

MDPI

Article

Shear Bond between Ultra-High Performance Fibre Reinforced Concrete Overlays and Normal Strength Concrete Substrates

Sara Javidmehr and Martin Empelmann *

iBMB, Division of Concrete Construction of TU Braunschweig, D-38106 Braunschweig, Germany; s.javidmehr@ibmb.tu-bs.de
* Correspondence: massivbau@ibmb.tu-bs.de; Tel.: +49-531-391-5409

Abstract: Strengthening or retrofitting of existing structures is a more sustainable and resource-efficient solution than replacing them with new constructions. To enhance the performance and effectiveness of strengthening works the use of high-performance materials is a promising method. Using ultra-high performance fibre reinforced concrete (UHPFRC) as supplementary concrete is one of such solutions leading to high structural resistance and better durability. For such UHPFRC overlays the shear bond resistance of the interface between the existing substrate, usually normal strength concrete (NSC), and the UHPFRC is a significant design aspect. This paper presents the results of push-off tests conducted on NSC-UHPFRC specimens, which were produced with different substrate treatment methods. Using different surface measurement techniques including the sand patch method and digital microscopy, the effects of substrate roughness and treatment method on shear bond behaviour and failure mechanisms are investigated, and the results are analysed with design approaches and further calculation models in the technical literature. Based on the results, the significance of considering roughness parameters and failure mode for the design of high-performance overlays is highlighted. Furthermore, the effectiveness of different substrate treatment methods is discussed and an effective treatment method is suggested.

Keywords: shear bond; UHPFRC; push-off test; tensile bond strength; concrete overlay; strengthening; existing infrastructures; digital microscopy; surface roughness

Citation: Javidmehr, S.; Empelmann, M. Shear Bond between Ultra-High Performance Fibre Reinforced Concrete Overlays and Normal Strength Concrete Substrates. *Sustainability* **2021**, *13*, 8229. https://doi.org/10.3390/su13158229

Academic Editors: Fausto Minelli, Enzo Martinelli and Luca Facconi

Received: 9 June 2021
Accepted: 19 July 2021
Published: 23 July 2021

Publisher's Note: MDPI stays neutral with regard to jurisdictional claims in published maps and institutional affiliations.

1. Introduction

With the growing demand for rehabilitation and strengthening works, the necessity for more efficient and sustainable strengthening strategies expands. One very promising option is the use of overlays made of high-performance materials. For concrete floors and decks of, e.g., bridges with deficient bending and shear resistance, the application of a thin, and thus weight-reduced, layer of ultra-high performance fibre reinforced concrete (UHPFRC), instead of ordinary overlays with the same load-bearing capacity, decreases the dead load of the strengthened structure significantly [1–6]. At the same time, due to the very compact microstructure of ultra-high performance concrete (UHPC), the toppings or overlays made of UHPFRC slow down the diffusion of corrosive substances plus water absorption and enhance the freeze-thawing resistance and abrasion resistance of structures, so that the overall durability of the structure is enhanced significantly [7–9]. By adding fibres, further advantageous properties such as reduced crack widths and improved shrinkage control can be used [10–14].

To achieve the best structural and economic benefits, the strengthened concrete members should show a nearly monolithic behaviour, which has to be assured by the shear bond strength of the interface [15]. The basic load-bearing mechanisms contributing to shear bond strength between concrete cast at different times (concrete-concrete bond) are defined in the shear-friction theory [16] and have been investigated experimentally and modelled in several theoretical approaches [17–28]. According to the shear friction theory, the following load-bearing mechanisms contribute to the shear bond strength of interfaces without the application of shear connectors:

- adhesion
- cohesion (mechanical interlocking)
- friction due to external compressive and clamping forces

Adhesion is caused by various chemical and physical mechanisms in microscale (e.g., covalent or ionic bonds, Van der Waals forces) and is dependent on the properties of the new and old concrete, and the quality of substrate, as well as wettability, permeability and moisture content of old concrete substrate [20,25]. The cohesion is activated after the decay of adhesion with a relative horizontal and vertical displacement of the interface and is mainly affected by the mechanical interlocking caused by micro- as well as macro-roughness (compare Figure 1) [26]. The friction in the contact area is activated in a macroscopic scale only in the presence of lateral compression, which is of secondary importance for flexural members with concrete overlays.

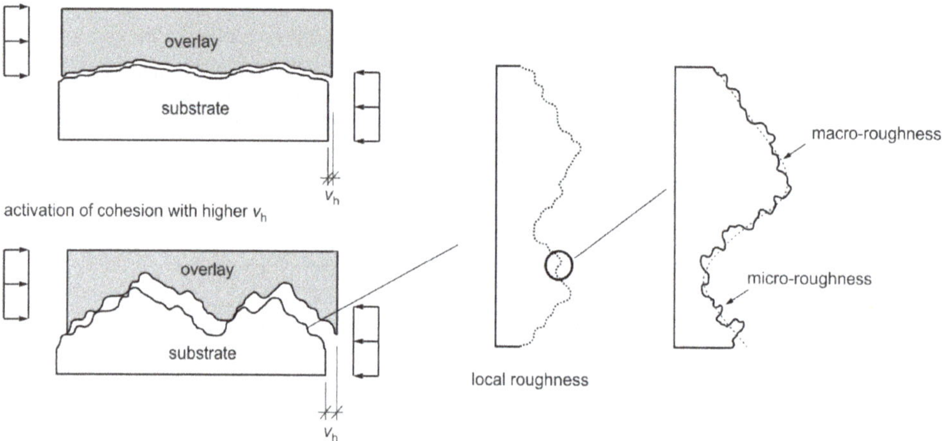

Figure 1. Adhesion and cohesion activation in different loading stages in a simplified grain model (**left**) by various horizontal displacements; micro- and macro-roughness of the real interface (**right**).

Existing research works explain the general shear bond mechanisms between normal strength concrete mixtures cast at different times (e.g., precast elements with cast-in-place concrete). Despite several research works, there is still no consensus on the effects of interface roughness on the adhesive and cohesive bond in both shear and tension. Regarding the shear bond between NSC and UHPFRC, a few experimental investigations are also available [22–25]. However, the possible interaction between adhesion and cohesion as well as their effects on the failure modes of NSC-UHPFRC interfaces are still unknown. Based on existing models and experiments, some calculation approaches exist, which are limited to NSC-NSC interfaces and suppose higher shear and adhesion bond with increasing interface roughness. The interface roughness has been classified in early works merely based on qualitative assessment. Currently, measured roughness parameters are used for this classification. However, higher-level quantitative approaches are required for a realistic and economic calculation of NSC-UHPFRC shear bond strength.

This paper tends to investigate the aforementioned points. First, a short overview of existing quantitative approaches for the calculation of shear bond strength is given. Afterwards, the experiments conducted by the authors on the shear bond between NSC and UHPFRC are presented that include different interface treatment methods and roughnesses.

The obtained experimental results are then compared with the calculative approaches and the resulted failure modes are discussed in terms of adhesion and cohesion. Finally, the possible advantages of UHPRFC overlays in combination with treatment methods and required interface roughness are analysed.

2. Calculation of Shear Bond Strength

2.1. Interface Roughness Parameters

Concrete interfaces are classified in EN 1992-1-1 (EC2) [29] and Model Code 2010 (MC2010) [30] based on roughness as "very smooth", "smooth", "rough" and "very rough". For the classification of interface roughness, the "peak to mean" roughness value R_t, determined using the sand patch method [31] is used in the mentioned standards and design code provisions (compare Table 1). With this method, a sand volume V of 25 cm^3 is distributed in a cylindrical form over the macro-texture surface. Using the measured diameter d (compare Figure 2), the "peak to mean" roughness value R_t is determined using Equation (1):

$$R_t = \frac{40 \cdot V}{\pi \cdot d^2} \quad (1)$$

Table 1. Defined surface roughness categories according to EC2 and MC2010.

Category	Peak to Mean Roughness Value R_t
very rough (e.g., using water jetting or grooving)	≥3.0 mm
rough (e.g., using water jetting or sand blasting)	≥1.5 mm
smooth	<1.5 mm
very smooth	not measurable

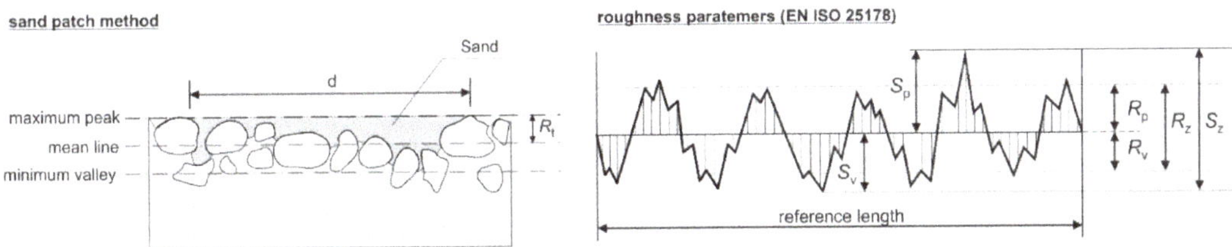

Figure 2. Defined roughness parameter R_t using sand patch method (**left**) and roughness parameters according to EN ISO 25178 (**right**).

An alternative, and more exact, method is using digital microscopy. The 3D-measured topology of the interface provides valuable data on the micro-texture of the interface and can be quantified afterwards using different depth and surface roughness parameters, e.g., following the definitions in EN ISO 25178 [32]. In this context, the maximum height S_z, the maximum peak height S_p and the maximum valley depth S_v (from the mean line) as well as mean peak to mean valley depth R_z are useful parameters (compare Figure 2).

Beside the height parameters, EN ISO 25178 defines also hybrid parameters such as the developed interfacial area ratio S_{dr}, which quantifies the increased contact area A_{net} in the interface between overlay and substrate in comparison to the shear plane area A_{sh} according to Equation (2).

$$S_{dr}\ [\%] = \frac{(A_{net} - A_{sh})}{A_{sh}} \times 100 \quad (2)$$

The height parameters of roughness are reasonable parameters for the definition of cohesion in a rough interface, whereas the area parameters, such as S_{dr}, help by the quantification of roughness effects on interface adhesive bond.

2.2. Design Approaches Based on Roughness Categories

For design of shear interface between the existing concrete substrate (NSC) and the ultra-high-performance concrete overlay (UHPFRC), the acting shear stress should be limited to the calculated ultimate shear resistance τ_{cal}. Several design codes provide equations based on shear friction theory, according to which the ultimate shear resistance τ_{cal} consists of adhesive bond and the mechanical interlocking for interfaces without lateral compression and without shear connectors. These two load-bearing mechanisms are expressed, e.g., in EC2 [29] and MC2010 [30] in Equation (3)

$$\tau_{cal} = c \cdot f_{ctd} \leq 0.5 \cdot \nu \cdot f_{cd} \quad (3)$$

where:
c coefficient depending on interface roughness
f_{ctd} design value of concrete tensile strength
v effectiveness factor of concrete
f_{cd} design value of concrete compressive strength

In EC2 and MC2010, the design value of concrete tensile strength f_{ctd} is determined for normal strength concrete as $\alpha_{ct} \cdot f_{ctk,0.05} / \gamma_c$ ($\alpha_{ct} = 0.85$ is a factor considering sustained load effects, $f_{ctk,0.05}$ corresponds to 5 % concrete tensile strength and $\gamma_c = 1.35$ is the partial safety factor). In the current draft of the new Eurocode 2 prEN 1992-1-1 [33] (prEC2), $\sqrt{f_{ctk,0.05}} / \gamma_c$ is used in Equation (3) instead of f_{ctd}. The suggested values for the coefficient c and the effectiveness factor v are summarized in Table 2.

Table 2. Coefficients for different surface roughness categories according to EC2, MC2010 and prEC2.

Category	Peak to Mean Roughness Value R_t	Coefficient c		Effectiveness Factor v		
		EC2/MC2010	prEC2	EC2	MC2010	prEC2
very rough (e.g., using water jetting or grooving)	≥3.0 mm	0.50	0.19	0.70		
rough (e.g., using water jetting or sandblasting)	≥1.5 mm	0.40	0.15	0.50	$0.55 \cdot (30/f_{ck})^{\frac{1}{3}} < 0.55$	0.25
smooth	<1.5 mm	0.20	0.075	0.20		
very smooth	not measurable	0.025	0.0095	0.0		

As shown in Table 2, the methods consider a stepwise increase of the shear bond strength values for different roughness categories and do not consider the quantified roughness explicitly. The changes in the considered value of tensile strength in the new Eurocode draft also show the missing consensus on the correlation between the shear bond strength and tensile strength of concrete. It is also worth noting that in all approaches, the lowest concrete strength should be considered for the calculation of τ_{cal}, if substrate and overlay are made of two different strength classes. Moreover, all approaches are only valid for concrete strength classes up to C100/115. Thus, the use of bond coefficients according to a higher specific adhesion of UHPFRC is not possible with the referred design approaches.

2.3. Calculation Approaches Using Roughness Parameters

2.3.1. Approach Proposed by Gohnert

Gohnert [34] was the first to propose a method which considers explicitly the difference between the mean peak height and the mean valley depth R_z (in mm) as a roughness parameter to determine the concrete-concrete bond strength τ_{cal} according to Equation (4):

$$\tau_{cal} = 0.209 \cdot R_z + 0.7719 \quad (4)$$

The proposed equation is based on precast elements made of normal strength concrete (compressive strength between 22.8 MPa and 56.2 MPa) with a ribbed interface produced by different methods at various production sites.

2.3.2. Approach Proposed by Santos and Júlio

A further calculation approach proposed by Santos and Júlio [35] considers the mean valley depth R_v (in mm) as a roughness parameter to define the coefficient c accordingly:

$$\tau_{cal} = c \cdot f_{ctd} = \frac{1.062 \cdot R_v^{0.145}}{\gamma} \cdot f_{ctd} \leq 0.5 \cdot f_{cd} \quad (5)$$

This empirical power function in Equation (5) correlates the roughness parameter R_v determined based on 2D profiles of interfaces prepared with different treatment methods in combination with a partial safety factor γ that is suggested to be $\gamma = 2.6$.

2.4. Overview of Existing Calculation Approaches for NSC-UHPFRC Shear Bond

As shown in the previous sections, the design approaches consider surface categories using a very simple method for the determination of macro-roughness. There are only a few approaches in the literature which include roughness parameters in the calculation of shear bond strength of the interface explicitly. Overall, the existing approaches are mostly validated on normal strength concrete overlays. The different contribution of adhesion and cohesion as well as their possible effects on the failure plane have not been investigated entirely until now. For UHP(FR)C overlay, there are some specifications in AFGC/Setra [36] as well as in SIA 2052 [37] prescribing interface design requirements that are not considered in the present paper.

In conclusion, for the design of UHPFRC overlays enhanced calculation approaches using interface roughness parameters are necessary, making the avoidance of shear connectors possible and also minimizing the interface treatment effort leading to a more rapid and resource-efficient strengthening. The following experimental investigations and evaluations of test results in this paper aim to investigate the above-mentioned points.

3. Experimental Investigation of NSC-UHPFRC Bond Strength

3.1. Test Specimens and Testing Programme

Within the experimental investigations in a research project [38] a total number of 31 tests on 3 monolithic reference specimens and 28 NSC-UHPFRC specimens were conducted. The reference push-off tests were made of normal strength concrete (NSC) with a concrete strength grade C30/37. The push-off tests for the investigation of NSC-UHPFRC bond were made of an NSC (C30/37) substrate, which was cast first. The NSC substrate was cured for one day under a plastic sheet in the mould and stored after demoulding in a controlled atmosphere for about 35 days. The testing programme included differently treated interfaces, which were: left-as-cast and brushed (B), sandblasted (S), water-blasted (J) and grooved (G). Figure 3 shows the observed surface topologies of (B), (S), (J) and (G) test series.

The specimens labelled with (B) were wire-brushed. For specimens labelled as (BD) the UHPFRC was cast against a dry wire-brushed substrate, whereas the substrates of all further test series are pre-wetted before casting. Through the sandblasting with a 0.4 mm sand the fine grain was exposed. In the water-jetting treatment, a water pressure of 2.500 bars was jetted on the interface from a distance of 25 cm. With the grooving treatment technique, a well-defined interface texture was produced (compare Figure 4).

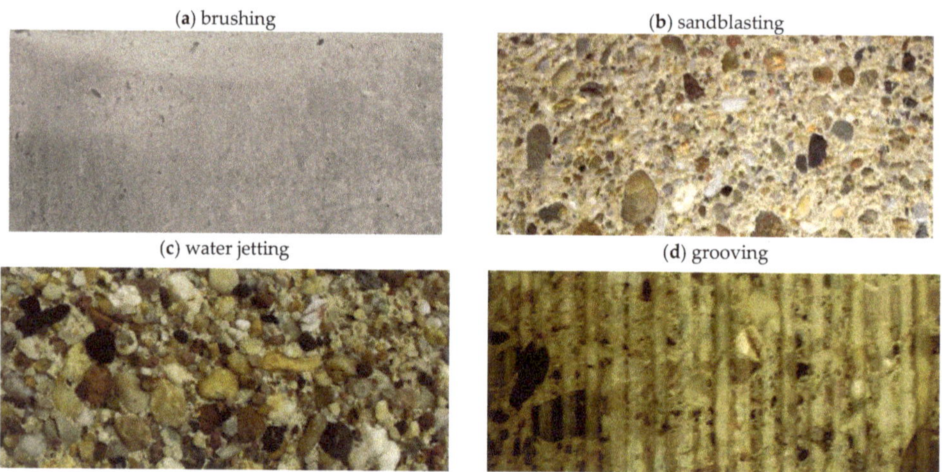

Figure 3. Comparison between the surface topology of the brushed (**a**), sandblasted (**b**), water-jetted (**c**) and grooved (**d**) substrates.

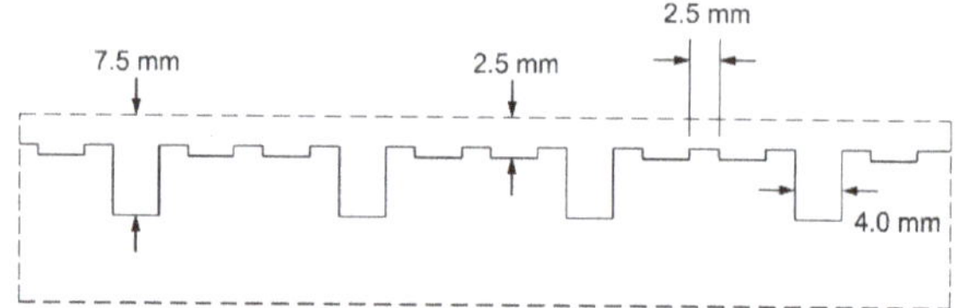

Figure 4. Interface geometry of grooved interface.

After substrate treatment and preparation, an overlay made of an UHPFRC (MQ4) with 1.25 Vol. % fibre content (Weidacon FM 0.19/13 with l_f/d_f = 13.0/0.19 mm ≈ 68) was added to the NSC substrate after 28 days (see Figure 5, left). Table 3 provides the mixture properties of substrate and overlay concrete. The flowability of the fresh NSC and UHPFRC was measured using slump test according to EN 12350-5 38, which was 56 cm for the NSC and 66.5 cm for the UHPFRC. The dimensions and reinforcement configuration of the test specimens are shown also in Figure 5.

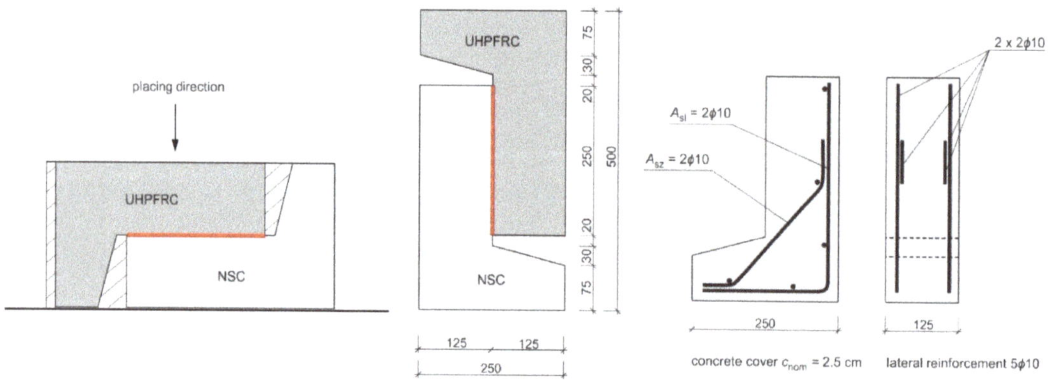

Figure 5. Casting and placing of UHPFRC, specimen dimensions in [mm] and reinforcement configuration of push-off test specimens.

Table 3. Concrete mixture of NSC and UHPFRC.

NSC C30/37		UHPFRC M4Q-1.25	
Components	**[kg/m³]**	**Components**	**[kg/m³]**
Cement CEM II / B-S 42.5N	360.0	Cement CEM I 52.5 R	795.4
Silica fume	50.0	Silica fume	168.6
Superplasticiser	1.1	Superplasticiser	24.1
Quartz sand 0/2 mm	706.0	Quartz powder	198.4
Quartz sand 2/8 mm	531	Quartz sand 0.125/0.5 mm	971.0
Quartz sand 8/16 mm	525	Steel fibres (Weidacon FM 0.19/13.0)	99.39
Water	185	Water	188

In addition to the test specimens, reference cylindrical specimens (Ø/h = 150/300 mm) were produced from the same batch for the investigation of material properties of hardened NSC and UHPFRC. The reference tests were conducted at the age of testing, which was 35 days for NSC and 7 days for UHPFRC [39]. The mean values of material properties are provided in Table 4 at the testing age of the push-off specimens.

Table 4. Material properties of hardened NSC and UHPFRC.

	NSC (Age of 35 Days)			UHPFRC			NSC-UHPFRC
Specime	**Splitting Tensile Strength [40]** $f_{ct,sp,NSC}$ **[N/mm²]**	**Compressive Strength [41]** $f_{c,NSC}$ **[N/mm²]**	**Young's Modulus [42]** $E_{c,NSC}$ **[N/mm²]**	**Compressive Strength** $f_{c,UHPFRC}$ **(Age of 7 Days)**	**Compressive Strength** $f_{c,UHPFRC}$ **(Age of 28 Days)**	**Young's Modulus** $E_{c,UHPFRC}$ **[N/mm²]**	**Tensile Bond Strength** $f_{ct,bond}$ **[N/mm²]**
monolithic	2.3	29.9	32,833	-	-	-	2.0
BD	2.9	35.3	30,000	101.0	154.9	43,133	-
B	2.7	35.3	29,700	119.5	157.9	45,500	-
S	2.7	34.6	27,367	106.0	-	42,833	2.2
J	2.8	36.2	33,833	101.8	145.8	41,567	2.7
G	3.0	38.1	34,800	115.5	161.8	43,900	3.0

It should be noted that the properties of the hardened UHPFRC in Table 4 were determined at the testing age of push-off experiments. For the strength classification of the used UHPFRC mixture, further compression tests were conducted at an age of 28 days.

Besides, tensile bond strength values of the NSC-UHPFRC interface $f_{ct,bond}$ were determined at the age of testing (35 days) using pull-off tests, which are listed in Table 4 for different substrate treatment methods. The plate samples of pull-off tests are illustrated in Figure 6. On each plate, 5 ring notches were drilled, the stamp was glued on each ring and a tensile load was applied with a pull-off tester. During the drilling of notches, the samples with brushed surfaces (B and BD test series) were damaged, unfortunately, so that no pull-off test results are available for these samples.

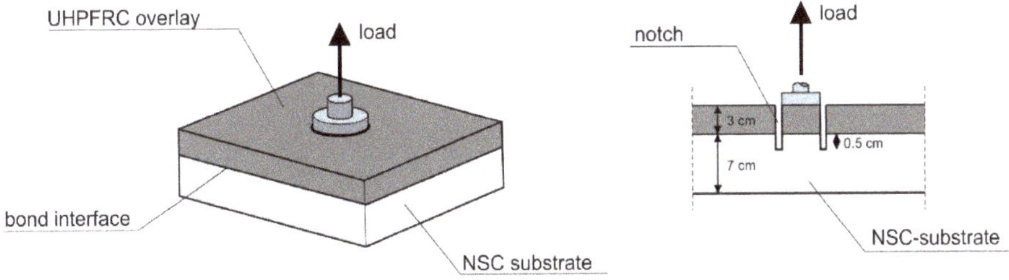

Figure 6. Dimensions and reinforcement configuration of pull-off test specimens.

3.2. Interface Quality and Quantification of Roughness

After interface treatment, each surface was inspected visually to detect possible damages or specimen features of each method (compare Figure 7). The inspections showed in some water-jetted specimens similar damages of aggregate and aggregate-cement paste bond in form of microcracks or aggregate splitting as depicted in Figure 7b.

Figure 7. Detailed imaging of a sandblasted interface (**a**) and a water-jetted interface (**b**).

For a quantified evaluation of interface roughness for different treatment methods the "peak to mean" roughness values R_t were determined using the sand patch method. Furthermore, the surface micro-texture was measured using a digital microscope and height as well as surface parameters, i.e., S_p and S_{dr}, were determined in accordance to EN ISO 25178 29 using the software MountainsLab® [43]. Each interface was measured in six measurement sections with a reference area of 6 × 5 mm. The mean value of surface roughness parameters for each surface treatment method is documented in Table 5.

Table 5. Roughness parameters of interfaces with different preparation methods.

Specimen	Peak to Mean Roughness Values R_t [mm]	Maximum Peak Height S_p [mm]	Mean Height S_a [mm]	Maximum Height S_z [mm]	Developed Interfacial Area Ratio S_{dr} [%]
BD	-	0.15	0.018	0.82	18.6
B	-	0.15	0.017	0.79	16.6
S	0.52	0.24	0.04	1.65	37.8
J	2.87	0.98	0.11	6.52	163.8
G	1.81	0.45	0.05	3.63	79.9

Figure 8 shows exemplarily the results of surface measurements using digital microscopy for different interface preparation methods.

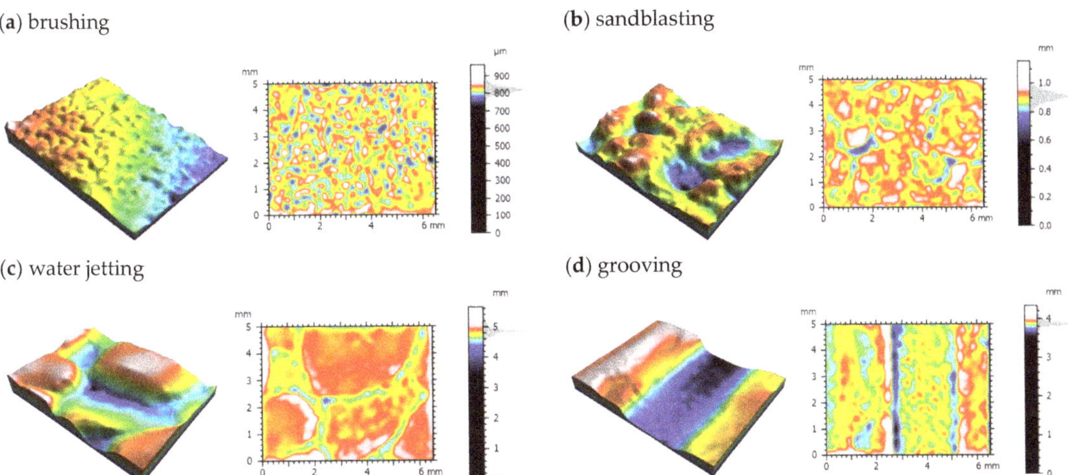

Figure 8. Representative measured surface topology with digital microscopy for the brushed (**a**), sandblasted (**b**), water-jetted (**c**) and grooved (**d**) substrates.

It should be noted that, because of the described measurement procedure (several measurements of one surface), the maximum values of peak and height correspond to the mean values despite using the definitions and notation of EN ISO 25178 in Table 5 ($S_v \approx R_v$; $S_z \approx R_z$). Moreover, most roughness parameters are shown to be correlated with each other, which is affected, e.g., by the microstructure of the substrate (especially the aggregate size). For the NSC substrate of the tested specimens, a correlation between the peak to mean roughness R_t determined using the sand patch method and the maximum peak height S_p is shown in Figure 9.

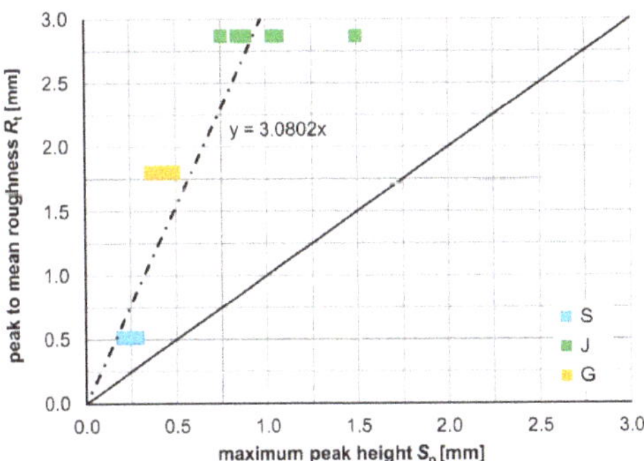

Figure 9. Correlation between peak to mean roughness value R_t and maximum peak height S_p (average value).

3.3. *Testing Procedure*

The load was applied displacement-controlled with a loading rate of 0.001 mm/s over steel load plates (w/t = 250 mm/20 mm). To avoid eccentric loading, the machine load was introduced over a spherical cap into the upper load plate. During the tests, the vertical and horizontal displacement of the joint as well as the strain of the NSC substrate and the UHPFRC overlay were monitored by LVDTs and strain gauges (compare Figure 10). The strain gauges were placed following the principal compressive stress trajectories in the NSC and UHPFRC.

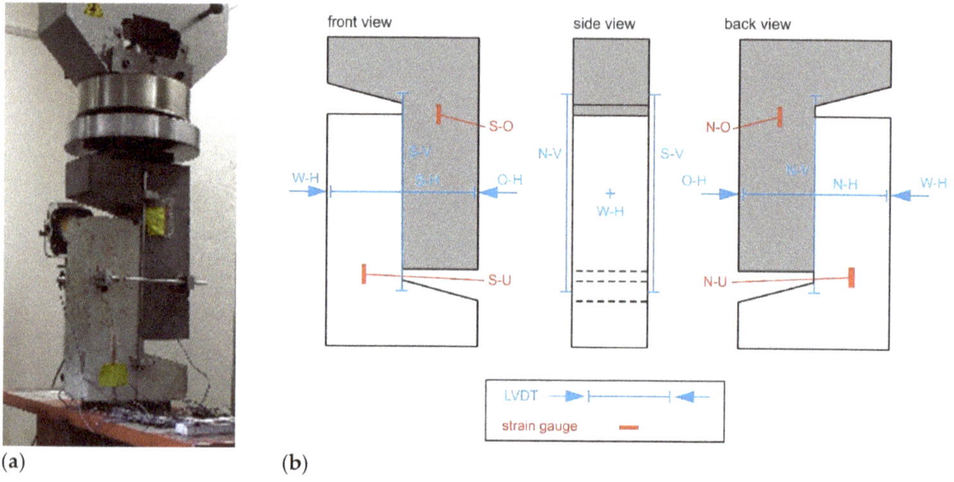

Figure 10. Setup of push-off tests (**a**), applied strain gauges and LVDTs to the test specimens (**b**).

4. Test Results

4.1. Tensile Bond Strength

Based on the discussion in Section 2.1, the increased roughness affects the effective bond surface, which can be considered with the surface parameter S_{dr}. In this context, a modified tensile bond strength $f^*_{ct,bond}$ is defined as:

$$f^*_{ct,bond} = \frac{f_{ct,bond}}{1 + S_{dr}/100} \quad (6)$$

The tensile bond strength values are compared in Table 6 with the uniaxial tensile strength f_{ct} ($0.9 \cdot f_{ct,sp}$) of the substrate. It should be noted that the uniaxial tensile strength f_{ct} is determined as $0.9 \cdot f_{ct,sp}$ according to the recommendation in [31]. The comparison shows that the bond strength is lower than the uniaxial tensile strength for all specimens and lower in the case of water-jetting compared to sand-blasted and grooved interfaces, which can be reasoned by the observed pre-damaged substrates of water-jetted interfaces (compare Figure 7). A further observation was the changing failure mode of pull-off tests with increasing surface roughness. Some grooved (G) and sandblasted interfaces (S) showed a mixed failure mode with the failure plane locating in both NSC and UHPC. This might be also an additional reason for the higher tensile bond strength of these specimens in comparison to water-jetted specimens (J) with the failure plane in NSC.

Table 6. Tensile bond strength of specimens determined in pull-off tests.

Specimen	Tensile Bond Strength $f_{ct,bond}$ [N/mm²]	Mod. Tensile Bond Strength $f^*_{ct,bond}$ [N/mm²]	$f^*_{ct,bond}/f_{ct}$ [-]
monolithic	2.0	2.0	1.0
S	2.2	1.59	0.76
J	2.7	1.05	0.5
G	3.0	1.60	0.76

4.2. Load-Slip Behaviour and Failure Mode

The monolithic push-off tests are considered as reference test specimens with the failure according to Figure 11 and the load-displacement behaviour shown in Figure 12.

Figure 11. Typical failure of the monolithic specimen.

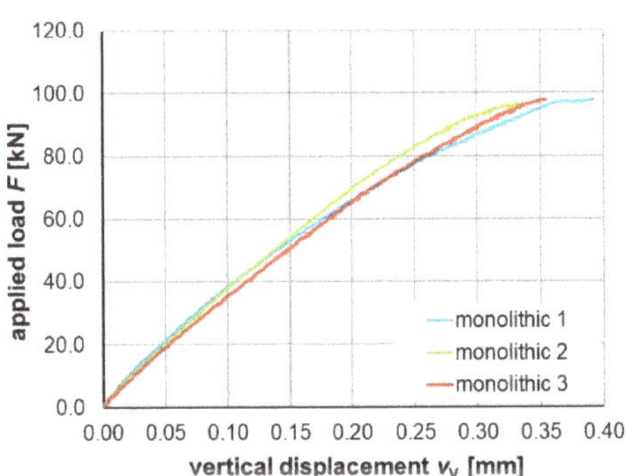

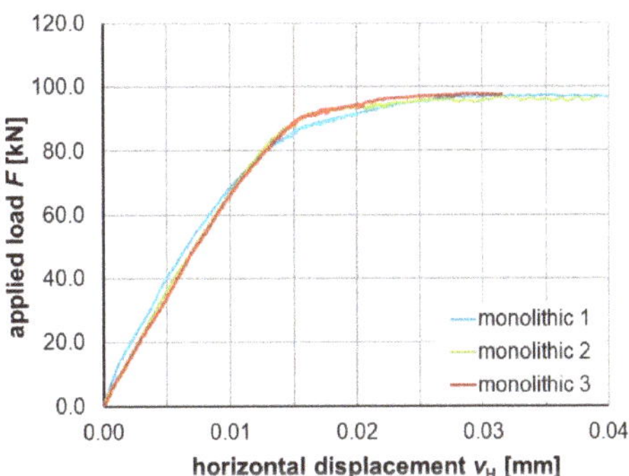

Figure 12. Load-vertical displacement (**left**) and load-horizontal displacement (**right**) of shear failure plane of monolithic specimens.

Considering the shear bond strength to be

$$\tau_{exp} = \frac{F_u}{A_{sh}} \quad (7)$$

the mean shear bond strength of monolithic specimens corresponds to

$$\tau_{exp} = \frac{97.5 \cdot 10^3}{31250} = 3.12 \text{ N/mm}^2 \quad (8)$$

that nearly equals $1.5 \cdot f_{ct} = 1.5 \cdot 0.9 \cdot f_{ct,sp} \approx 1.5 \cdot 0.9 \cdot 2.38 = 3.11 \text{ N/mm}^2$.

Similar to tests on monolithic specimens, the push-off tests on NSC-UHPFRC specimens also show an almost rigid bond-slip behaviour up to a load threshold depending on joint treatment and preparation method. In this state, the interfacial horizontal displacement (measured difference between the measured values with LVDTs W-H and O-H) is very small and the compressive strain measured in NSC and UHPFRC (strain gauge SO and SU as well as NO and NU) have almost similar values. After a certain load threshold, the lower compressive strain (SU and NU) increases and a higher increase rate of relative horizontal displacement of the interface is distinguished hereafter. Figures 13 and 14 show the vertical and horizontal load-displacement curves.

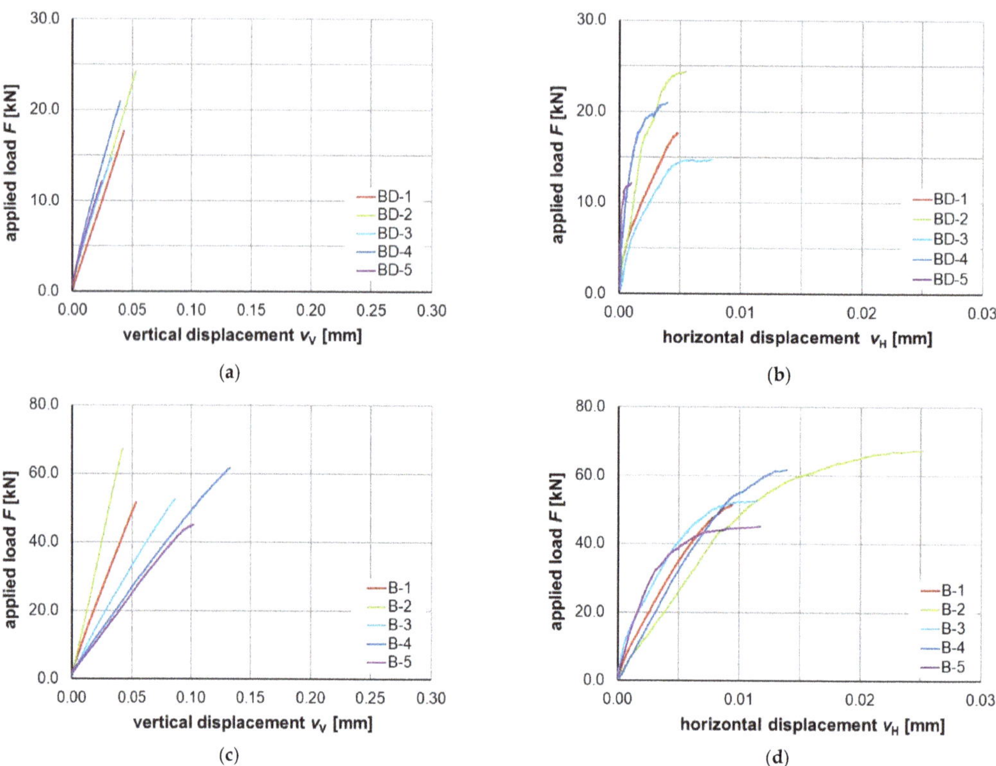

Figure 13. Load-displacement of the shear interface with brushed dry (**a**,**b**) and brushed wet (**c**,**d**) interface treatment.

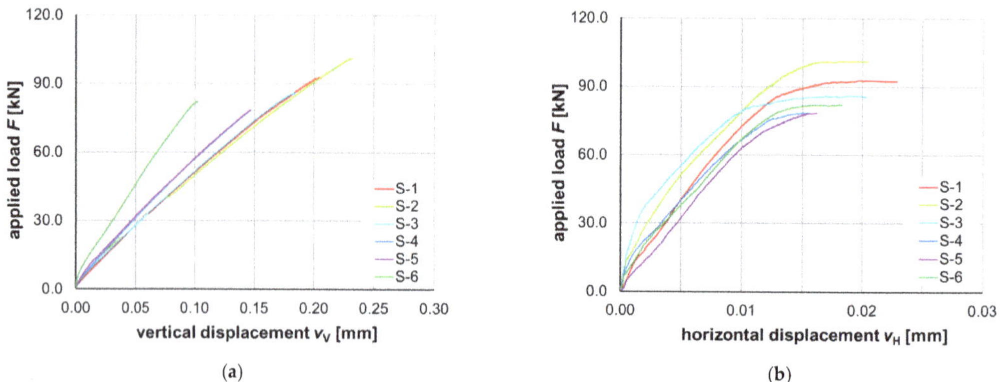

Figure 14. *Cont.*

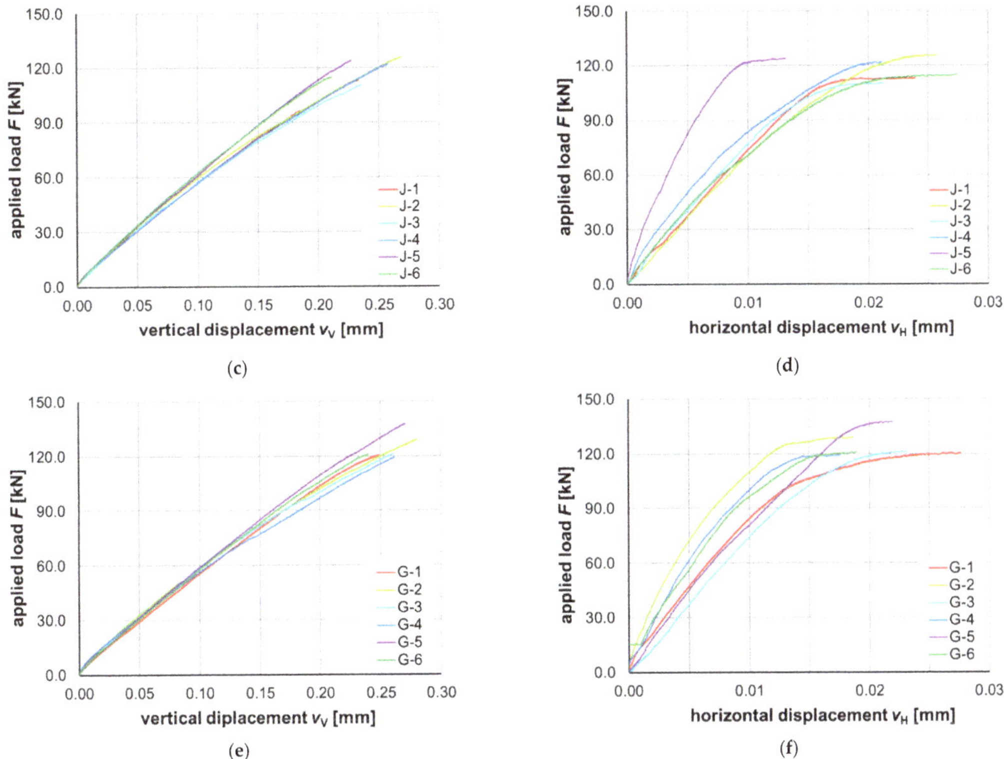

Figure 14. Load-displacement of the shear interface of sandblasted (**a**,**b**), water-jetted (**c**,**d**) and grooved (**e**,**f**) interface treatment.

The vertical displacement-force diagrams show an almost linear behaviour of interface up to failure irrespective of joint treatment method or roughness. In general, the vertical displacement at failure state is around 10 times of the horizontal displacement. Based on the horizontal displacement-force diagrams, it can be concluded that the failure of specimens with higher interface roughness is accompanied by a higher relative horizontal displacement of the interface prior to failure, which correlates with the relative displacement in the interface, but also includes the difference between lateral deformation of NSC and UHPFRC parts of the specimen. As predicted, with increasing roughness of the interface the ultimate load increases. However, the failure plane of specimens varies based on the interface roughness. For (BD) and (B) tests a pure adhesive failure was observed, whereas (S) and (G) specimens showed a mixed failure mode with failure plane approaching through NSC specimens in some regions and also an adhesive failure of some other regions (compare Figure 15). From a maximum peak height S_p of about 4.5 mm, the failure plane is predominantly located in the NSC substrate (cohesive failure).

Figure 15. Representative failure planes of the brushed (**a**), sandblasted (**b**), water-jetted (**c**) and grooved (**d**) substrates.

5. Evaluation of Test Results and Discussion

To investigate the suitability of current calculation approaches for the determination of NSC-UHPFRC shear bond strength, the experimental shear bond strength τ_{exp} values are first evaluated using the design code approaches in Section 2.2 (compare Figure 15). The design approaches of EC2/MC2010 and prEC2 are used to determine the calculated shear bond strength τ_{cal} with mean tensile strength values and without safety factors. Using the design approaches, no investigation was possible for the tests with very smooth substrates (B and BD).

As shown in Figure 16 and based on the statistical quantities mean value m, variation of coefficient v and standard deviation s of the model prediction accuracy (a normal distribution is assumed), the existing design methods of EC2/MC2010 as well as of prEC2 are too conservative for the obtained test results by the authors, in which the results of newly proposed prEC2 are slightly more conservative.

Noting that the design approaches are not validated for NSC-UHPFRC shear bond, a further reason for the underestimated shear bond strength are the uncertainties regarding the correlation between roughness parameters and shear bond strength. In this background the stepwise increase of the shear bond strength assumed in design codes and associated to a joint category (compare Table 2) is shown in Figure 17 (dotted lines). The shear bond strength values obtained in the current tests show a certain correlation with the categories, but are not feasible, especially for the categories "very smooth" and "smooth".

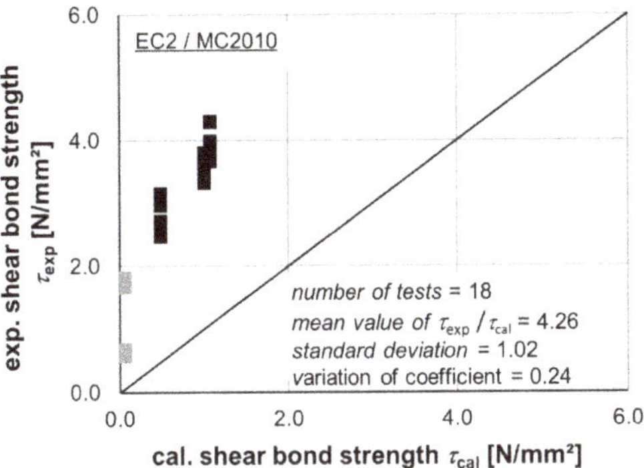

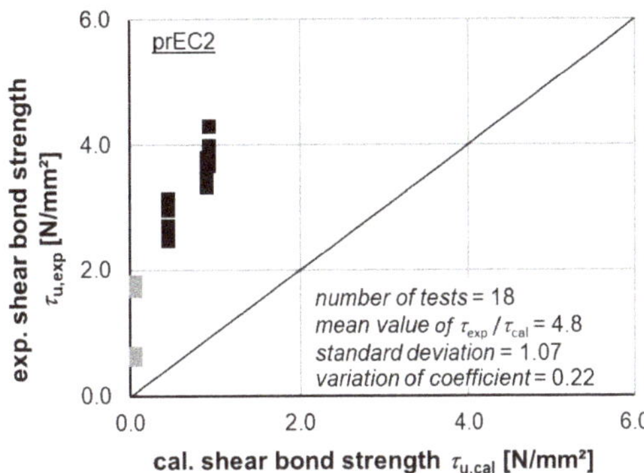

Figure 16. Experimental shear bond strength vs. predicted shear bond strength with the design approach of EC / MC2010 (**left**) and predicted shear bond strength with the design approachof prEN (**right**).

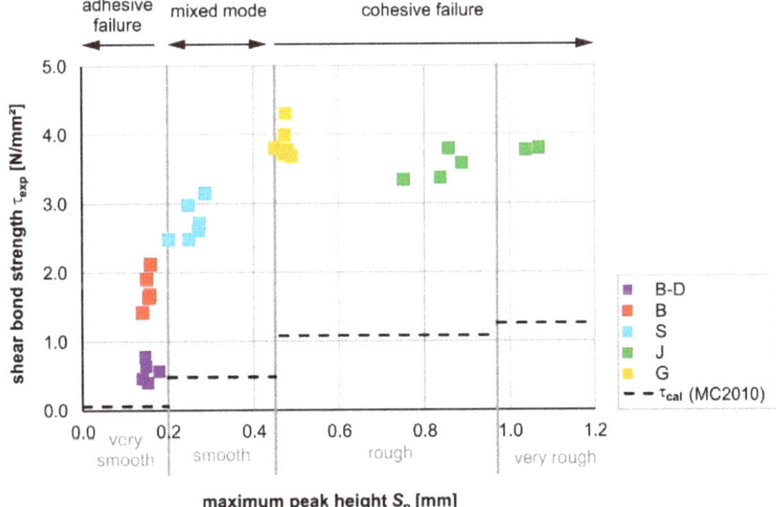

Figure 17. Correlation between roughness value maximum peak height S_p and the shear bond strength τ_{exp} of tests in comparison to roughness categories and prediction of design approach in MC2010.

In this context, it should be noted, that the test results show that the minimum shear bond strength was reached by specimens with an adhesive failure and the maximum value was reached in case of cohesive failure, indicated in Figure 16, and allocated principally to the joint category of the design codes. In addition, the effect of fibres in tests with a mixed-mode failure could not be specified directly and requires further investigations.

Based on Figure 17, it can be further observed that a linear or power correlation between maximum peak height S_p and the shear bond strength τ_{exp} is more representative, which reaches a certain plateau (at a shear bond strength value $\tau_{exp} \approx 3.3$ N/mm^2) observed as a cohesive failure mode with the failure plane locating in NSC substrate. In fact, the results make apparent that for rough and very rough interfaces the cohesive failure is the dominant failure mode and should be considered as an upper bound of shear bond strength for NSC-UHPFRC interfaces. That means that above a certain roughness no further increase in the load-bearing capacity is obtained. Therefore, using efficient treatment methods that produce a well-defined interface texture, such as the grooving method is suggested for NSC-UHPFRC interfaces.

Furthermore, the results show that the interface topology and substrate properties cannot be merely defined based on the roughness parameters. The interfaces treated with water jetting reach a higher roughness, although the NSC-UHPFRC bond strength of such interfaces is slightly lower than the grooved interfaces. This is assumedly caused by the treatment-induced microcracks and aggregate splitting in the substrate (compare Figure 7) and gives further evidence for the advantageousness of grooving as a treatment method.

In order to establish a more convincing calculation approach, the interface texture has to be considered more in detail. Therefore, the linear approach of Gohnert and the power function of Santos & Julio (without safety factors) are implemented in Figure 18. The comparison shows that the approaches of Gohnert and Santos & Julio have a better prediction accuracy than the design approaches in Figure 15 (mean values of the prediction accuracy factors of the model with the mean values $m = 1.8$ and $m = 2.1$, respectively, instead of $m = 4.26$ and $m = 4.8$, respectively), with the linear correlation proposed by Gohnert as the best fit. Nevertheless, since the enhanced approaches are validated for the NSC-NSC bond, they are still too conservative for the evaluation of the NSC-UHPFRC shear bond strength. Nevertheless, the comparison clearly confirms that better results can be reached by an explicit consideration of interface roughness in the determination of the NSC-UHPFRC shear bond strength.

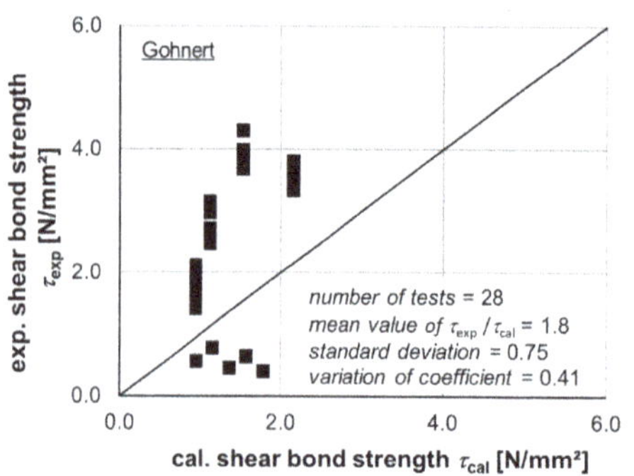

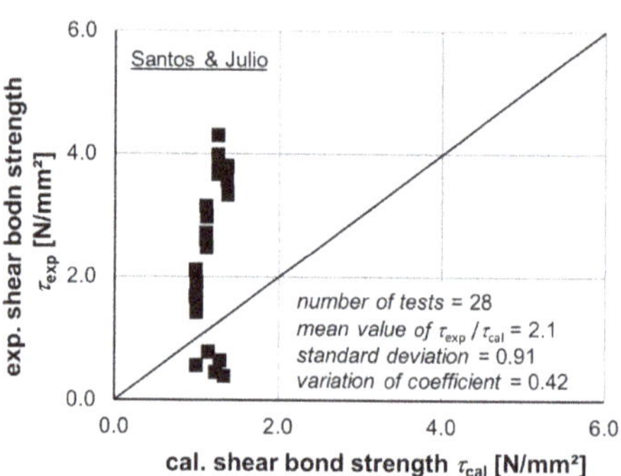

Figure 18. Experimental shear bond strength vs. predicted shear bond strength with the calculative approach of Gohnert (**left**) and predicted shear bond strength with the calculative approach of Santos & Julio (**right**).

It is worth noting that the comparison of the design approaches (Figure 16) with the further calculation approaches (Figure 18) merely intends to show the difference between an implicit and explicit consideration of roughness parameters, respectively. Although the approaches are used with mean concrete tensile strength values and without safety factors, the safety factors cannot be verified in this study. For safe shear bond approaches, further studies are needed, which have to cover various influencing factors such as general concrete mixture parameters, interface treatment and preparation as well as shrinkage and time-dependent behaviour of concrete in a separate study.

6. Conclusions and Outlook

This paper presented the results of a research study on NSC-UHPFRC interfacial bond strength. Within the conducted push-off tests, the shear bond strength between NSC interfaces with different treatment methods and surface roughness have been determined. Additionally, the observed failure modes were used to signify the main load-bearing mechanism. Based on the results of the experimental investigations and the discussion with design approaches and calculation models from the technical literature, the following general statements can be made:

- The failure type in the shown experiments varied based on the treatment method of the interface between adhesive failure, cohesive failure (cohesive) and a mixed failure mode (adhesive/cohesive). The minimum shear bond strength was reached by specimens with an adhesive failure and the maximum value was reached in case of a cohesive failure.
- If the joint surface is prepared by water jetting, very rough joint surfaces can be achieved. However, the increase in maximum bond strength was disproportionate to the roughness increase. The specimens failed with a cohesive failure mode in the NSC substrate.
- The joint preparation using grooving showed the best results concerning the load-bearing capacity of the joint. Therefore, grooving seems to be an effective and gentle preparation method for the preparation of the joint surface before the application of UHPC-overlays. It appeared that the high energy applied to the surface during water jetting led to microcracking and damaging of the substrate.
- The investigations showed that for NSC-UHPFRC interfaces the cohesive failure mode in NSC substrate has to be considered as an upper bound for the design of UHPFRC overlays.
- The calculation of shear bond strength based on roughness categories in design codes appears too crude for a UHPFRC overlay concrete to exploit the potential of the UHPFRC and should be reconsidered.
- By evaluation of tests results, the approaches with a linear or power function between interface roughness parameter and shear bond strength of the interface showed better results in comparison to design approaches. However, as all calculation methods are derived empirically for NSC-NSC interfaces, the shear bond strength was underestimated in all cases.
- Based on the observed failure modes, the effect of the fibres was not fully understood. It appeared that in the case of cohesive failure modes there is no significant influence on the shear bond strength, however, in the case of mixed failure mode and the post-peak behaviour there is a certain influence.

Furthermore, the investigations show the advantageousness of using novel and more detailed methods such as digital microscopy or laser scanning, which reveal important information on the texture of the interface. The determined interface roughness parameters (height and area parameters) were used for a better interpretation of the test results. Such parameters can be used to enhance the calculation approaches by considering the interface explicitly. However, for suitable refined approaches, further investigations on the correlation between failure mode, adhesive bond strength and interface roughness are compulsory. Also, the effects of steel fibres in tests with a mixed failure mode should be further studied.

Overall, the research study shows that strengthening of the existing structures with UHPFRC overlays is a very promising method for several reasons. Using more profound calculation approaches for UHPFRC overlays, strengthenings can reach high load-bearing capacities and a nearly monolithic structural behaviour with very little treatment of the joint surfaces. Furthermore, the strengthening process is more sustainable as well as time- and resource-efficient, as the joint treatment and installation effort of shear connectors can be reduced.

Author Contributions: Conceptualization, M.E. and S.J.; methodology, M.E.; validation, S.J. and M.E.; formal analysis, S.J.; investigation, S.J. and M.E.; resources, M.E.; data curation, S.J.; writing—original draft preparation, S.J.; writing—review and editing, M.E.; visualization, S.J.; supervision, M.E.; project administration, M.E.; funding acquisition, M.E. All authors have read and agreed to the published version of the manuscript.

Funding: The investigations presented in this paper are parts of a research project conducted at the iBMB, Division of Concrete Construction of the TU Braunschweig within the research project "Shear load transfer in UHPFRC-NSC joints", funded by the Deutscher Ausschuss für Stahlbeton (DAfStb,

German Committee of Reinforced Concrete)–project number V 504. The Open Access Publication Funds of Technische Universität Braunschweig funded the publication of the paper.

Institutional Review Board Statement: No application.

Informed Consent Statement: No application.

Data Availability Statement: No application.

Acknowledgments: The authors appreciate the financial and advisory support of the Deutscher Ausschuss für Stahlbeton (DAfStb, German Committee of Reinforced Concrete. We acknowledge also the support by the Open Access Publication Funds of Technische Universität Braunschweig.

Conflicts of Interest: The authors declare no conflict of interest. The funders had no role in the design of the study, in the collection of analyses and in the interpretation of data; in writing of the manuscript, and in decision to publish the results.

References

1. Brühwiler, E.; Bastien Masse, M. Strengthening the Chillon viaducts deck slabs with reinforced UHPFRC. In Proceedings of the IABSE Conference—Structural Engineering: Providing Solutions to Global Challenges, Geneva, Switzerland, 23–25 September 2015; pp. 1171–1178.
2. Pelke, E.; Jaborek, A.; Berger, D.; Brühwiler, E. Überführungsbauwerk der L3378 bei Fulda-Lehnerz—Erster Einsatz von UHPC in Deutschland im Straßenbrückenbau, Teil 1: Projektentwicklung und Baudurchführung. *Beton-und Stahlbetonbau* **2018**, *113*, 831–841. [CrossRef]
3. Hadl, P.; delle Pietra, R.; Hoang, K.H.; Pilch, E.; Tue, N.V. Application of UHPC as road bridge topping within the adaptation of an existing conventional bridge to an integral abutment bridge in Austria. *Beton-und Stahlbetonbau* **2015**, *110*, 162–170. [CrossRef]
4. Methner, R.; Müller, R. Bridge repair, strengthening and sealing with Ultra High Performance Concrete (UHPC)—Experience from three Swiss major bridges. *Beton-und Stahlbetonbau* **2019**, *114*, 126–133. [CrossRef]
5. Orgass, M.; Klitsch, B.; Wißler, M.; Tauscher, F.; Dehn, F. Overpass bridge of the L 3378 road near Fulda-Lehnerz. First application of UHPC in road bridge construction in Germany. Part 2. *Beton-und Stahlbetonbau* **2018**, *113*, 821–830. [CrossRef]
6. Adam, V.; Bielak, J.; Will, N.; Hegger, J. Experimental investigations regarding strengthening of bridge deck slabs with textile reinforced concrete. *Beton-und Stahlbetonbau* **2020**, *115*, 952–961. [CrossRef]
7. Vande Voort, T.L.; Suleiman, M.T.; Sritharan, S. *Design and Performance Verification of UHPC Piles for Deep Foundations;* Center for Transportation Research and Education, Iowa State University: Ames, IA, USA, 2008.
8. Tayeh, B.A.; Abu Bakar, B.H.; Mergat Johari, M.A.; Voo, J.L. Mechanical and permeability properties of the interface between normal concrete substrate and ultra-high performance fibre concrete overlay. *Constr. Build. Mater.* **2012**, *36*, 538–548. [CrossRef]
9. Zanotti, C.; Talukdar, S.; Banthia, N. Astate-of-the-art on concrete repairs and some thoughts on ways to achieve durability in repairs. In *Infrastructure Corrosion and Durability—A Sustainability Study*; OMICS Group eBooks: Constanta, Romania, 2014.
10. Empelmann, M.; Oettel, V.; Cramer, J. Calculation of crack width of concrete components reinforced with steel fibres and reinforcing bars. *Beton-und Stahlbetonbau* **2020**, *115*, 136–145. [CrossRef]
11. Minelli, F.; Plizzari, G. Derivation of a simplified stress-crack width law for Fibre Reinforced Concrete through a revised round panel test. *Cem. Concr. Compos.* **2015**, *58*, 95–104. [CrossRef]
12. Cuenca, E.; Conforti, A.; Monfardini, L.; Minelli, F. Shear transfer across a crack in ordinary and alkali activated concrete reinforced by different fibre types. *Mater. Struct.* **2020**, *53*, 24. [CrossRef]
13. Banthia, N.; Chokri, K.; Ohama, Y.; Mindess, S. Fibre-reinforced cement based composites under tensile impact. *Adv. Cem. Based Compos.* **1994**, *1*, 131–141. [CrossRef]
14. Banthia, N.; Bindiganavile, V.; Azhari, F.; Zanotti, C. Curling control in concrete slabs using fibre reinforcement. *J. Test. Eval.* **2014**, *42*, 390–397. [CrossRef]
15. Empelmann, M.; Henke, V.; Sender, C. *Strengthening of Flexural Members with a Post-Installed Concrete Overlay (in German). Final Report;* Fraunhofer IRB Verlag: Stuttgart, Germany, 2009.
16. Randl, N. Investigations on the load transmission between old and new concrete with different joint roughnesses (in German: Untersuchungen zur Kraftübertragung zwischen Alt- und Neubeton bei Unterschiedlichen Fugenrauigkeiten). Ph.D. Thesis, Universität Innsbruck, Innsbruck, Austria, 1997.
17. Lenz, P. Concrete-Concrete bond potentials for shear joints (in German: Beton-Beton-Verbund Potenziale für Schubfugen). Ph.D. Thesis, Technische Universität München, München, Germany, 2012.
18. Zilch, K.; Müller, A. *Experimentelle Untersuchung zum Ermüdungstragverhalten von unbewehrten Schubfugen an nachträglich ergänzten Betonbauteilen. DAfStb Research Report;* Technische Universität München: München, Germany, 2004.
19. Santos, P.M.D.; Julio, E.N.B.S. Interface shear transfer on concrete composite members. *ACI Struct. J.* **2014**, *111*, 1–10.
20. Erhard, D.; Chorinsky, G. *Repair of Concrete Floors with Polymer Modified Cement Mortars. Adhesion between Polymers and Concrete;* Springer US: Boston, MA, USA, 1986; pp. 230–234.

21. Gowripalan, N.; Gilbert, R.I. *Design Guidelines for Ductal Prestressed Concrete Beams-Design Guide*; School of Civil and Environmental Engineering, The University of New South Wales: Sydney, NSW, Australia, 2000.
22. Jang, H.-O.; Lee, H.-S.; Cho, K.; Kim, J. Experimental study on shear performance of plain construction joints integrated with ultra-high performance concrete (UHPC). *Constr. Build. Mater.* **2017**, *152*, 16–23. [CrossRef]
23. Farzad, M.; Shafieifar, M.; Azizinamini, A. Experimental and numerical study on bond strength between conventional concrete and Ultra High-Performance Concrete (UHPC). *Eng. Struct.* **2019**, *186*, 297–305. [CrossRef]
24. Zhang, Y.; Zhang, C.; Zhu, Y.; Cao, J.; Schao, X. An experimental study: Various influence factors affecting interfacial shear performance of UHPC-NSC. *Constr. Build. Mater.* **2020**, *236*, 117480. [CrossRef]
25. Valikhani, A.; Jaberi Jahromi, A.; Mantawy, I.M.; Aziziamini, A. Experimental evaluation of concrete-to-UHPC bond strength with correlation to surface roughness for repair application. *Constr. Build. Mater.* **2020**, *238*, 117753. [CrossRef]
26. Randl, N.; Steiner, M.; Peyerl, M. Hochfester Aufbeton zur Tragwerksverstärkung. Teil 1: Kleinkörperversuche. *Beton-und Stahlbetonbau* **2020**, *115*, 106–116. [CrossRef]
27. Walraven, J.C. Fundamental analysis of aggregate interlock. *ASCE J. Struct. Div.* **1981**, *107*, 2245–2270. [CrossRef]
28. Santos, P.M.D.; Julio, E.N.B.S. A state-of-the-art review on shear-friction. *Eng. Struct.* **2012**, *45*, 435–448. [CrossRef]
29. EN 1992-1-1: Eurocode 2: Eurocode 2. Design of Concrete Structures-Part 1-1: General Rules and Rules for Buildings; European Committee for Standardization CEN. 2004. Available online: https://www.phd.eng.br/wp-content/uploads/2015/12/en.1992.1.1.2004.pdf (accessed on 15 July 2021).
30. *Fib—International Federation for Structural Concrete (2013): Fib Model Code for Structures*; Ernst & Sohn: Berlin, Germany, 2010.
31. Kaufmann, N. Das Sandflächenverfahren. *Straßenbautechnik* **1971**, *24*, 131–135.
32. ISO 25178-2. *Geometrical Product Specifications (GPS)—Surface Texture: Areal—Part 2: Terms, Definitions and Surface Texture Parameters*; ISO: Geneva, Switzerland, 2012.
33. prEN 1992-1-1 Eurocode 2: Design of concrete structures—Part 1-1: General rules—Rules for buildings. *Bridges Civ. Eng. Struct. Rev.* **2020**, *11*, 16.
34. Gohnert, M. Horizontal shear transfer across a roughened surface. *Cem. Concr. Compos.* **2003**, *25*, 379–385. [CrossRef]
35. Santos, P.M.D. Júlio ENBS. Factors affecting bond between new and old concrete. *ACI Mater. J.* **2011**, *108*, 449–456.
36. AFGC. *Ultra High Performance Fibre-Reinforced Concretes—Recommendations*; Association Française de Génie Civil: Paris, France, 2013.
37. *SIA 2052: Ultra-Hochleistungs-Faserbeton (UHFB)—Baustoffe, Bemessung und Ausführung*; Schweizerischer Ingenieur und Architektenverein: Zürich, Switzerland, March 2016. Available online: http://shop.sia.ch/normenwerk/ingenieur/sia%202052/d/D/Product (accessed on 15 July 2021).
38. Empelmann, M.; Javidmehr, S.; Zühlsdorf, M.-R.; Wilkening, M.; Oettel, V. Shear Load Transfer in UHPFRC-NSC Joints. Final Report of DAfStb-Project No. V 504. 21. 20 March. (in press)
39. *EN 12350-5: Testing Fresh Concrete. Part 5: Flow Table Test. CEN/TC 104*; Deutsches Institut für Normung e.V.: Berlin, Germany, 2009.
40. *EN 12390-6 Testing harDened Concrete. Tensile Splitting Strength of Test Specimens, CEN/TC 104/DIN*; Deutsches Institut für Normung e.V.: Berlin, Germany, 2009.
41. *EN 12390-3: Testing Hardened Concrete. Compressive Strength of Test Specimens, CEN/TC 104/DIN*; Deutsches Institut für Normung e.V.: Berlin, Germany, 2009.
42. *EN 12390-13: Testing Hardened Concrete. Determination of Secant Modulus of Elasticity in Compression, CEN/TC 104/DIN*; Deutsches Institut für Normung e.V.: Berlin, Germany, 2009.
43. MountainsLab® Premium: Surface Imaging, Analysis and Metrology Software. Digital Surf, License Number DS-0D475CF1. Available online: www.digitalsurf.com (accessed on 15 July 2021).

Article

Assessment of a Municipal Solid Waste Incinerator Bottom Ash as a Candidate Pozzolanic Material: Comparison of Test Methods

Flora Faleschini [1,2,*], Klajdi Toska [1], Mariano Angelo Zanini [1], Filippo Andreose [1], Alessio Giorgio Settimi [2], Katya Brunelli [2] and Carlo Pellegrino [1]

1 Department of Civil, Environmental and Architectural Engineering, University of Padova, 35131 Padova, Italy; klajdi.toska@dicea.unipd.it (K.T.); marianoangelo.zanini@dicea.unipd.it (M.A.Z.); filippo.andreose@unipd.it (F.A.); carlo.pellegrino@unipd.it (C.P.)
2 Department of Industrial Engineering, University of Padova, 35131 Padova, Italy; alessiogiorgio.settimi@unipd.it (A.G.S.); katya.brunelli@unipd.it (K.B.)
* Correspondence: flora.faleschini@dicea.unipd.it or flora.faleschini@unipd.it; Tel.: +39-049-827-5585

Abstract: New generations of green concretes are often consuming large amounts of industrial waste, as recycled or manufactured aggregates and alternative binders substituting ordinary Portland cement. Among the recycled materials that may be used in civil engineering works, construction and demolition waste (C&DW), fly ashes, slags and municipal solid waste incinerator bottom ashes (MSWI BA) are those most diffused, but at the same, they suffer due to a large variability of their properties. However, the market increasingly asks for new materials capable of adding some specific features to construction materials, and one of the most interesting is the pozzolanic activity. Hence, this work deals with an experimental study aimed at assessing the technical feasibility of using an industrial waste comprised largely of MSWI BA, with small quantities of C&DW and electric arc furnace slag (EAFS), in green cement-based mixtures (cement paste and mortars). The aim of the work is to achieve the goal of upcycling such waste and avoiding its disposal and landfilling. Particularly, the test methods for assessing the pozzolanic activity of this waste are discussed, analyzing the efficacy of indirect methods such as the strength activity index (SAI), the conductivity test and the efficiency factor (k), together with a direct method based on lime consumption.

Keywords: mortars; MSWI bottom ash; pozzolanic activity; supplementary cementing materials; sustainability

Citation: Faleschini, F.; Toska, K.; Zanini, M.A.; Andreose, F.; Settimi, A.G.; Brunelli, K.; Pellegrino, C. Assessment of a Municipal Solid Waste Incinerator Bottom Ash as a Candidate Pozzolanic Material: Comparison of Test Methods. *Sustainability* **2021**, *13*, 8998. https://doi.org/10.3390/su13168998

Academic Editors: Fausto Minelli, Enzo Martinelli and Luca Facconi

Received: 7 May 2021
Accepted: 3 August 2021
Published: 11 August 2021

1. Introduction

Pozzolanic materials have been largely adopted in construction materials, since the Roman ages, when pozzolanic systems were developed to realize the *opus caementicium* using the volcanic ashes from the area close to Pozzuoli, Naples, from which they take their name. Probably, the adoption of these kinds of materials has even an older origin, as it is believed that pozzolanic concrete was used in Mesoamerica too, in the period between 1100 and 850 B.C. More recently, pozzolanic materials have found large application in concrete aiming to realize blended cement mixes, with the twofold objective of reducing cement environmental footprint and costs. The definition of a pozzolan material can be found in reference [1], as "a siliceous and aluminous material which, in itself, possesses little or no cementitious value but which will, in finely divided form in the presence of moisture, react chemically with calcium hydroxide at ordinary temperature to form compounds possessing cementitious properties". According to such definition, it is well recognized that pozzolans may have varying sources, being both natural and man-made, and the mechanisms behind their interaction with cement might differ significantly [2–4]. As a result, the intensity of the pozzolanic reaction may differ depending on the materials. Furthermore, pozzolans can also be considered as supplementary cementing materials (SCMs), a category of materials that include those compounds contributing to the properties of hardened concrete through both hydraulic and pozzolanic activity.

SCMs are often used in the construction industry to minimize ordinary Portland cement consumption in cement-based materials, such as concrete and mortar mixes. Some advantages are the ability to make concrete mixtures more economical, reduce permeability, increase strength, or influence other properties, both in the hardened and fresh state [5]. Further, their use significantly lowers the environmental footprint of concrete. It is indeed worth recalling that ordinary Portland cement is responsible for 866 kg CO_2/t of clinker [6]. This amount is due to the process of CO_2 release via the calcination of carbonate minerals in the kiln feed, which accounts for about 60% of the CO_2 release, whereas the remaining 40% is associated to the combustion of the fuels used to heat the kiln feed [6]. This high carbon footprint may be lowered only by adopting strict environmental policies within the same cement industry (e.g., by means of energy efficiency, alternative fuels, biomasses, clinker substitution), developing alternative cements, favoring carbon capture and sequestration, and lastly, adopting SCMs.

The most common SCMs are pulverized fly ashes (PFA), ground granulated blast furnace slags (GGBFS) and silica fume (SF), whose consumption has greatly increased in recent years [7–11] and whose availability in the near future is questionable. Indeed, according to the Paris agreement, EU countries will no longer invest in coal power plants after 2020, thus posing a serious risk in the future supply of all their by-products, including PFA and SF [12]. Accordingly, many researchers have turned their attention to finding alternative materials to be used as SCMs, focusing on industrial or agriculture waste. Within the first group, it is worth mentioning rock-wool waste [13], electric arc furnace dust (EAFD) [14] and other steel slags, i.e., ladle furnace slag (LFS) [15–19], glass waste [20], co-combustion fly ashes [21] and municipal solid waste incineration (MSWI) bottom ash [22]; in the second, we can mention sugarcane bagasse ash [23] and rice husk ash [24,25]. When dealing with these materials, there are two main challenging objectives that need to be reached: ensuring that the waste is stable, both from a chemical and volumetric point of view, and that it is able to develop a pozzolanic behavior.

On one side, indeed, the leachability of heavy metals is a key concern when managing these materials, and thus weathering processes are often recommended before their reuse as secondary building material. For instance, weathering of MSWI BA for a period of 1–3 months is typically adopted, aiming at allowing oxidation, carbonation, neutralization of pH, dissolution and precipitation reactions to occur and chemically stabilize the ash, and this will reduce the solubility of the main toxic elements which might be released into the environment [26]. Such a topic is also worthy of being analyzed as it is directly linked to the classification of such materials (i.e., to define if it should be considered as a waste or a by-product). Particularly, if the "end-of-waste" status of inert waste [27] is achieved, such a classification allows simpler authorization processes than full and ordinary ones asked of waste treatment plants.

On the other side, technical feasibility of the waste use is of fundamental importance. For such scope, it is worth recalling that a material can be classified as a SCM if it contributes to the properties of hardened concrete through the development of hydraulic of pozzolanic reactions, in such a way to ensure some target properties, according to both EN 450-1 [28] and ASTM C618 [29]. According to international regulations and literature, there is a consensus that the activity of most SCMs is linked to some main parameters: the content of active silica, fineness, specific surface area, water to powder ratio, curing temperature and alkalinity of the pore solution [30].

To assess the reactivity of SCMs and thus if a material displays pozzolanic activity, several tests have been proposed in literature, which may be divided mainly into two classes: direct and indirect. In the first group, the pozzolanic activity of a material can be analyzed through a Frattini test and simplified saturated lime (SL) [31,32], which are the most well-known. These methods consist of a direct evaluation of the consumption of $Ca(OH)_2$ through X-ray diffraction (XRD), thermo-gravimetric analysis (TGA) or chemical titration. They differ on the kind of solution that the candidate pozzolanic materials are added to, being in the former case a solution containing Portland cement and in the latter saturated lime water.

In the second group of test methods, i.e., the indirect ones, a property of the concrete where the SCM is added is analyzed and compared to that of ordinary concrete. Particularly, it is possible to analyze the strength activity index (SAI), which measures the relative compressive strength ratio between the SCM-concrete and the ordinary one [28]. Further test methods and indexes have been derived in literature to evaluate the reactivity of SCMs, and it is worth citing the conductivity test (an indirect method), proposed by Luxan et al. [33], and the definition of the efficiency factor k of a SCM [34], which is defined as the part of the SCM in a pozzolanic concrete, which can be considered as equivalent to Portland cement, having the same properties as concrete without SCM.

However, according to several works carried out on different waste materials and SCMs, such tests do not always correlate with each other [35]. Furthermore, the roadmap for identifying if a material can be suitably used as a pozzolanic material is very complex, and often, at least some indicators typically used for this characterization fail [36]. Reasons for the observed discrepancies in the results obtained with different methods were associated to various causes, including: the amount of entrapped air in the mortars used to evaluate the SAI due to the use of admixtures; the adoption of blended cements, having different particles fineness; and uncertainties in the absolute amount of $Ca(OH)_2$ when adopting lime saturated test methods [37]. Donatello et al. [35], who analyzed comparatively the pozzolanic activity of incinerator sewage sludge ash (ISSA), coal fly ash (FA), metakaolin (MK), silica fume (SF) and silica sand (SS) through both direct and indirect test methods, recommended using a combination of these tests to provide a robust evaluation of the reactivity of a potential pozzolanic material. Particularly, they found that the SAI index and Frattini test methods correlated better and were tightly controlled methods.

The above context shows that prior to proceeding into a complex and long process for qualifying a recycled material as a SCM or as an industrial pozzolan, it is fundamental to verify if it is able to display pozzolanic activity. This work is developed with the following aims: to compare different test methods to assess the pozzolanic potential of a waste material; to evaluate if the raw material displays pozzolanic potential for being activated through a further process; to identify the optimum cement substitution ratio. Thus, the reactivity of an industrial waste, used in its raw state, is analyzed through different indirect methods, namely the SAI index, k-value and conductivity test. Further, the Frattini test is also carried out as a direct test method. The material selected as a candidate pozzolanic material is an industrial waste, obtained from the treatment of a municipal solid waste incinerator bottom ash (MSWI BA), blended with a minor content of electric arc furnace slag (EAFS) and construction and demolition waste (C&DW). Particularly, MSWI BA production in Italy is very high: in 2016, more than 10^6 kg of MSWI BA were produced, and among them, 80% comes from plants located in North Italy [38]. The results allow us to discuss the correlation between the different test methods adopted, and in the specific case for this industrial waste, to suggest the best technological process for improving the reactivity of the analyzed material.

2. Materials and Methods

2.1. Raw Waste Material: Characterization

The industrial waste is a blended mix of MSWI BA, EAF slag and C&DW. This blend is produced in a waste treatment plant in Italy with varying grading fractions, being an all-in (0–31 mm), a fine (0–4 mm) and a coarse (4–16 mm) fraction, the application of which is intended for civil engineering purposes with non-structural properties. Particularly, the (0–4 mm) fraction contains more than 95% of MSWI bottom ash, and it is shown in Figure 1. In this study the waste is used in a 0–4 mm and 0–1 mm grading, the latter being sorted from the raw material fine fraction, without any further treatment except the mechanical sorting and a weathering of three-to-six months at atmospheric conditions carried out at the treatment facility.

Figure 1. MSWI BA (0–4 mm).

MSWI BA density (s.s.d.) is 2285 kg/m^3, whereas the chemical composition obtained through XFR is listed in Table 1, together with the composition of cement type CEM II/A-LL 42.5R, used as a reference. The XRF test method has been often adopted in literature when characterizing the composition of MSWI BA [39–41]. The main constituents are Ca, Si, Al, Fe and Mg oxides; particularly, SiO_2 + Al_2O_3 + Fe_2O_3 content is about 42.4%, which is below the limit proposed by EN450-1 for fly ashes [28], and that proposed by ASTM C618-19 [29] for natural pozzolans, but it is close to the latest limits proposed for fly ashes.

Table 1. Chemical composition of MSWI BA and cement.

	MgO (%)	Al_2O_3 (%)	SiO_2 (%)	P_2O_5 (%)	SO_3 (%)	K_2O (%)	CaO (%)	TiO_2 (%)	Cr_2O_3 (%)	MnO (%)	Fe_2O_3 (%)	CuO (%)	ZnO (%)	PbO (%)
MSWI BA	8.13	12.55	21.76	2.56	4.80	0.91	36.90	1.31	0.57	0.39	8.09	0.59	1.24	0.16
Cement	2.38	4.79	19.71	0.10	2.95	1.03	65.46	0.21	-	0.04	3.28	-	-	<0.3

Additionally, to complete the characterization of this material for its application as a SCM, the authors are focusing also on environmental safety issues, through leaching and ecotoxicological tests, that may be adopted to verify the potential presence of harmful substances. Some preliminary results on this aspect can be found in [42], showing that a low risk exists when the material is tested in monolithic mortar and concrete specimens.

XRD tests were carried out using a Siemens/Bruker D5000 Diffractometer, with CuKα radiation and operation conditions of 40 keV and 30 mA, on a pulverized sample of the material. The XRD pattern is shown in Figure 2, identifying as the most relevant crystalline phases quartz, calcite, gehlenite, magnetite, wuestite, mayenite, larnite and then calcium sulfate tetrahydrate. Quartz, calcite and magnetite are often found in MSWI BA in large quantities [43,44], whereas the other constituents generally appear less abundantly. It is worth mentioning the presence of mayenite and gehlenite, which in literature were found in activated MSWI, obtained after thermal treatment [45,46]. Particularly, mayenite (C12A7) is widely used in calcium aluminate cements as a minor phase, and even in geopolymers to improve setting and early-age strength. It is well recognized that this mineral could lead to an improvement in the hydraulic reactivity of cements, especially at young ages [47,48]. However, recent studies demonstrated that rapid hydration of C12A7 may cause flocculation in the system, leading to the formation of regions with local defects within their mineralogic structure and, thus, to possible weakening of the paste matrix in cement-based materials [49]. Lastly, between 25° and 35° 2θ it is possible to identify the main hump in the spectrum, representing an amorphous phase.

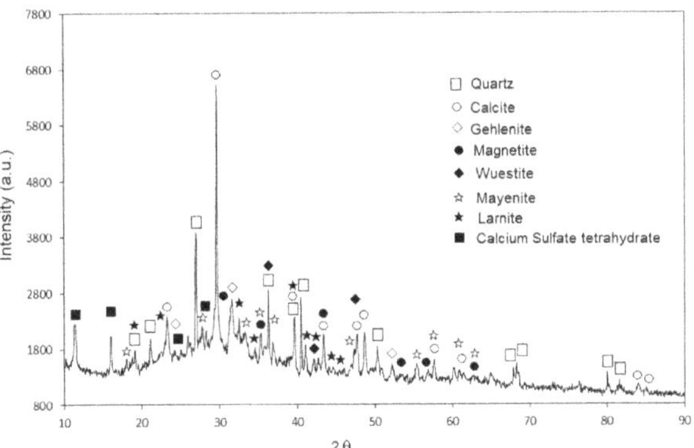

Figure 2. XRD pattern of the MSWI BA.

2.2. Methods

In this work, the reactivity of the raw waste material is analyzed through different methods. Among the indirect tests to evaluate the pozzolanic activity, the evaluation of the strength activity index (SAI) and the efficiency k-factor is performed, together with the rapid conductivity test. Further, the Frattini test is carried out to evaluate CaO consumption, thus being considered a direct test method. For this scope, varying mortar mixtures were created and tested under compressive and flexural strength tests; in addition, cement pastes were realized to evaluate the setting time evolution and their mechanical properties.

2.2.1. Mortar Specimens—Mechanical Tests

Overall, twelve mortar mixes were realized and tested, four being the reference (labelled as "Mix Ref") and eight being the experimental ones where MSWI BA replaces cement at a 20%w ratio. It is worth noting that the proportioning followed the direct weight replacement (DWR) instead of the direct volume replacement (DVR) method, i.e., a certain amount of MSWI BA replaces the same amount of Portland cement in weight. Among the MSWI BA mixes, four were realized with the raw (0–4 mm) grading, named with the letter "F", and four with the (0–1 mm) particle fractions, named with the letters "EF". For casting all the mortars, the same cement type was used, classified as a CEM II-A/LL 42.5-R, with rapid strength gain. Tap water without any deleterious materials was used for mortar realization, together with a natural sand (0–4 mm) and a water reducing agent, added in all the mortar mixtures except for those realized with the highest water/cement (w/c) ratio. The addition of the plasticizer allowed the mortars to have a fluid workability, even if the mixes containing the MSWI BA had a slightly reduced flow compared to the references. Indeed, it is worth recalling that BA typically has a high water demand, but the addition of the plasticizer can control it well. Table 2 shows the features of the analyzed mortar mixes.

Other than compressive strength at 28 days, flexural strength was also evaluated at the same age. For the reference and "Mix EF" samples, the same tests were carried out also at a longer age, i.e., after 56 days. For the tests, 40 × 40 × 160 mm prismatic mortar samples were casted, demolded after two days, maintained under controlled humidity and temperature conditions (20 ± 2 °C, >95% relative humidity) and tested under a three-point bending test in a 25 kN capacity universal loading machine; with the two parts of the samples remaining after the test, the compressive strength test was carried out in a 600 kN capacity universal loading machine.

Table 2. Mix design of mortars (in kg/m^3).

	Water	Cement	Water/Binder	Natural Sand	MSWI BA (0–4 mm)	MSWI BA (0–1 mm)	Plasticizer
Mix Ref 1	315	525	0.6	1575	-	-	-
Mix E1	315	420	0.6	1575	105	-	-
Mix EF1	315	420	0.6	1575	-	105	-
Mix Ref 2	266.7	533	0.5	1600	-	-	5.33
Mix E2	266.7	426.7	0.5	1600	106.67	-	5.33
Mix EF2	266.7	426.7	0.5	1600	-	106.67	5.33
Mix Ref 3	240	533	0.45	1600		-	5.33
Mix E3	240	426.7	0.45	1600	106.67	-	5.33
Mix EF3	240	426.7	0.45	1600	-	106.67	5.33
Mix Ref 4	220	550	0.4	1650	-	-	5.50
Mix E4	220	440	0.4	1650	110	-	5.50
Mix EF4	220	440	0.4	1650	-	110	5.50

2.2.2. SAI index

Strength activity index (SAI) values of the studied MSWI BA were calculated according to ASTM C618 [29], which defines SAI as the ratio of the compressive strength of the 20% SCM mortar to that of a control mortar. SAI should not be less than 70% after 28 days to classify a material as pozzolanic. In this work, other than the SAI for compressive strength, the same index was also evaluated for the flexural strength, both at 28 and 56 days of curing (the latter only for "Mix EF" samples).

2.2.3. Efficiency k-Factor

The *k*-value of the MSWI BA mortars is estimated here using the Δw concept, which is described in detail in the work conducted by Babu and Rao [50] and Schiessl and Hardtl [51]. For such scope, the results in terms of compressive strength evaluated at 28 days of the mortar mixes are used. The *k*-value is hence defined in such a way that the w/c ratio of the reference mix and the ratio $[w/(c + k{\cdot}MSWI)]$ of the mix with pozzolan material are the same, given a fixed strength level. Recall that the second ratio has the meaning of water to "effective" cementitious materials. For such scope, it is first necessary to obtain the compressive strength evolution vs. w/c (water/cement) or vs. w/b (water/overall binder as cement + MSWI BA) ratio, based on the experimental results. The strength model used is based on Abrams relation, where σ_c is the compressive strength, w is the water content, b is the overall binder content and μ_1 and μ_2 are the regression coefficients:

$$\sigma_c = \mu_1 \cdot (w/b)^{\mu_2} \tag{1}$$

Comparing the curves of the pozzolanic mortars and that of the reference mix, Δw can be expressed as:

$$\Delta w = (w/c) - (w/b) \tag{2}$$

In the above relation, the water to effective cementitious materials ratio $w/(c + k{\cdot}MSWI)$ should substitute the w/c term, and the water to overall binder content $w/(c + MSWI)$ substitutes w/b. Then, the *k*-value can be calculated via Equation (3):

$$k = w/(MSWI \cdot (\Delta w + w/(c + MSWI))) - c/MSWI \tag{3}$$

2.2.4. Conductivity Test

The test is based on the procedure developed by Luxán et al. [33], which aims to experimentally assess the compensated conductivity of a calcium hydroxide (CH) saturated solution, to which the potential pozzolanic material is added, over time which is applicable to natural products (about 120s). This method is recognized as an indirect test method characterized by a high rapidity, which however might suffer from some weaknesses, most in terms of the presence of soluble salt in non-natural pozzolans which typically deposit on the bottom of the solution.

The test is carried out as follows: first, 200 mL of a CH saturated solution prepared with distilled water is soaked under controlled temperature conditions (40 ± 1 °C) and the electrical conductivity is measured using a WTW MultiLine P4 Universal Meter. Then, the electrical conductivity measure is repeated after 120s from the addition of 5 g of the MSWI BA to the solution, here used in the (0–1 mm) fraction. The variation of the pH and of the conductivity indicates the reaction of dissolved $[Ca]^{2+}$ and $[OH]^{-}$ ions.

2.2.5. Frattini Test

The Frattini test method is considered as a direct evaluation of the pozzolanic activity according to the European standards. The procedure follows the work of Baki et al. [52], which consists of the preparation of a 20 g sample made of cement (80%) + MSWI BA (20%). The initial sample of MSWI BA is extracted from the (0–1 mm) fraction. This sample is then added to 100 mL distilled water at 40 °C and vigorously soaked for 20 s, after which it is placed in an electric oven at 40 °C for four days. After, it is filtered under vacuum conditions and then analyzed via titration to quantify both $[OH]^{-}$ and $[Ca]^{2+}$ ions, the former using 0.1 mol/L HCl solution and five drops of methyl-orange indicator, the latter with 0.03 mol/L EDTA solution and Pattond and Reeders indicator. The last titration is performed after a pH correction to achieve a pH value of 12.5 ± 0.2. The same test was carried out for a sample made using cement only and another where 20% of the cement was replaced by natural sand.

2.2.6. Cement Paste Specimens—Setting Times and Mechanical Properties

Seven cement pastes (one reference and six with MSWI BA) were realized to evaluate both initial and final setting times and also compressive strength at 28 days. Cement pastes were realized with a fixed w/b ratio equal to 0.3 and a varying cement replacement ratio (adopting the DWR method), up to 50%w. MSWI BA was used in the (0–1 mm) fraction.

Setting time was evaluated using Vicat apparatus, which measures paste resistance to the penetration of a needle under a load of 300 g. The time elapsed between zero and the instant at which the distance between the needle and the baseplate is 6 ± 3 mm is taken as the initial set time. Instead, the final setting time was considered as the time elapsed between zero and the instant at which the needle penetrates the paste to a maximum depth of 3 mm. Instant zero is considered from the moment when mixing water is added to the mixture. Set tests were carried out in a room with relative humidity and temperature of 54 ± 2.0% and 19 ± 1.0 °C, respectively.

Compressive strength tests on hardened cement pastes were carried out on 50 mm side cubic specimens, at 28 days of curing in a room with relative humidity and temperature of 95 ± 5.0% and 20 ± 2.0 °C, respectively. Tests were carried out in a 600 kN capacity universal loading machine.

3. Results and Discussion

3.1. SCM-Mortars and SAI Values

Table 3 lists the mechanical properties of the analyzed mortars in terms of hardened density (ρ) and compressive (f_c) and flexural strength (f_{cf}), after 28 days for all the samples and also after 56 days for the reference and "Mix EF" samples. Values refer to the average results of at least three samples.

Table 3. Mechanical properties of the mortars (ave. = average results; st. dev. = standard deviation).

	28 Days			56 Days		
	ρ (kg/m^3)	f_c (MPa)	f_{cf} (MPa)	ρ (kg/m^3)	f_c (MPa)	f_{cf} (MPa)
Mix Ref 1 (ave.)	2174	26.87	5.55	2250	42.62	7.50
(st. dev.)	9	0.91	0.23	8	1.34	0.19
Mix E1 (ave.)	2052	10.33	3.37	-	-	-
(st. dev.)	39	0.38	0.38	-	-	-
Mix EF1 (ave.)	2046	17.12	5.55	2066	20.28	5.09
(st. dev.)	21	0.16	0.46	22	0.83	0.76
Mix Ref 2 (ave.)	2195	32.29	6.78	2214	44.89	8.34
(st. dev.)	20	1.45	0.05	4	0.70	0.23
Mix E2 (ave.)	2049	22.74	4.71	-	-	-
(st. dev.)	9	0.94	0.13	-	-	-
Mix EF2 (ave.)	2082	23.70	6.05	2137	28.48	5.71
(st. dev.)	6	0.62	0.13	5	2.56	0.27
Mix Ref 3 (ave.)	2193	42.86	7.94	2285	47.78	8.65
(st. dev.)	31	0.71	0.19	26	0.64	0.35
Mix E3 (ave.)	1990	22.66	5.27	-	-	-
(st. dev.)	24	1.14	0.21	-	-	-
Mix EF3 (ave.)	2175	29.00	6.91	2160	29.12	6.16
(st. dev.)	28	1.72	0.24	27	2.95	0.12
Mix Ref 4 (ave.)	2240	43.50	7.38	2208	50.38	8.85
(st. dev.)	30	2.45	0.70	31	1.82	4.86
Mix E4 (ave.)	2164	33.86	6.73	-	-	-
(st. dev.)	28	0.71	0.08	-	-	-
Mix EF4 (ave.)	2133	29.58	6.81	2105	28.22	5.14
(st. dev.)	11	0.42	0.33	27	0.33	0.21

Results highlight that the hardened density is lower when the MSWI BA partially substitutes cement, despite the grading used, than in the reference mortars. Due to long-term hydration, hardened density increases with time for all the analyzed mixes. Further, substituting 20% of cement with the MSWI BA has a severe impact on both compressive and flexural strength, even if this last parameter is affected in a slighter way, despite the grading of the ash added. The reasons for strength losses can be argued to be the replacement of a stronger material (cement matrix) with a weaker one (MSWI BA) and the increase in the pore fraction of the concrete, due to the ash particles reduced fineness compared to that of cement particles. Additionally, the DWR method adopted here might have an impact too, as it influences the water demand of the mix. Further, the impact of the substitution is less influential for those mortars realized with the (0–1 mm) MSWI BA, rather than for the mixes made with the coarser fraction of this material; the only exception applies for the mixes with a low w/b ratio, which have comparable strength both when MSWI is used in the (0–4 mm) and (0–1 mm). Instead, mixtures realized with (0–4 mm) MSWI BA show a higher strength loss when a low w/b ratio is used.

Table 4 lists the values of the SAI, both for the compressive and flexural strength ratios, evaluated for all the mixtures containing the MSWI BA. Analyzing the SAI for compressive strength, the target value of 0.7 at 28 days is exceeded only for Mix E2 and E4 when using the coarse ash and for Mix EF2 when using the fine fraction of the ash. However, when using the fine MSWI BA, the SAI_c index is generally higher (average value *a. v.* = 0.68; standard deviation *st. dev.* = 0.04) than when using the coarse ash (*a. v.* = 0.60; *st. dev.* = 0.177). Particularly, on average, the (0–1 mm) fraction allows limited strength losses, showing a maximum decrease of 37% compared to the reference mix, and the variability of results is limited. Instead, when the ash is used in the coarse fraction, the variability of the results is larger, as demonstrated by the high *st. dev.* value of the results.

Table 4. SAI values (for compressive and flexural strength).

	28 Days								56 Days			
	Mix E1	Mix EF1	Mix E2	Mix EF2	Mix E3	Mix EF3	Mix E4	Mix EF4	Mix EF1	Mix EF2	Mix EF3	Mix EF4
$SAI_{,c}$	0.38	0.63	0.70	0.73	0.53	0.68	0.78	0.68	0.48	0.63	0.61	0.56
$SAI_{,f}$	0.64	1.00	0.69	0.89	0.66	0.87	0.91	0.92	0.68	0.68	0.71	0.58

Values of SAI_c at 56 days were evaluated for the "Mixes EF", and they are lower than those for 28 days; particularly, the *a. v.* decreases to 0.57, with a *st. dev.* = 0.07. This result is due to a more pronounced strength gain of the reference mixes over time than in those containing the MSWI BA, as shown in Figure 3, which displays the compressive strength for reference and "Mixes EF" mortars. Indeed, "Mixes EF" display less strength increase than reference mixes, despite the w/b ratio adopted.

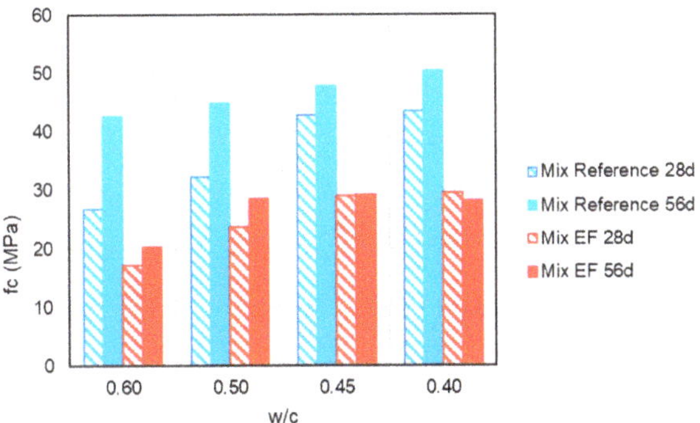

Figure 3. Compressive strength of reference and MSWI BA (0–1 mm) mortars at 28 d and 56 d.

Concerning instead SAI_f at 28 days, the best result is obtained in Mix E4 for the mortars realized with the coarse ash, whereas for the fine ash mortars Mix EF1 behaves best, displaying the same flexural strength as the reference mortar. As discussed for the compressive strength parameter, when the ash is used in the (0–1 mm) grading, the average SAI_f values is higher (*a. v.* = 0.92), with a small variability of the results (*st. dev.* = 0.05), than when it is used in the (0–4 mm) grading (*a. v.* = 0.72; *st. dev.* = 0.13). Furthermore, at the highest maturation age, SAI_f values of "Mixes EF" decrease, similarly to SAIc, with an *a. v.* = 0.66 and *st. dev.* = 0.06.

According to the above results, it is possible to derive compressive strength evolution as a function of the w/b ratio according to Equation (1), as shown in Figure 4a–c, respectively for reference, MSWI BA (0–4 mm) and MSWI BA (0–1 mm) mortars. Regression equations have a high R^2 value, demonstrating the goodness of the fitting relations. According to these regressions, it is possible to observe that at the lowest w/b value, mortars realized with the coarser fraction of the MSWI BA provide higher strength; conversely, at the highest w/b ratio, the mortars realized with the finest MSWI BA fraction suffer less strength losses than the counterparts with the coarser ash. Indeed, the slope of "Mix E" is steeper than that of "Mix EF". According to reference [53], where the influence of the w/b ratio on strength development of pozzolanic mortars was studied, the overall water content available in a mix for binder hydration impacts also pozzolan reactivity. Particularly, the compressive strength of mortars due to pozzolanic reaction increases with the w/b ratio: this suggests that probably the (0–1 mm) fraction, which performs sufficiently well also at high w/b ratios, has also a positive filling effect due to its reduced size compared to the coarser fraction.

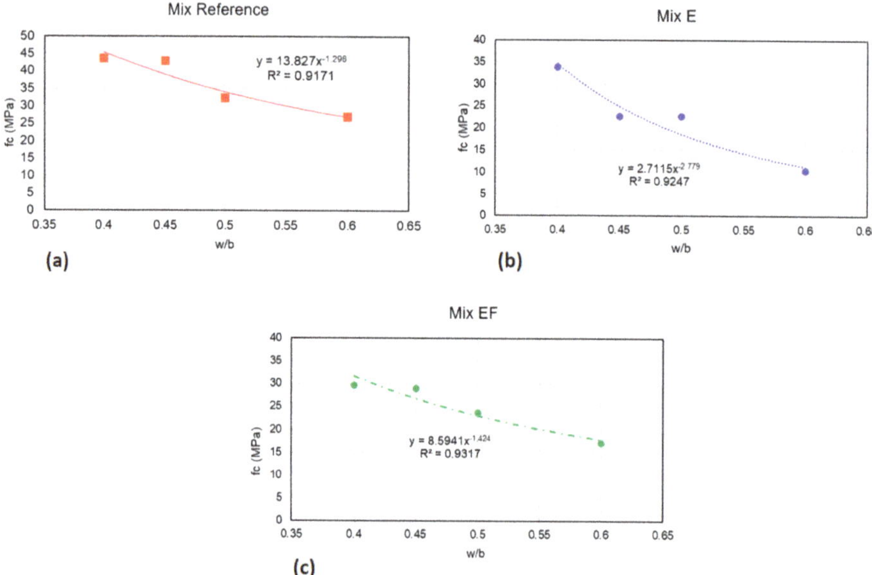

Figure 4. Compressive strength evolution as a function of w/b ratio for: (**a**) reference; (**b**) MSWI BA (0–4 mm); and (**c**) MSWI BA (0–1 mm) mortars.

3.2. k-Value of MSWI BA Mortars

The efficiency of the mortars is evaluated using the Δw method, obtained starting from the equations shown in Figure 4. Here, the k-value is determined for each w/b ratio of the (0–4 mm) MSWI BA, and it is plotted in Figure 5. Results highlight how the best efficiency is obtained for mortars having the lowest w/b ratio, and those values are similar to those typical of PFA and GGBFS, but lower than SF. Indeed, according to reference [34], k-values of supplementary cementing materials used at 25%w (with cement at 75%w) range from 0.1 up to 1.4 at 28 days, for a reference mortar realized with $w/b = 0.5$ and aggregate/cement ratio = 3. The highest values refer to high-calcium fly ash of high sulfur content and low-calcium fly ash, whereas the lowest values are typical of nickel slag, Milos earth and diatomaceous earth. However, it is worth observing that the efficiency of MSWI BA is almost null when the w/b ratio is high.

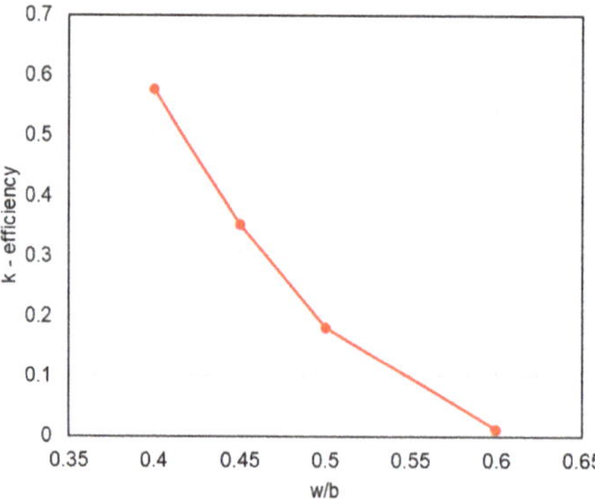

Figure 5. MSWI BA (0–4 mm) efficiency in mortars at 28 days.

3.3. Conductivity Test

The first measure of the solution without any ash addition revealed a conductivity value of 14 mS/cm, whereas the second measure, at 120 s from the addition of the MSWI BA, was about 7.30 mS/cm. The measure was repeated in triplicate, showing similar results. According to the classification provided by Luxán et al. [33], when the variation of the conductivity in this time window exceeds 1.2 mS/cm, the material can be classified as pozzolanic. However, it should be recalled that several studies indicated this test might provide approximated results [54], as it does not consider the presence of soluble salts in non-natural pozzolans, and thus should not be used as a conclusive test method.

3.4. Frattini Test

The test provides the values of the $[Ca]^{2+}$ and $[OH]^{-}$ oxides, which decrease in the solution as a consequence of the calcium hydroxide depletion after the pozzolanic reaction. Such values are then plotted in a graph together with the lime saturation curve: the experimental values measured here lay below this curve, indicating that the material can be considered active, and thus, it displays pozzolanic activity. The result is illustrated in Figure 6, together with the values typically displayed by other pozzolanic materials, i.e., metakaolin, PFA and silica fume [35]. Further, Figure 6 displays the results of the tested cement, used to realize the mortars, which is a cement CEM II/A-LL 42.5 R type, including low percentages of limestone. It is worth mentioning that the values provided by the cement are located over the red curve (i.e., the lime saturation curve), as expected. Another point is also plotted, which indicates a 20g sample made of cement (80%) + sand (20%), which is located above the lime saturation curve, properly indicating that no pozzolanic reaction took place.

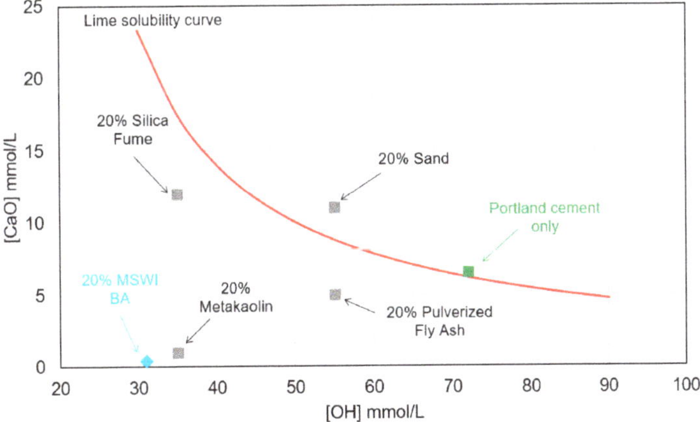

Figure 6. Results of the Frattini test (in light blue, MSWI BA).

3.5. Setting Times and Mechanical Properties of Cement Pastes

Seven cement pastes were tested to evaluate the initial and final setting time. Results are listed in Table 5, where the time values are expressed in minutes. It is worth highlighting how cement replacement with a large amount of MSWI BA reduces significantly the initial and final setting time, with a decrease of about −70% and −62% respectively, at 50%w replacement ratio. Conversely, at a low substitution rate, i.e., at 10%w, the initial setting time increases by about 40%, with few reductions as regards the final setting (−10%). The results obtained here can be justified according to the mineralogic composition of the raw MSWI BA, which displays some relevant XRD peaks corresponding to mayenite, a mineral that hydrates very quickly and is responsible for rapid-strength gain in some calcium-aluminate cements [47,55,56]. Mayenite is a crystalline phase that has been detected in other bottom ashes too, e.g., in references [38,57], and in ladle furnace slags [58,59]. Particularly in the second case, its presence has been associated to a reduction in the setting times of pastes [60,61] and flash set phenomena [48], which typically hinder strength development at a higher age because of the formation of a hard and relatively impermeable shell around the slag particles [62].

Table 5. Setting time of cement pastes with MSWI BA.

	Initial Setting (min)	Final Setting (min)
Reference	205	400
MSWI BA 5%w	245	395
MSWI BA 10%w	285	360
MSWI BA 15%w	255	345
MSWI BA 20%w	205	325
MSWI BA 25%w	135	255
MSWI BA 50%w	65	155

Table 6 lists instead the results of compressive strength tests carried out on the same cement paste mixes at 28 days: it is possible to clearly see how the mechanical strength is severely affected when high replacement ratios are used. Indeed, for a 50%w replacement ratio, about 50% strength loss is displayed. However, when small amounts of MSWI BA are added as a subsitute for cement, the strength does not change significantly, and for up to a 10% replacement ratio, even a slight strength enhance is instead observed. Such results almost agree with those of setting times: indeed, it is argued that replacing huge amounts of cement with the MSWI BA induces flash set phenomena, due to the rapid hydration of mayenite. This process leads the other particles with little available water for longer hydration and strength gain, thus inducing the severe strength loss observed. Instead, when the amount of the substitution is low (i.e., about 10%), the overall content of mayenite is not sufficient to induce any flash set, and thus, the water is available for hydration of the cement particles, and possible further pozzolanic activities can take place. Other reasons why the pastes exhibit such strength loss at replacement ratios equal to or higher than 15% might be linked to both the physical and chemical properties of MSWI BA, e.g., the weaker nature of the ash than cement, the different water demand of the particles and its less fineness.

Table 6. Compressive strength of cement pastes with MSWI BA (average values ± standard deviation).

	Compressive Strength f_c (MPa)
Reference	48.71 ± 0.60
MSWI BA 5%w	49.21 ± 1.33
MSWI BA 10%w	49.00 ± 1.61
MSWI BA 15%w	38.14 ± 0.36
MSWI BA 20%w	36.08 ± 1.03
MSWI BA 25%w	34.89 ± 1.05
MSWI BA 50%w	23.62 ± 1.23

4. Conclusions

This work shows the results of an experimental campaign aimed at addressing the pozzolanic activity of a grossly grounded MSWI BA. Different test methods were used, both direct and indirect, to evaluate if their adoption allows the same judgement to be obtained. Even though the material is classified as a non-pozzolanic material, due to the insufficient fineness and the lower $SiO_2 + Al_2O_3 + Fe_2O_3$ content than that typically required by the main codes, it has been demonstrated that it owns some pozzolanic potentials. Indeed, some amorphous structures and the positive outcomes of almost all the tests carried out here demonstrate that the material may be further treated to ensure an increase of its pozzolanic ability. However, the presence of mayenite in the mineralogic composition of the material may compromise the ability to realize cement-based materials with a high replacement ratio of this MSWI BA, due to the occurrence of a possible flash set. Such phenomena could avoid the longer hydration of cement particles, and thus, it may hinder achievement of the required compressive strength. At this stage, a cement replacement ratio up to 10% seems feasible to not compromise the strength gain. A further study aimed at assessing the microstructure and morphology of mortars realized with this MSWI BA may help to understand better the reactions governing the hardening phase of this material.

Author Contributions: Conceptualization, F.F. and K.B.; methodology, F.F. and K.B.; formal analysis, F.F. and F.A.; investigation, F.F., K.T., K.B., M.A.Z., F.A. and A.G.S.; resources, F.F. and C.P.; data curation, K.T. and F.A.; writing—original draft preparation, F.F. and M.A.Z.; writing—review and editing, F.F., K.T. and M.A.Z.; supervision, F.F.; project administration, C.P.; funding acquisition, F.F. and C.P. All authors have read and agreed to the published version of the manuscript.

Funding: The APC was funded by the University of Padova, Department of Civil, Environmental and Architectural Engineering—BIRD.

Data Availability Statement: Data are available at corresponding author request.

Acknowledgments: The authors would like to acknowledge Eng. Sabrina Pastore for her help during the experimental campaign.

Conflicts of Interest: The authors declare no conflict of interest.

References

1. *ASTM C125-20. Standard Terminology Relating to Concrete and Concrete Aggregates*; ASTM International: West Conshohocken, PA, USA, 2020.
2. Uzal, B.; Turanli, L.; Mehta, P.K. High-volume natural pozzolan concrete for structural applications. *ACI Mater. J.* **2007**, *104*, 535.
3. Sata, V.; Jaturapitakkul, C.; Kiattikomol, K. Influence of pozzolan from various by-product materials on mechanical properties of high-strength concrete. *Constr. Build. Mater.* **2007**, *21*, 1589–1598. [CrossRef]
4. Alnahhal, M.F.; Alengaram, U.J.; Jumaat, M.Z.; Alqedra, M.A.; Mo, K.H.; Sumesh, M. Evaluation of industrial by-products as sustainable pozzolanic materials in recycled aggregate concrete. *Sustainability* **2017**, *9*, 767. [CrossRef]
5. Siddique, R.; Khan, M.I. *Supplementary Cementing Materials*; Springer Science Business Media: Berlin, Germany, 2011.
6. Barcelo, L.; Kline, J.; Walenta, G.; Gartner, E. Cement and carbon emissions. *Mater. Struct.* **2014**, *47*, 1055–1065. [CrossRef]
7. Cheriaf, M.; Rocha, J.C.; Pera, J. Pozzolanic properties of pulverized coal combustion bottom ash. *Cem. Concr. Res.* **1999**, *29*, 1387–1391. [CrossRef]
8. Chindaprasirt, P.; Rattanasak, U.; Jaturapitakkul, C. Utilization of fly ash blends from pulverized coal and fluidized bed combustions in geopolymeric materials. *Cem. Concr. Compos.* **2011**, *33*, 55–60. [CrossRef]
9. Nath, P.; Sarker, P.K. Effect of GGBFS on setting, workability and early strength properties of fly ash geopolymer concrete cured in ambient condition. *Constr. Build. Mater.* **2014**, *66*, 163–171. [CrossRef]
10. Yang, H.M.; Kwon, S.J.; Myung, N.V.; Singh, J.K.; Lee, H.S.; Mandal, S. Evaluation of strength development in concrete with ground granulated blast furnace slag using apparent activation energy. *Materials* **2020**, *13*, 442. [CrossRef] [PubMed]
11. Wu, Z.; Khayat, K.H.; Shi, C. Changes in rheology and mechanical properties of ultra-high performance concrete with silica fume content. *Cem. Concr. Res.* **2019**, *123*, 105786. [CrossRef]
12. Pellegrino, C.; Faleschini, F.; Meyer, C. Recycled materials in concrete. In *Developments in the Formulation and Reinforcement of Concrete*; Woodhead Publishing: Sawston, UK, 2019; pp. 19–54.
13. Silva, K.D.C.; Silva, G.C.; Natalli, J.F.; Mendes, J.C.; Silva, G.J.B.; Peixoto, R.A.F. Rock wool waste as supplementary cementitious material for portland cement-based composites. *ACI Mater. J.* **2018**, *115*, 653–661. [CrossRef]
14. Da Silva Magalhães, M.; Faleschini, F.; Pellegrino, C.; Brunelli, K. Cementing efficiency of electric arc furnace dust in mortars. *Constr. Build. Mater.* **2017**, *157*, 141–150. [CrossRef]
15. Parron-Rubio, M.E.; Perez Garcia, F.; Gonzalez-Herrera, A.; Oliveira, M.J.; Rubio-Cintas, M.D. Slag substitution as a cementing material in concrete: Mechanical, physical and environmental properties. *Materials* **2019**, *12*, 2845. [CrossRef] [PubMed]
16. Papayianni, I.; Anastasiou, E. Effect of granulometry on cementitious properties of ladle furnace slag. *Cem. Concr. Compos.* **2012**, *34*, 400–407. [CrossRef]

17. Manso, J.M.; Rodriguez, Á.; Aragón, Á.; Gonzalez, J.J. The durability of masonry mortars made with ladle furnace slag. *Constr. Build. Mater.* **2011**, *25*, 3508–3519. [CrossRef]
18. Anastasiou, E.K. Effect of high calcium fly ash, ladle furnace slag, and limestone filler on packing density, consistency, and strength of cement pastes. *Materials* **2021**, *14*, 301. [CrossRef] [PubMed]
19. Santamaria, A.; Ortega-Lopez, V.; Skaf, M.; García, V.; Gaitero, J.J.; San-José, J.T.; González, J.J. Ladle furnace slag as cement replacement in mortar mixes. In Proceedings of the Fifth International Conference on Sustainable Construction Materials and Technologies, Kingston University, London, UK, 14–17 July 2019; pp. 14–17.
20. Bignozzi, M.C.; Saccani, A.; Barbieri, L.; Lancellotti, I. Glass waste as supplementary cementing materials: The effects of glass chemical composition. *Cem. Concr. Compos.* **2015**, *55*, 45–52. [CrossRef]
21. Faleschini, F.; Zanini, M.A.; Brunelli, K.; Pellegrino, C. Valorization of co-combustion fly ash in concrete production. *Mater. Des.* **2015**, *85*, 687–694. [CrossRef]
22. Tang, P.; Chen, W.; Xuan, D.; Zuo, Y.; Poon, C.S. Investigation of cementitious properties of different constituents in municipal solid waste incineration bottom ash as supplementary cementitious materials. *J. Clean. Prod.* **2020**, *258*, 120675. [CrossRef]
23. Ribeiro, D.V.; Morelli, M.R. Effect of calcination temperature on the pozzolanic activity of Brazilian sugar cane bagasse ash (SCBA). *Mater. Res.* **2014**, *17*, 974–981. [CrossRef]
24. Antiohos, S.K.; Papadakis, V.G.; Tsimas, S. Rice husk ash (RHA) effectiveness in cement and concrete as a function of reactive silica and fineness. *Cem. Concr. Res.* **2014**, *61*, 20–27. [CrossRef]
25. Cosentino, I.; Restuccia, L.; Ferro, G.A.; Tulliani, J.M. Type of materials, pyrolysis conditions, carbon content and size dimensions: The parameters that influence the mechanical properties of biochar cement-based composites. *Theor. Appl. Fract. Mech.* **2019**, *103*, 102261. [CrossRef]
26. Chimenos, J.M.; Fernandez, A.I.; Nadal, R.; Espiell, F. Short-term natural weathering of MSWI bottom ash. *J. Hazard. Mater.* **2000**, *79*, 287–299. [CrossRef]
27. Pivato, A.; Pasetto, M.; Faleschini, F.; Beggio, G.; Hennebert, P. Research to industry and industry to research. *Detritus* **2019**, *8*, I–III.
28. British Standards Institution. *EN 450-1:2012. Fly Ash for Concrete. Definition, Specifications and Conformity Criteria*; CEN: Brussels, Belgium, 2012.
29. ASTM C618. *Standard Specification for Coal Fly Ash and Raw or Calcined Natural Pozzolan for Use in Concrete*; ASTM International: West Conshohocken, PA, USA, 2019.
30. Papadakis, V.G.; Tsimas, S. Supplementary cementing materials in concrete: Part I: Efficiency and design. *Cem. Concr. Res.* **2002**, *32*, 1525–1532. [CrossRef]
31. British Standards Institution. *EN 196-5:2011. Methods of Testing Cement. Pozzolanicity Test for Pozzolanic Cement*; CEN: Brussels, Belgium, 2011.
32. ASTM C311/C311M-18. *Standard Test Methods for Sampling and Testing Fly Ash or Natural Pozzolans for Use in Portland-Cement Concrete*; ASTM International: West Conshohocken, PA, USA, 2018.
33. Luxan, M.D.; Madruga, F.; Saavedra, J. Rapid evaluation of pozzolanic activity of natural products by conductivity measurement. *Cem. Concr. Res.* **1989**, *19*, 63–68. [CrossRef]
34. Papadakis, V.G.; Antiohos, S.; Tsimas, S. Supplementary cementing materials in concrete: Part II: A fundamental estimation of the efficiency factor. *Cem. Concr. Res.* **2002**, *32*, 1533–1538. [CrossRef]
35. Donatello, S.; Tyrer, M.; Cheeseman, C.R. Comparison of test methods to assess pozzolanic activity. *Cem. Concr. Compos.* **2010**, *32*, 121–127. [CrossRef]
36. Bumanis, G.; Vitola, L.; Stipniece, L.; Locs, J.; Korjakins, A.; Bajare, D. Evaluation of Industrial by-products as pozzolans: A road map for use in concrete production. *Case Stud. Constr. Mater.* **2020**, *13*, e00424. [CrossRef]
37. Thorstensen, R.T.; Fidjestol, P. Inconsistencies in the pozzolanic strength activity index (SAI) for silica fume according to EN and ASTM. *Mater. Struct.* **2015**, *48*, 3979–3990. [CrossRef]
38. Inkaew, K.; Saffarzadeh, A.; Shimaoka, T. Modeling the formation of the quench product in municipal solid waste incineration (MSWI) bottom ash. *Waste Manag.* **2016**, *52*, 159–168. [CrossRef]
39. Schollbach, K.; Alam, Q.; Caprai, V.; Florea, M.V.A.; Van der Laan, S.R.; Van Hoek, C.J.G.; Brouwers, H.J.H. Combined characterization of the MSWI bottom ash. In Proceedings of the Thirty-Eighth International Conference on Cement Microscopy, Lyon, France, 17–21 April 2016; pp. 74–84.
40. Zhang, T.; Zhao, Z. Optimal use of MSWI bottom ash in concrete. *Int. J. Concr. Struct. Mater.* **2014**, *8*, 173–182. [CrossRef]
41. Caprai, V.; Gauvin, F.; Schollbach, K.; Brouwers, H.J.H. MSWI bottom ash as binder replacement in wood cement composites. *Constr. Build. Mater.* **2019**, *196*, 672–680. [CrossRef]
42. Pivato, A. Analysis of chemical and ecotoxicological data from leaching tests of inert wastes performed for the assessment of end of waste status. In Proceedings of the SIDISA 2021—XI International Symposium on Environmental Engineering, Turin, Italy, 29 June–2 July 2021.
43. Lancellotti, I.; Ponzoni, C.; Bignozzi, M.C.; Barbieri, L.; Leonelli, C. Incinerator bottom ash and ladle slag for geopolymers preparation. *Waste Biomass Valorization* **2014**, *5*, 393–401. [CrossRef]
44. Tang, P.; Florea, M.V.A.; Spiesz, P.; Brouwers, H.J.H. Application of thermally activated municipal solid waste incineration (MSWI) bottom ash fines as binder substitute. *Cem. Concr. Compos.* **2016**, *70*, 194–205. [CrossRef]

45. Qiao, X.C.; Tyrer, M.; Poon, C.S.; Cheeseman, C.R. Novel cementitious materials produced from incinerator bottom ash. *Resour. Conserv. Recycl.* **2008**, *52*, 496–510. [CrossRef]
46. Mazouzi, W.; Kacimi, L.; Cyr, M.; Clastres, P. Properties of low temperature belite cements made from aluminosilicate wastes by hydrothermal method. *Cem. Concr. Compos.* **2014**, *53*, 170–177. [CrossRef]
47. He, Z.; Li, Y. The influence of mayenite employed as a functional component on hydration properties of ordinary Portland cement. *Materials* **2018**, *11*, 1958. [CrossRef]
48. Adesanya, E.; Sreenivasan, H.; Kantola, A.M.; Telkki, V.V.; Ohenoja, K.; Kinnunen, P.; Illikainen, M. Ladle slag cement—Characterization of hydration and conversion. *Constr. Build. Mater.* **2018**, *193*, 128–134. [CrossRef]
49. Babu, K.G.; Rao, G.S.N. Efficiency of fly ash in concrete with age. *Cem. Concr. Res.* **1996**, *26*, 465–474. [CrossRef]
50. Schiessl, P.; Hardtl, R. Efficiency of Fly Ash in Concrete Evaluation of Ibac Test Results. In *Technical Report of Institut für Bauforschung*; RWTH Aachen University: Aachen, Germany, 1991.
51. Baki, V.A.; Nayır, S.; Erdoğdu, Ş.; Ustabaş, İ. Determination of the pozzolanic activities of trachyte and rhyolite and comparison of the test methods implemented. *Int. J. Civil. Eng.* **2020**, *18*, 1053–1066. [CrossRef]
52. Hashemi, S.S.G.; Mahmud, H.B.; Djobo, J.N.Y.; Tan, C.G.; Ang, B.C.; Ranjbar, N. Microstructural characterization and mechanical properties of bottom ash mortar. *J. Clean. Prod.* **2018**, *170*, 797–804. [CrossRef]
53. Sata, V.; Tangpagasit, J.; Jaturapitakkul, C.; Chindaprasirt, P. Effect of W/B ratios on pozzolanic reaction of biomass ashes in Portland cement matrix. *Cem. Concr. Compos.* **2012**, *34*, 94–100. [CrossRef]
54. Velázquez, S.; Monzó, J.M.; Borrachero, M.V.; Payá, J. Assessment of the pozzolanic activity of a spent catalyst by conductivity measurement of aqueous suspensions with calcium hydroxide. *Materials* **2014**, *7*, 2561–2576. [CrossRef]
55. Sriwong, C.; Phrompet, C.; Tuichai, W.; Karaphun, A.; Kurosaki, K.; Ruttanapun, C. Synthesis, microstructure, multifunctional properties of mayenite Ca 12 Al 14 O 33 (C12A7) cement and graphene oxide (GO) composites. *Sci. Rep.* **2020**, *10*, 1–19.
56. Qiao, X.C.; Ng, B.R.; Tyrer, M.; Poon, C.S.; Cheeseman, C.R. Production of lightweight concrete using incinerator bottom ash. *Constr. Build. Mater.* **2008**, *22*, 473–480. [CrossRef]
57. Manso, J.M.; Ortega-López, V.; Polanco, J.A.; Setién, J. The use of ladle furnace slag in soil stabilization. *Constr. Build. Mater.* **2013**, *40*, 126–134. [CrossRef]
58. Skaf, M.; Ortega-López, V.; Fuente-Alonso, J.A.; Santamaría, A.; Manso, J.M. Ladle furnace slag in asphalt mixes. *Constr. Build. Mater.* **2016**, *122*, 488–495. [CrossRef]
59. Henríquez, P.A.; Aponte, D.; Ibáñez-Insa, J.; Bizinotto, M.B. Ladle furnace slag as a partial replacement of Portland cement. *Constr. Build. Mater.* **2021**, *289*, 123106. [CrossRef]
60. Adolfsson, D.; Robinson, R.; Engström, F.; Björkman, B. Influence of mineralogy on the hydraulic properties of ladle slag. *Cem. Concr. Res.* **2011**, *41*, 865–871. [CrossRef]
61. Wang, Y.; Ni, W.; Suraneni, P. Use of ladle furnace slag and other industrial by-products to encapsulate chloride in municipal solid waste incineration fly ash. *Materials* **2019**, *12*, 925. [CrossRef]
62. Le Saoût, G.; Lothenbach, B.; Hori, A.; Higuchi, T.; Winnefeld, F. Hydration of Portland cement with additions of calcium sulfoaluminates. *Cem. Concr. Res.* **2013**, *43*, 81–94. [CrossRef]

Article

Innovative Design Concept of Cooling Water Tanks/Basins in Geothermal Power Plants Using Ultra-High-Performance Fiber-Reinforced Concrete with Enhanced Durability

Salam Al-Obaidi [1,2,*], Patrick Bamonte [1], Francesco Animato [3], Francesco Lo Monte [1], Iacopo Mazzantini [3], Massimo Luchini [3], Sandra Scalari [3] and Liberato Ferrara [1]

1 Department of Civil and Environmental Engineering, Politecnico di Milano, 20133 Milan, Italy; patrick.bamonte@polimi.it (P.B.); francesco.lo@polimi.it (F.L.M.); liberato.ferrara@polimi.it (L.F.)
2 Roads and Transportations Engineering Department, University of Al-Qadisiyah, Diwaniyah 58001, Iraq
3 Enel Green Power, 00198 Rome, Italy; francesco.animato@enel.com (F.A.); iacopo.mazzantini@enel.com (I.M.); massimo.luchini@enel.com (M.L.); sandra.scalari@enel.com (S.S.)
* Correspondence: salammaytham.alobaidi@polimi.it

Abstract: The structure presented in this paper is intended to be used as a prototype reservoir for collecting water coming from the cooling tower of a geothermal plant, and is primarily designed to compare the performance of different materials (traditional reinforced concrete and Ultra-High-Performance Fiber-Reinforced Concrete (UHPFRC)) as well to assess the performance of different structural solutions (wall with constant thickness versus wall provided with stiffening buttresses). In the absence of specific code provisions, given the novelty of the UHPFRC used, the main properties used for the design were determined through a dedicated experimental campaign (tensile/flexural properties and shrinkage). The main focus of the design was on the Serviceability Limit States, more specifically the requirements regarding water tightness. Given the rather simple structural layout, especially in the compartments where no stiffening buttresses are present, linear elastic analysis was used to determine the internal actions. The nonlinear behavior ensuing from the peculiar tensile constitutive response of the material was taken into account locally, in order to determine the stress level, the depth of the compression zone and the crack width. The performance was finally compared with the reference compartment (made with ordinary reinforced concrete), through on-site observations and measurements.

Keywords: durability; UHPFRC; water-retaining structures; aggressive environment

Citation: Al-Obaidi, S.; Bamonte, P.; Animato, F.; Lo Monte, F.; Mazzantini, I.; Luchini, M.; Scalari, S.; Ferrara, L. Innovative Design Concept of Cooling Water Tanks/Basins in Geothermal Power Plants Using Ultra-High-Performance Fiber-Reinforced Concrete with Enhanced Durability. *Sustainability* **2021**, *13*, 9826. https://doi.org/10.3390/su13179826

Academic Editors: Fausto Minelli, Enzo Martinelli and Luca Facconi

Received: 7 July 2021
Accepted: 29 August 2021
Published: 1 September 2021

1. Introduction

Reinforced concrete structures under Extremely Aggressive Exposure (EAE) are known to suffer from durability and time-dependent problems, which require continuous repair actions that are critical regarding the economic impact and service disruption problems (especially in the case of strategic infrastructures, including harbors, highways or facilities for the production of energy) [1]. Typical pathologies in reinforced concrete structures under EAE conditions consist of excessive corrosion of the steel reinforcement, accompanied by cracking and spalling of the concrete cover. Corrosion is typically initiated by altering the pH environment of the concrete, particularly in the vicinity of the steel reinforcement (concrete cover) or by localized chloride attack to the reinforcement. This mechanism may occur, e.g., in concrete structures that are in contact with extremely aggressive waters (industrial chemical water, marine or deicing salts environments). Following the prescriptions to design a structure with a specific service life under extremely aggressive exposure conditions, designers are obligated to focus on the quality of the concrete cover to ensure its ability to protect the reinforcement from corrosion, at least prior to reaching the design target service life. Furthermore, reinforced concrete structures in industrial applications are also severely prone to corrosion and erosion due to the harsh environment created by the chemical treatment and processing methods. Among the aggressive exposure conditions, the XA3 exposure class (highly aggressive chemical environment) has been reported in many industrial and chemical applications as well as in one of the case studies selected as the topic of the present paper: the basins for a cooling tower in geothermal power plants owned and run by Enel Green Power in different sites in Tuscany (Italy), which have suffered from cracking, loss of waterproofing layers, as well as abrasion and erosion of the concrete due to the exposure to the geothermal water, actually containing different types of aggressive chemicals, including chloride and sulfate ions [2]. Since these degradation mechanisms required quick maintenance works in order to maintain the plants in operation, repairing of the degraded concrete layers, sealing of the joints between the bed and walls, adding non corroded construction elements and re-applying the bituminous coating have been conducted at the deteriorated parts of the artefacts to extend their service life. However, some parts could not be retrofitted due to limited maintenance time.

Currently available design codes and related standards specify, depending on the exposure class and required service life, the structural design criteria, such as the crack-width limits, as well as detail the criteria, including minimum cover thickness, and material criteria, including type of cement and w/c content, to meet the durability requirements specifically limiting the ingress of chemical aggressive and to provide a design service life of 50 to 100 years. As a matter of fact, the aforementioned deterioration scenarios may also appear in the early stages of the structure's operational life, likely due to the inherent characteristics of ordinary reinforced concrete materials and/or lack of experience during the construction. Therefore, the use of advanced alternative materials and of the corresponding construction and design methodologies need to be implemented in the design and construction of structures and infrastructures, with specific focus on extremely aggressive environments.

Currently, in both industry and academia, there is significant driving momentum in the development and application of advanced cement-based materials in the construction and retrofitting of structures and infrastructures, to provide them with enhanced durability and hence extended service life and overall improved environmental and economic sustainability. In this respect, Ultra-High-Performance Concrete (UHPC) or Ultra-High-Performance Fiber-Reinforced Concrete (UHPFRC), with its highly compact matrix and ability to control and distribute the induced damage into multiple and narrow cracks, representing its most resilient features, is the material to be implemented for applications demanding superior durability in real structural service conditions, which, for cement-based material, is necessary to be considered in the cracked state [3,4].

In addition, the peculiar composition of UHPFRCs, featured by a high content of cement and of supplementary cementitious materials with a low water/binder ratio, is highly conducive to autogenous healing of the cracks [5–7], which can be furthermore stimulated through the addition of crystalline mineral additions, mineral admixtures and other tailored constituents, including SAPs [7–10]. The improvement of mechanical performance achieved through the synergy of the mix composition and the presence of steel fibers in further interaction with the aforementioned enhanced durability and self-healing features may make the development and application of UHPFRCs able to improve the environmental footprint, thanks to optimized use of material, and significantly extend the structure service life in extremely aggressive environments [11–13]. Moreover, thanks to their recently demonstrated high recyclability potential, this category of material is also able to bring a breakthrough innovation into the concept of longer structural service life [14].

As a matter of fact, current solutions for new concrete constructions in EAE have so far seldomly implemented on a massive scale advanced cement-based construction materials, such as Ultra-High-Performance (Fiber-Reinforced) Concrete (UHPC/UHPFRC), mainly because of a lack of standards and of technical awareness by most designers and contractors. Having said that, to date, the claimed superior durability of UHPC/UHPFRC has been almost exclusively demonstrated by means of laboratory specimens, focusing mainly on the uncracked state. To contribute in filling this gap between laboratory testing and real-scale structural applications, the EC H2020-funded ReSHEAience project aims to test and validate new design concepts through long-term monitoring of six proof-of-concept pilot structures, selected as representative of the cutting-edge economy sectors such as green energy, blue growth and conservation of reinforced concrete heritage [15]. The focus of this work is on the pilot constructed within the framework of ReSHEALience project and located in Enel Green Power geothermal power plant in the municipality of Chiusdino in Tuscany (Italy), namely, a geothermal cooling tower water collection basin. The basin is operating in parallel with the actual basins of the plant. The main pathologies in the basins containing the geothermal water are cracking produced by expansion (sulphate attack), cement paste dissolution (leaching) and/or efflorescence (acid attack), together with reinforcement corrosion triggered by the chlorides also present in the water. In the project framework outlined above, the geothermal basin of the Chiusdino plant has been redesigned to resist the mechanical actions and chemical attack as well. With the aim of implementing a signature durability-based design of concrete structures, which through the incorporation of degradation mechanisms into structural design algorithms [16] will contribute to move from a prescriptive to performance-based design approach [17–21], three types of structural concrete solutions have been designed and constructed. They consist of the following:

(a) A 100 mm-thick cast-in-place ordinary reinforced concrete wall.
(b) A 60 mm-thick cast-in-place UHPC wall.
(c) A solution consisting of 30 mm-thin precast UHPC slabs "supported" by 200 $\times$ 200 mm^2 cast-in-place UHPC columns/buttresses.

The aim of this paper is to analyze each structural solution both at the serviceability and ultimate limit states, based on the mechanical properties of the materials experimentally identified through a tailored experimental campaign [22]. Moreover, dedicated tests are also being conducted on the basin to validate the proof of concept behind each structural system, as also affected by the different construction methods implemented on the construction site. The general aim of the work is to pave the way toward a performance-based design approach for long-term durability of UHPFRC structures, also in the context of the currently ongoing international code-writing and standardization efforts.

2. Description of the Pilot

Within the framework of the ReSHEALience project, the main aim is to upgrade the concept of UHPC/UHPFRC to Ultra-High-Durability Concrete (UHDC) by implementing macro- and nanoscale constituents in strain-hardening steel fiber-reinforced cementitious material [15]. The purpose of this "tailoring" the UHDC material concept is to enhance the durability in real structural service scenarios, which explicitly include the presence of "micro-cracks", as further explained, as a natural "status" of the material into its service state, thus reducing the need and frequency for maintenance and the related costs, and contribute to wholesome extension of the service life of structures subjected to extremely aggressive environments. The most distinguished characteristics of the UHDC materials employed in this project is the incorporation into the mix-design of selected nano-constituents, namely, alumina nano-fibers and cellulose nano-crystals, which, besides contributing to enrich the microstructure of the concrete matrix and preventing or reducing the transport of aggressive substances in the uncracked state [23–25], also provide functionalities such as self-curing [20,21] and, in main synergy with crystalline admixtures, stimulate autogenous healing [26]. Moreover, as reported in the literature [27–30] and proven by [23], the signature mechanical tensile behavior of UHPFRC is characterized by a stable multiple cracking process, which spreads the localized damage into multiple tiny cracks up to a relatively higher deformation level as compared not only to plain and reinforced concrete but even to ordinary fiber-reinforced concrete (FRC). This unique feature may allow and require tailored structural concepts to be designed for its full exploitation, not merely confined to the reduction of structural element thickness. This step is fundamental in order to tackle the challenges brought in by the newly developed materials and to formulate a holistic durability-based structural design approach, which will be validated through the innovative structure concepts designed, built and tested, of which the tank to collect and contain the geothermal water dealt with in this paper is a landmark representation in the framework of the project [31].

The structure consists of a 300 mm-thick foundation slab made of ordinary reinforced concrete, and is divided into three compartment basins by means of vertical partitional walls. The in-plan dimensions of the whole tank are approximately 22.50 × 7.00 m^2. The three basins are characterized by the same in-plan dimensions (7.50 × 7.00 m^2), but differ in materials and structural typology: one in ordinary reinforced concrete (RC) and the other two in Ultra-High-Durability Concrete (UHDC). The vertical walls, which characterize the three compartments, have different thicknesses, depending on the material used and the geometry of the structural elements (Figure 1).

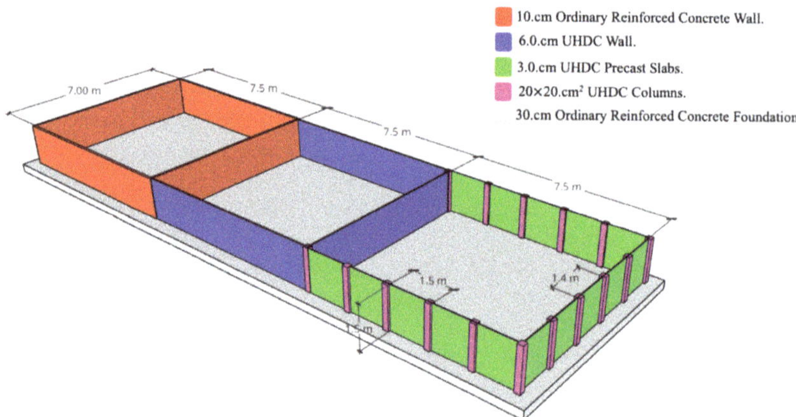

Figure 1. Geometrical and material description of the basins.

Basins 1 and 2 consist of cast-in-place continuous walls with a constant thickness, respectively, equal to 100 mm for the ordinary reinforced concrete Basin 1 and 60 mm for the UHDC Basin 2. Basin 3, on the other hand, consists of 30 mm-thick precast slabs, stiffened by cast-in-place vertical buttresses with a square cross section (200 × 200 mm^2) also made of UHDC, and spaced at a mutual distance of 1.50 m on the long side and 1.40 m on the short side. The dimensions chosen for the buttresses are in line with those needed in similar structures to ensure adequate support for the wooden superstructure.

When the picture in Figure 2a was taken (20 February 2020), the structure was almost completed, with only some of the columns of Basin 3 still to be cast. Figure 2b shows the structure completed and filled with geothermal water during one of the validation tests, performed on 20 April 2021.

Figure 2. Pictures of the structure, taken on 13 February 2020 (**a**) and on 20 April 2021 when a full load test was performed filling the three basins (**b**).

3. Materials

3.1. Reinforced Concrete (Basin 1)

For the foundation slab and the four walls made with the ordinary reinforced concrete traditional solution, concrete C25/30 was used, characterized by the following mechanical properties (EN 1992-1-1 [32]):

- Characteristic cylindrical compressive strength: f_{ck} = 25 MPa.
- Average direct tensile strength: f_{ctm}= 2.6 MPa.
- Average indirect tensile strength (in bending): f_{cfm} = 3.1 MPa.
- Instantaneous modulus of elasticity: E_{cm} = 31 GPa.

The values of the average direct and indirect tensile strength as well as of the Young's modulus of elasticity correspond to the concrete class chosen, as indicated in the Italian Standard (NTC 2018 [33]), and are in line with those reported in EN 1992-1-1 Table 3.1 [32]. Nonetheless, a Schmidt hammer (rebound hammer) tester was used to evaluate the concrete compressive stress of the reinforced concrete walls and the foundation. The results indicated average values of 28–37 MPa, as converted to characteristic cylindrical values.

3.2. UHDC (Basins 2 and 3)

The UHDC mix used for the other two compartments, whose mix design and chemical compositions of the employed cement and slag are shown in Tables 1 and 2, respectively, is characterized by the presence of dispersed fiber reinforcement, to the advantage of the overall mechanical behavior, which allowed structural solutions without the need of a specific traditional reinforcement. The straight brass-plated fibers, with the characteristic length and diameters being 20 mm and 0.22 mm, respectively, provide the minimum tensile strength of 2400 MPa. In consideration of the peculiarities of the material, a specific experimental campaign of mechanical characterization was performed in order to investigate the post-cracking behavior, especially focusing on the multi-cracking stage [34].

All the mechanical tests were performed after at least 90 days after casting, in order to limit the possible influence brought in by the delayed hydration of the slag. In the curing period, all specimens were stored in controlled environmental conditions (RH = 90%, T = 20 °C).

Different testing methods have been carried out in order to investigate the multiple cracking phenomena experienced by the UHDC under tensile and flexural loads, as shown in Figure 3. Indirect tensile tests have been performed employing the so-called Double Edge Wedge Splitting (DEWS) test methodology, able to provide in a straightforward manner the tensile stress–COD (Crack Opening Displacement) relationship of the material. Moreover, flexural 4-Point Bending Tests (4PBT) were also performed on the deep and thin beams, as below:

- Deep beams—DB ($L \times b \times h = 500 \times 100 \times 100$ mm^3);
- Thin beams—TB ($L \times b \times h = 500 \times 100 \times 25$ mm^3).

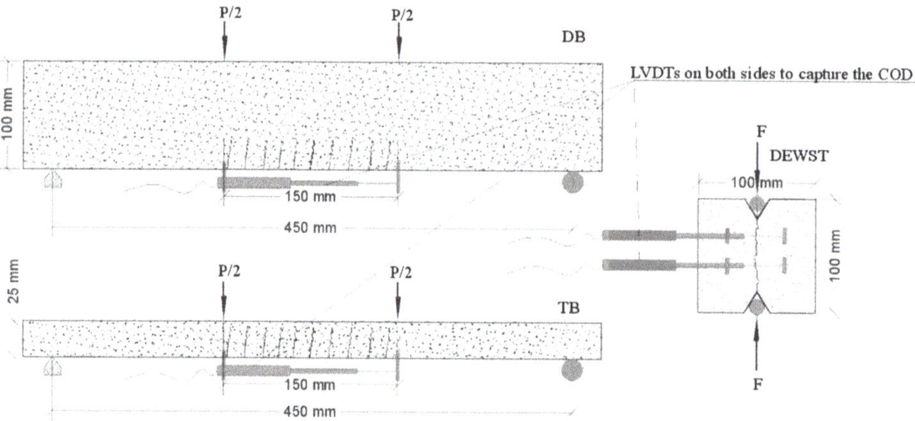

Figure 3. Different types of mechanical tests to evaluate the tensile behavior of UHDC: 4PBT under flexural load P, for the deep beam (DB) and thin beam (TB), and an indirect tensile test, Double Edge Wedge Splitting (DEWST), under load F.

The 4PBT on thin beams were adapted to verify the effects of fiber orientation in a "structural" specimen with the same thickness employed in the construction of the basin walls.

In Figure 4, the experimental nominal flexural stress σ_n (black curves) is plotted as a function of the Crack Opening Displacement (COD) for the reference mix, as measured across a 150 mm-wide central zone from the 4-Point Bending Tests on deep (Figure 4a) and thin beams (Figure 4b). The nominal stress is evaluated as follows:

$$\sigma_n = \frac{PL/6}{bh^2/6} = \frac{PL}{bh^2} \quad (1)$$

where P is the total load applied (N), L is the distance between the two supports (450 mm for both DB and TB), and b and h are the specimen width and height in mm, respectively. In the same plots, the response simulated employing the tensile stress–COD constitutive response calibrated on the basis of the DEWS tests [35] is shown (pink curves). In addition, for design purposes, the red curves show the results obtained assuming an elastic–perfectly plastic constitutive law, with the direct tensile strength equal to 7.0 MPa; up to a COD approximately equal to 2.0 mm, the numerical results are in good agreement with the experimental results as described in details in [22], where all the mechanical parameters below were calibrated.

As a consequence, the following mechanical properties were assumed (including a dispersion coefficient γ_d = 1.25 as per prescriptions contained in *fib* Model Code 2010):

- Average direct tensile strength: f_{ctm} = 7.0/1.25 = 5.6 MPa.
- Average indirect tensile strength (in bending): f_{cfm} = 12.0/1.25 = 9.6 MPa.
- Instantaneous modulus of elasticity: E_{cm} = 41.7 GPa.

Table 1. The Ultra-High-Durability Concrete (UHDC) mix components and their proportions.

Constituents	XA-CA	XA-CA + ANF	XA-CA-CNC
CEM I 52,5 R (kg/m^3)	600	600	600
Slag (kg/m^3)	500	500	500
Water (liter/m^3)	200	200	200
Steel fibers Azichem Readymesh 200® (kg/m^3)	120	120	120
Sand 0–2 mm (kg/m^3)	982	982	982
Superplasticizer Glenium ACE 300 ® (liter/m^3)	33	33	33
Crystalline Admixture Penetron Admix ® (kg/m^3)	4.8	4.8	4.8
Alumina nanofibers NAFEN® * (kg/m^3)	-	0.25	-
Cellulose nanofibrils Navitas® * (kg/m^3)	-	-	0.15

* % by weight of cement.

Table 2. Chemical composition of the employed cement and slag.(LOI: loss on ignition @1000 °C).

Oxide (wt.%)	CaO	SiO_2	Al_2O_3	MgO	SO_3	Fe_2O_3	TiO_2	Mn_2O_3	K_2O	Na_2O	Other	LOI
PC	59.7	19.5	4.9	3.3	3.4	3.5	0.2	0.1	0.8	0.2	0.4	2.5
BFS	39.2	38.9	10.2	6.4	1.3	0.4	0.6	0.3	0.5	0.8	0.3	1.2

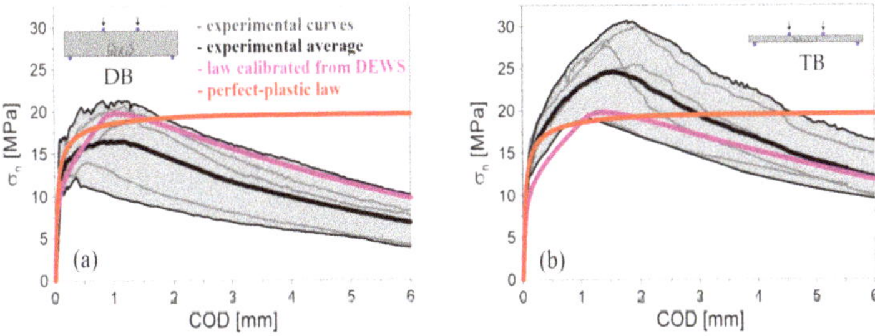

Figure 4. Maximum nominal stress in bending as a function of the Crack Opening Displacement for (**a**) deep beams (DB) and (**b**) thin beams (TB).

4. Structural Analysis

The design of the walls that make up the structure was carried out with reference to the maximum water height (1.50 m), using simple design equations based on a cantilever structural scheme for Basins 1 and 2, and on a slab clamped along three edges for Basin 3 (Figure 5), where the fixity of three edges is guaranteed by the monolithic connection with the buttresses along the two vertical edges and with the foundation slab along the bottom edge, as illustrated in Figure 2.

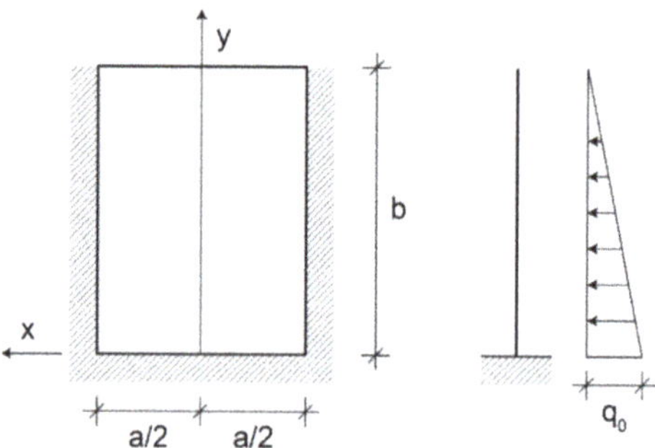

Figure 5. Structural scheme assumed for the design of Basins 1, 2 and 3.

The loads acting on the structure are the self-weight of the structural members, since no particular finishes are foreseen, and the hydrostatic pressure is due to the water contained inside the tank.

The self-weight is calculated on the basis of the specific weight of the materials and the dimensions of the structural elements. The specific weight values assumed are the following:

- Ordinary reinforced concrete: γ_{RC} = 25 kN/m^3.
- Fiber-reinforced cementitious composite: γ_{UHDC} = 25 kN/m^3.

The hydrostatic pressure of the water, acting on the vertical walls and on the bottom of the tank, was calculated by assuming a value equal to γ_w = 10 kN/m^3 for the specific weight of the water.

Two different load conditions were considered, for the serviceability (SLS) and ultimate (ULS) conditions, respectively. In the SLS, the height of the water table, as per the design provisions, is assumed at 1.30 m; at the ULS, an overflow of the water is assumed, and the maximum water height (h_w) compatible with the bearing capacity of the three basins is determined.

A summary of the design process is shown in Table 3, where a and b are the dimensions of the wall considered (see Figure 5; where a/b > 1.5, a one-way cantilever layout was assumed, indicated with symbol ∞ in Table 2), f_{cfm} is the indirect flexural tensile strength, $M_{max.}$ is the maximum bending moment along the y-axis, and $t_{min.}$ is the minimum wall thickness required to keep the maximum normal stress below the indirect flexural tensile strength.

Table 3. Design of the wall thickness of the three basins.

Basin	1 (RC)	2 (UHDC)	3 (UHDC)
a (m)	∞	∞	1.50
b (m)	1.50	1.50	1.50
a/b	∞	∞	1.00
f_{cfm} (MPa)	3.10	9.60	9.60
M_{max} (kNm/m)	5.63	5.63	1.01
t_{min} (mm)	104	59	25

A more refined evaluation of the stresses of the structural members of the tank, including the expected stress concentrations at the corners, was then carried out by means of a three-dimensional finite elements linear elastic model developed with the commercial software MIDAS Gen 2012. The FE model consists of 300 mm-thick horizontal plate elements for the foundation slab and vertical plate elements of different thickness to represent the external and internal walls. The buttresses in the third basin have been modelled using beam elements with a square section (200 × 200 mm^2). A three-dimensional view of the model is shown in Figure 6, where the geometric properties of the plate and beam elements are highlighted with different colors. One of the slabs on each side of Basin 3 has been modelled as 20 mm thick in order to assess the limits of employment of the developed UHDC material.

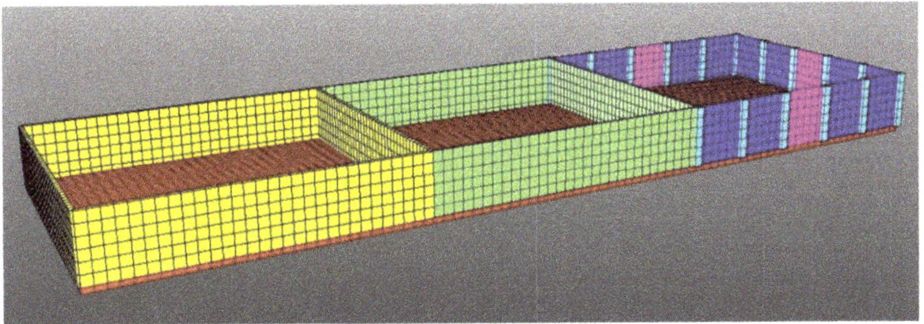

Figure 6. Three-dimensional view of the model used for finite element analysis and geometry of the plate elements (thicknesses: red = 300 mm; yellow = 100 mm; green = 60 mm; blue = 30 mm; magenta = 20 mm) and beams (light blue: section 200 × 200 mm^2).

The interaction between the foundation slab and the soil was accounted for through the Winkler model, with a constant that represents the link (supposedly linear elastic) between the pressure and the corresponding displacement. For the characteristics of the soil, reference was made to a detailed description relating to a similar structure located in the same plant. The characteristics of the soil referred to in the aforementioned report were assumed, due to the proximity between the two sites. In particular, for the purpose of modelling the contact between the foundation bed and the ground, a value of the subgrade stiffness equal to 30 MN/m^3 was assumed.

The mechanical and physical properties of the materials for the purposes of structural analysis, i.e., elastic modulus and specific weight, were introduced by defining two materials through the appropriate functions of the code. The analysis linear elastic, tensile and compressive strengths, as well as the reinforcement present in the reinforced concrete elements, did not affect the results.

Given the simple structural scheme that characterizes Basins 1 and 2 (i.e., cantilever wall), attention is focused only on the most significant results obtained for Basin 3, where walls of different thicknesses (20 and 30 mm) were adopted, as explained above. As expected, given the values shown in Table 2, in the 30 mm-thick walls the values of the maximum tensile stress are lower than the value of the indirect tensile strength during bending, f_{ctf} = 9.60 MPa (Figure 7a); on the contrary, the value of f_{cfm} (9.60 MPa) is exceeded in the walls with a thickness equal to 20 mm (Figure 7b).

The fact that that the tensile strength in bending is exceeded in limited portions of the structure should, however, pose no problems, considering that (a) the structure possesses sufficient redundancy to compensate for the limited (and localized) cracking phenomena; and (b) one of the primary objectives of the structure is to provide information about the behavior of UHDC in the cracked state.

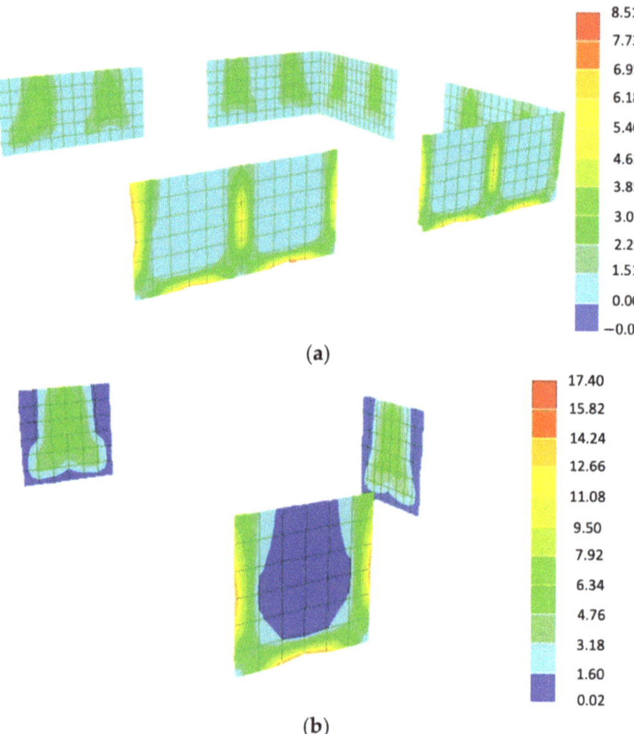

Figure 7. Basin 3: (**a**) maximum tensile stress on the walls with a thickness = 30 mm; (**b**) maximum tensile stress on the walls with a thickness = 20 mm *all units are in MPa.*

4.1. Analysis in Service Conditions (SLS)

The analysis at the Serviceability Limit States of the basins built with UHDC (second and third basin) and ORC (the first basin) was carried out by means of a sectional approach, where the capability of the UHDC material to redistribute the stresses thanks to the contribution of the fibers is explicitly accounted for (Figure 8). For the first basin, the section is not cracked under the service level of water. To this end, compatibility is ensured by assuming that plane sections remain plane, the compression stress distribution is assumed to be linear, while the tensile stresses are assumed to have an elastic–perfectly plastic distribution. As previously mentioned, the maximum tensile stress is limited to the value worked out by Lo Monte and Ferrara [34], namely, 5.60 MPa (coefficient of dispersion included). By enforcing equilibrium, under the maximum bending moment expected in the SLS (M_{SLS}), the strain distribution along the thickness can be determined, and more specifically the maximum tensile strain $\varepsilon_{t,max}$. The expected maximum crack width w_{max} is then worked out by assuming the characteristic length equal to the wall thickness t; therefore, $w_{max} = \varepsilon_{t,max} \times t$.

Table 4 summarizes the results of the SLS design, in terms of tensile and compressive stresses and expected crack widths. Notably, even in the presence of cracking, the estimated crack widths are either very small (Basins 2 and 3b) or nil (Basin 3a), thereby indicating reduced leakage likelihood. In addition, the maximum stress in compression is very low, considering that the UHDC developed attains values of compressive strength that can easily exceed 100 MPa. Thanks to the stress redistribution brought by the behavior of the plate theory [36], sections 3a and 3b show a very low values in terms of acting moments compared the other sections where the cantilever one-way behaviors were dominated.

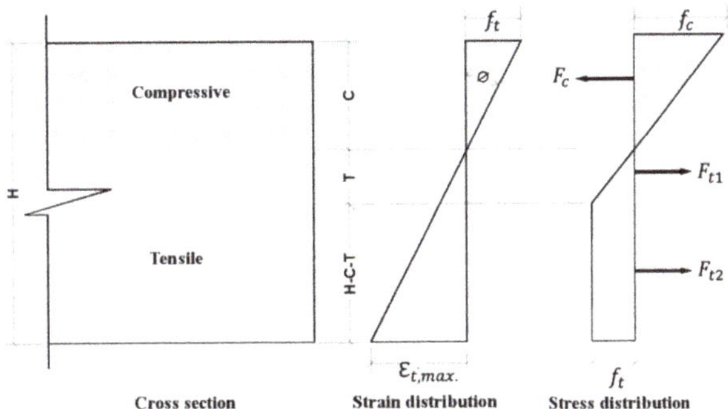

Figure 8. Stress and strain distribution across the thickness of the UHDC section.

Table 4. Evaluation of the state of stress and strain in the r/c and UHDC basins at SLS.

Basin	1	2	3a	3b
h (mm)	100	60	30	20
M_{SLS} (kNm/m)	3.67	3.67	0.67	0.67
$\sigma_{c,max}$ (MPa)	2.17	6.13	4.45	12.86
$\varepsilon_{t,max}$ (‰)	0.006	0.14	0.10	0.39
w_{max} (μm)	-	8	-	8

4.2. Analysis in Ultimate Conditions (ULS)

As previously mentioned, the ultimate limit state condition is evaluated in terms of the maximum height of the water (h_w), assuming overflow to take place, on the basis of the bending capacity Mu of the walls. To this end, a yield line analysis was carried out for the walls assuming a simple cantilever mechanism (global collapse) and a more complex (local collapse) mechanism for the walls provided with buttresses (Figure 9). Clearly, the safety of the walls is sufficient if the maximum height of the water h_w is larger than the height of the wall (b = 1.50 m), since overflows are unlikely to occur.

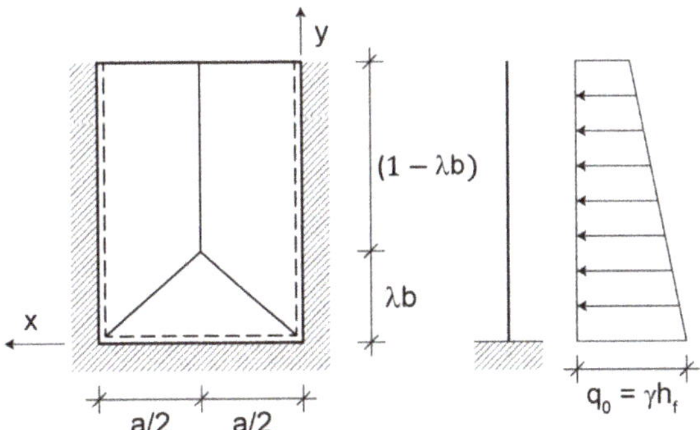

Figure 9. Local collapse mechanism assumed for Basin 3.

The maximum height of the water corresponding to the collapse mechanism shown in Figure 9 can be determined analytically by minimization of the following expression (as a function of the unknown parameter λ):

$$h_w = \frac{\frac{12M_u}{\gamma ab}\left(\frac{a}{b}\cdot\frac{1}{\lambda} + \frac{2b}{a} + \frac{4b}{a}\lambda\right) + \frac{5}{2}b + \frac{\lambda^2 b}{4}}{3-\lambda} \tag{2}$$

where M_u is the ultimate resisting moment in the assumed yield line mechanism section.

The results of the yield line analysis are summarized in Table 5. It is worth noting that for the 100 mm-thick reinforced concrete wall as well as for the UHDC walls with a thickness equal to both 60 and 30 mm, the governing collapse mechanism (i.e., characterized by the lower value of h_w) is the global mechanism, while for the assumed lowest value of the thickness the minimum value is obtained for the local mechanism. In all cases, however, $h_w > h$ (Figure 10).

Table 5. Evaluation of the maximum water height, h_w, in the UHDC basins at ULS.

Basin	1	2	3a	3b
t (mm)	100	60	30	20
M_u (kNm/m)	14.0	10.08	2.52	1.12
Mechanism	Global	Global	Global	Local
h_w (m)	2.63	1.90	2.55	2.28

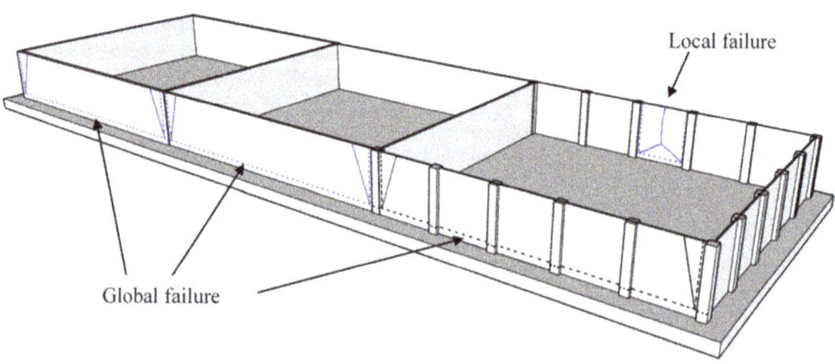

Figure 10. Illustration of local and global collapse mechanisms of the basins.

4.3. Computational Nonlinear Analysis

In order to validate the design assumptions and calculations highlighted above, a nonlinear analysis for representative "portions" of each of the three basins the pilot structure consists of was undertaken, also in view of the experimental validation currently on-going, whose preliminary results are going to be shown hereafter. Details of the analyzed elements and representative, related positions in the pilot and cross sections are shown in Figure 11.

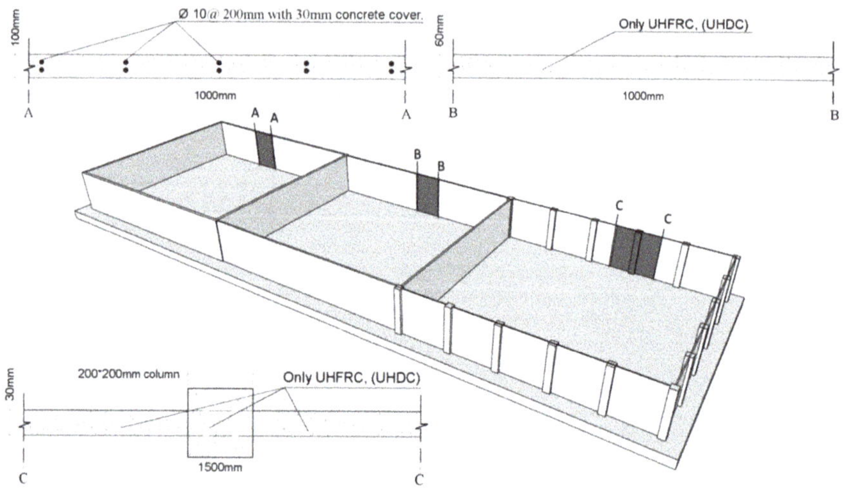

Figure 11. Sectional layout of the geothermal water basin walls assumed for nonlinear analysis.

4.3.1. Description of the Analysis Method

To perform the nonlinear analysis of each element, the fiber method was used, widely recognized as an efficient approach to perform the moment–curvature analysis of the reinforced concrete elements. Instead of evaluating the axial and flexural rigidity by finding the neutral axis corresponding to each strain increment, which is rather a time-consuming approach, the secant axial and flexural rigidity can be evaluated as $S_a = \frac{P}{\varepsilon_a}$, $S_b = \frac{M}{\varnothing}$, in order to calculate S_a and S_b corresponding to M and P.

The deformation vector $\begin{bmatrix} \varepsilon_0 & \varnothing \end{bmatrix}^t$ must be obtained first. A sectional tangent stiffness $[K_t]$ links an infinitesimal load vector $\begin{bmatrix} dP & dM \end{bmatrix}^t$ and an infinitesimal deformation vector $\begin{bmatrix} d\varepsilon & d\varnothing \end{bmatrix}^t$, as detailed in Equations (3) and (4).

$$dP = \frac{\partial P}{\partial \varepsilon} d\varepsilon + \frac{\partial P}{\partial \phi} d\phi \tag{3}$$

$$dM = \frac{\partial M}{\partial \varepsilon} d\varepsilon + \frac{\partial M}{\partial \phi} d\phi \tag{4}$$

$$\begin{bmatrix} dP \\ dM \end{bmatrix} = [K_t] \begin{bmatrix} d\varepsilon \\ d\varnothing \end{bmatrix} \tag{5}$$

Using this approach, and by dividing the concrete section into a sufficient number of fibers, strains in the middle of the strips can be iteratively imposed to find the solution. The Newton–Raphson method has been used to find the convergence of the solution with respect to the applied axial load. For a given axial strain and curvature, the strains in the middle of the strips and steel bars are evaluated as

$$\varepsilon_{strip} = \varepsilon_o + \phi * y_{strip} \varepsilon_{steel} = \varepsilon_o + \phi * y_{steel}$$

Then, using the assumed concrete stress–strain constitutive laws, which will be specified hereafter for each of the investigated cases, stresses were evaluated at the center of each steel and concrete layers (strip) and, through suitable integration of the stresses over the concrete and steel fibers, the resultant axial force and bending moment were finally calculated.

4.3.2. Reinforced Ordinary Concrete Section (the First Basin)

The first basin was constructed using ordinary reinforced concrete, with a cross section as shown in Figure 12. Given the boundary condition and the applied load, the overall behavior of the wall is represented by a 1.5 m-long cantilever beam fixed at the base of the basin. The section shown in Figure 12 below has been analyzed for the moment–curvature relationship using the procedure described above. To proceed with the nonlinear sectional analysis, the material constitutive laws described in Figure 13 were implemented for concrete and steel bars, as described in the following section. The unconfined compressive stress–strain relationship described by Mander et al. [37] was adopted (Figure 13a), while the bilinear softening tensile stress–strain relationship shown in Figure 13b was used to describe the tensile behavior of the concrete [38]. Elastic–perfectly plastic steel behavior, shown in Figure 13c, is used to describe the constitute law of the steel bars.

A MATLAB-R2020a code was written for the computational nonlinear analysis. The code was programmed to stop and report the failure when the maximum compressive concrete strain in the extreme fiber reaches the ultimate compressive strain, $\varepsilon_{cu} = -3.5\%$.

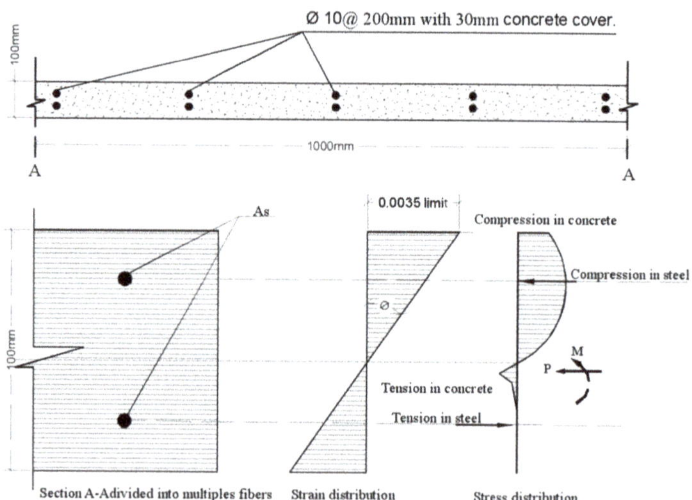

Figure 12. Ordinary reinforced concrete section analysis.

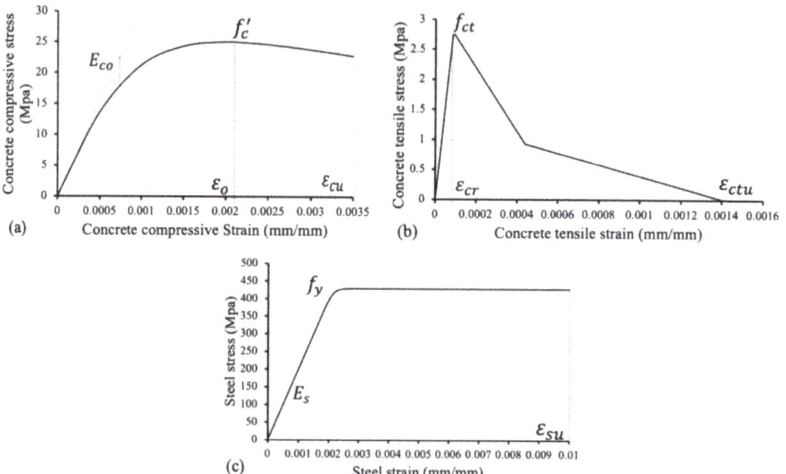

Figure 13. Materials constitutive relationships: (**a**) concrete compressive relationship, (**b**) concrete tensile relationship, and (**c**) steel compressive and tensile relationship.

4.3.3. UHDC Wall (Second Basin)

The same analysis procedure described previously was adapted for the second basin where the section consists of only UHPC material (Figure 14a). By implementing the concrete tensile stress–strain relationship adopted from Lo Monte and Ferrara (2020) [22], Figure 14b, the tensile behavior was identified. For the compressive stress–strain relationship, linear elastic behavior will be assumed with a modulus of elasticity of 41,700 MPa. The analysis of the section was programmed to terminate when the tensile strain at the extreme strip reaches the strain level of 0.015.

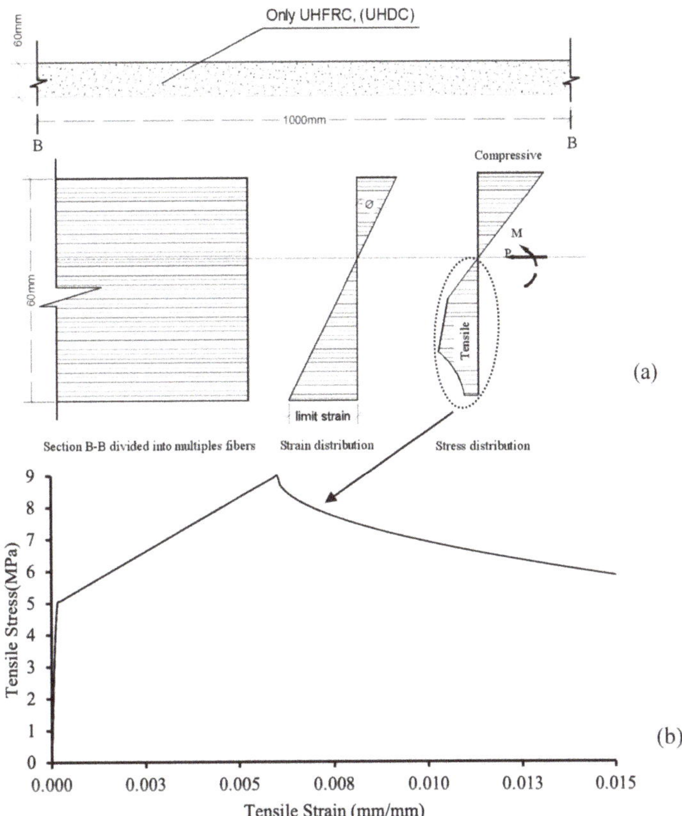

Figure 14. UHDC/UHPC sectional analysis by means of the fibers method (**a**), and the UHPC/UHDC tensile constitutive laws in term of strain (**b**).

4.3.4. UHDC Precast Slab Supported by UHDC Columns

The third cell of the basin is characterized by the 30 mm-thick precast UHPC/UHDC panels connected to each other through cast-in-place UHPC/UHDC columns where holes were made in the precast slabs along the vertical edges to increase the interlocking when the cast-in-place concrete of 200 × 200 mm^2 columns enters the holes and joins them together. For the connection of the system with the foundation, a groove was made in the foundation; then, the lower edge of precast elements, which also accommodates holes as the side edges, was inserted in the groove and the UHPC/UHDC concrete was poured into the groove to connect the foundation with the precast elements. However, the groove width was not sufficient to efficiently interlock with the lower part of the cast-in-place columns. Therefore, partial fixity or pin-end conditions were assumed during the analysis of the cross section, as shown in Figure 15, where, due to the applied load and expected behavior, the compression stress will be anticipated only in part of the column section and the precast slab has to equalize the compression force by performing in only tension behavior.

The results of the moment–curvature analysis results for all the sections are reported in Figure 16.

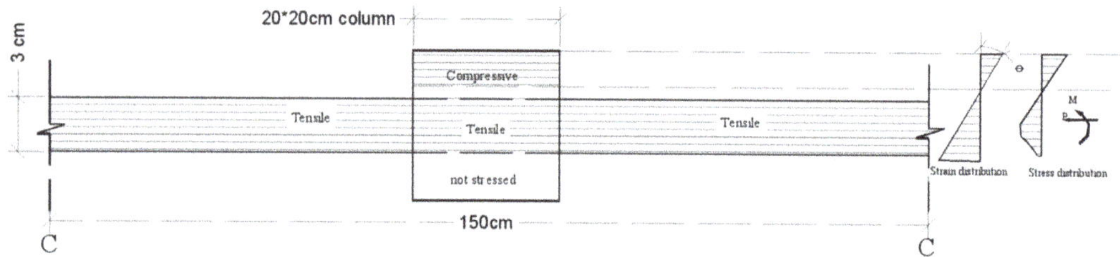

Figure 15. UHPC/UHDC precast panels supported by a UHPC/UHDC cast-in-place column.

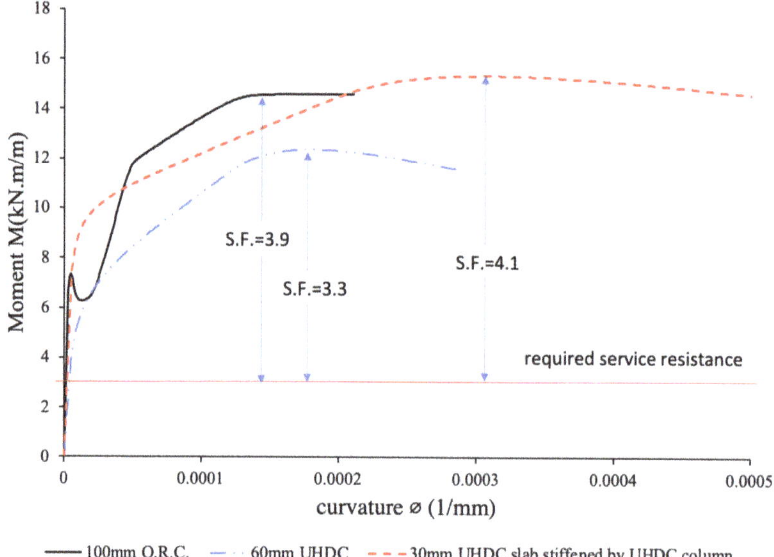

Figure 16. Moment–curvature relationship for the different employed sections.

4.4. Moment–Displacement Analysis

For the investigated structural elements, representative of the "current" development of the walls of the three basins, the maximum top edge deflection was also calculated by integrating the moment of area of the curvature diagram for each moment step, as shown in the formula below and illustrated in Figure 17.

$$\Delta = \sum_{i=1}^{curvature\ points} \left(\frac{\varnothing_i * x_i + \varnothing_{i+1} * x_{i+1}}{2} \right) * dx(i) \text{ (sum of trapezoidal moment of area), where}$$

x_i = distance from the position of xi to the wall's tip;
x_{i+1} = distance from the position of $xi + 1$ to the wall's tip;
$\varnothing_i$ = curvature corsponding to point x_i;
$\varnothing_{i+1}$ = curvature corsponding to point x_{i+1};
Δ = maximum displacement at the critical section.

A MATLAB code was employed to numerically integrate the curvature diagram for each step by creating loops to virtually create the span length and divided it into multiple segments (600–700 segments) based on the points of integration. Then, using the formula described above, the moment–displacement curves were obtained for each case. The moment–curvature graphs reported in Figure 16 were used to produce the moment–displacement graphs shown in Figure 18.

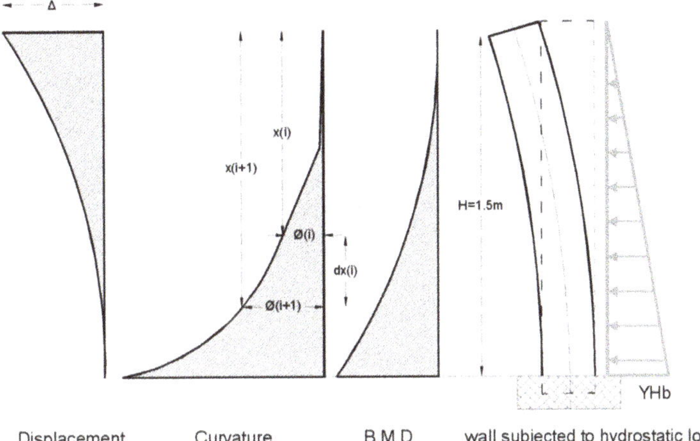

Figure 17. Illustration of displacement analysis using the moment of curvature area method.

Figure 18 compares the structural performance in terms of the water height–displacement/crack-width level for the three different structural element types (100 mm ordinary reinforced concrete wall, 60 mm cast in place UHDC wall and 30 mm precast UHDC panels joined by a 200 × 200 mm^2 cast-in-place UHDC wall). Moreover, the crack width, depth of compression zone and, where applicable, tensile strain of the longitudinal reinforcement bars are also reported for the three different sections.

Interestingly, it can be noticed that under a service water level around 130 cm, all the sections are not cracked, which satisfies the minimum requirements of the serviceability limit states for water-retaining walls for crack widths, and the maximum displacement at the tip point, which did not exceed 4 mm.

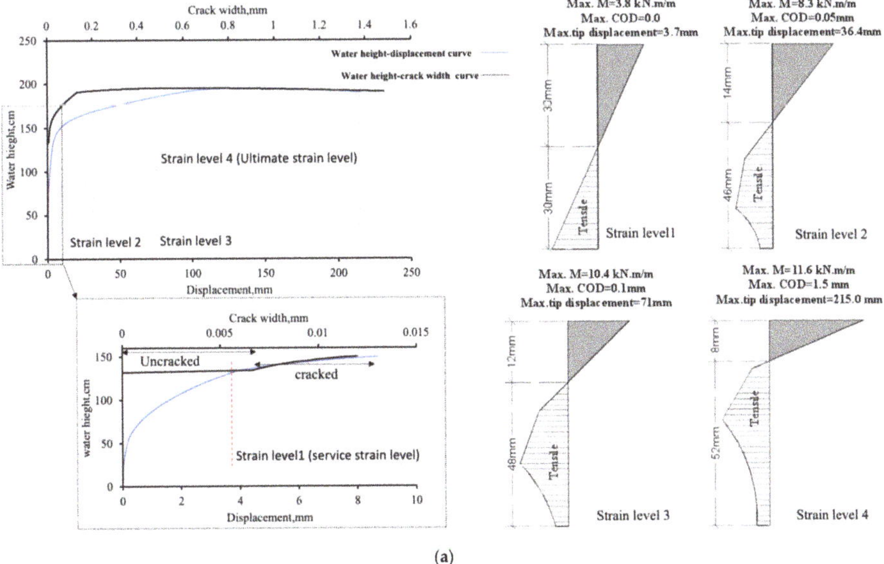

(a)

Figure 18. *Cont.*

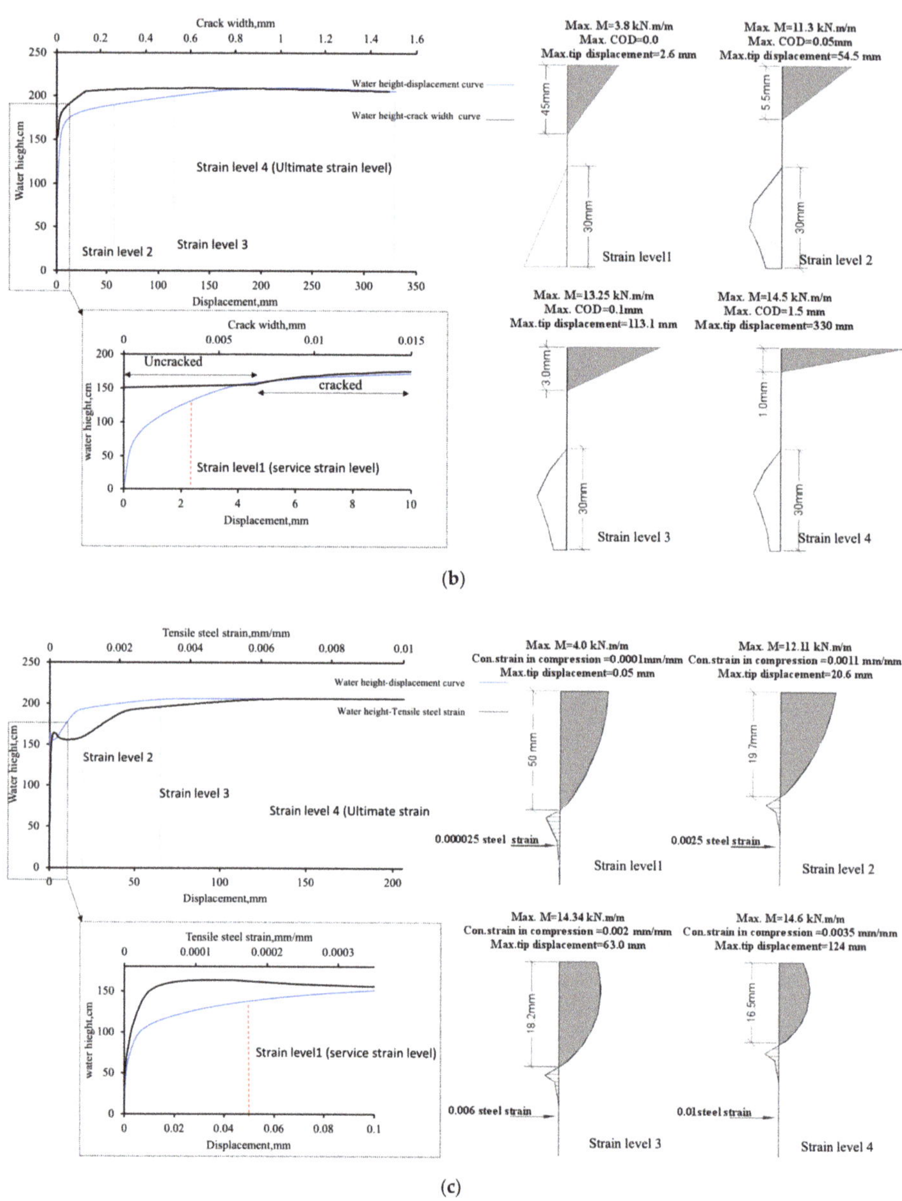

Figure 18. Sectional and structural analysis results for the three different structural element concepts: (**a**) 60 mm UHDC wall, (**b**) 30 mm UHDC slabs stiffened by a 20 × 20 cm^2 UHDC column, and (**c**) 100 mm ORC section wall.

Finally, looking into the ultimate limit states by gradually increasing the curvature and the strains of the tensile zone, the sections performed very well to resist a water level 1.5 times higher than the service water level even for the thin sections implemented with UHDC materials, though this would correspond to an unrealistically high displacement of the top edge of the wall.

This interestingly confirms the importance of the Serviceability Limit State conditions as governing the design of structures made of Ultra-High-Performance Concrete, in this case also incorporating hybrid (ordinary steel + fibers) reinforcement and hence the need to develop and validate reliable design tools for the correct prediction of the structural behavior under the aforementioned conditions, including the cracked state, and for the crack width calculation.

5. Validation Tests on the Real Structure (EGP Pilot)

Within the framework of the ReSHEALience project, several validations of the six proof-of-concept pilot structures have been planned to investigate the durability and assess the reliability of the structural design concepts of the pilots designed and built using the formulated and tested UHDC materials. For the geothermal water basin structure built by the project partner Enel Green Power, the carried-out validation tests, performed in collaboration with the Politecnico di Milano team, consisted of non-destructive monitoring of concrete quality via a rebound hammer and Ultrasonic Pulse Velocity tests, steel fiber dispersion survey and water-height versus displacement monitoring during filling and emptying of the basins. These also included steel strain monitoring via strain gauges placed during the construction in Basin 1 (ordinary reinforced concrete structure), concrete strain monitoring via strain gauges placed on the upper surface of concrete elements and reinforcement corrosion potential assessment via sensors installed during the construction process.

5.1. Steel Fibre Dispersion Survey

Steel fiber dispersion was surveyed for specific precast and cast-in-place UHPC/UHDC elements, as illustrated in Figure 19 below. The magnetic method proposed and validated by Ferrara et al. [39] was adopted in this survey for its robustness and easiness to handle on the work site. The method uses a probe with sensors spaced at 160 mm, which create a magnetic field, sensitive to the magnetic properties of the steel fibers aligned within it, thus resulting in variation in the measured inductance when the sensor is leaned in contact to the surface of the basin wall, along a different specified direction. The results of the measurements were recorded via a MATLAB script. A detailed description of the method and its calibration can be found in [39–41].

Results qualitatively plotted in Figure 19 indicate that fiber dispersion is quite homogenous for the pre-cast elements (b), which were casted horizontally. On the other hand, for the cast-in-place elements wall (a) and the column (c), some segregation of the steel fibers occurred in the top layers; this observation implies that the precast application of UHPC should be preferred, particularly for vertical structural elements where vertical casting may jeopardize the distribution of the steel fibers and affect the performance of the structural elements even prior to the application of the load.

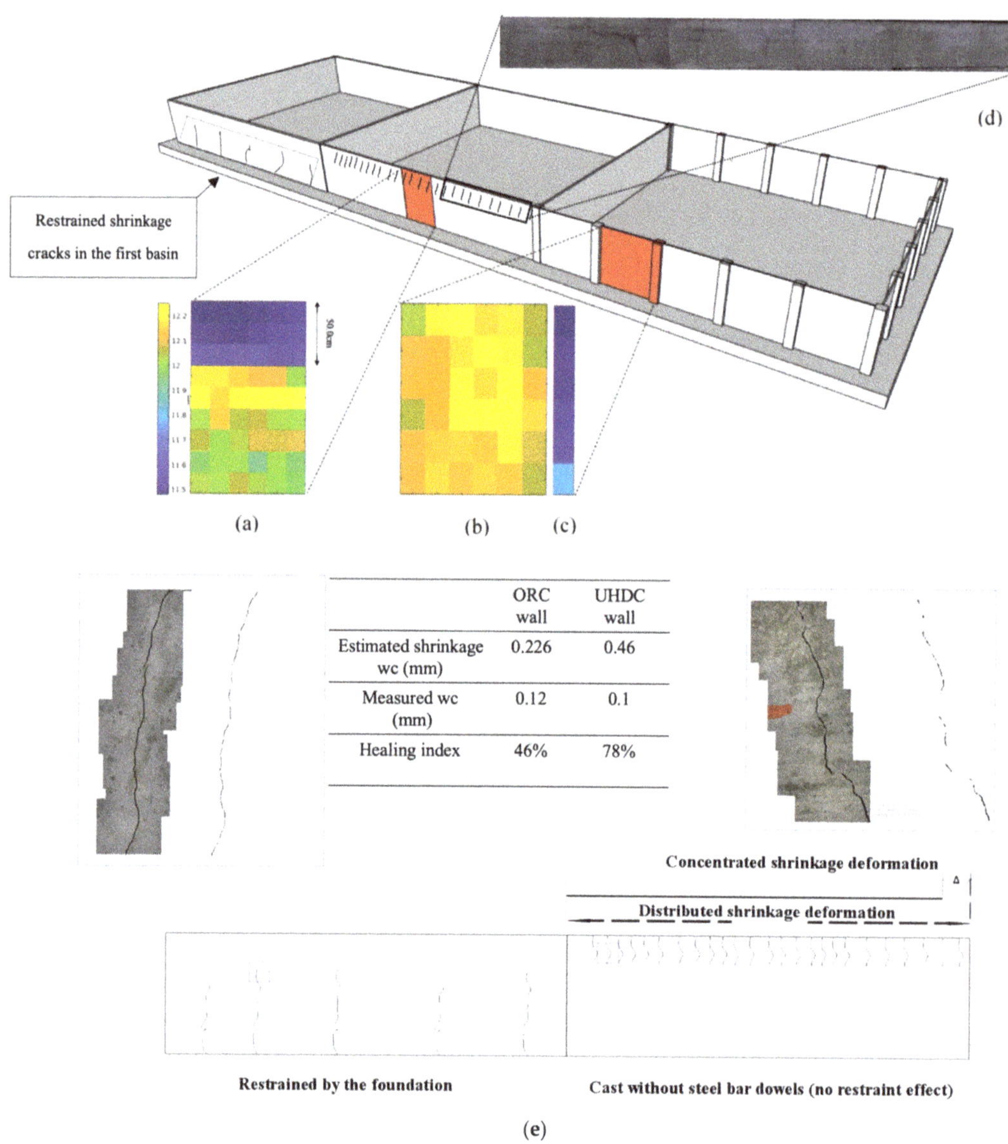

	ORC wall	UHDC wall
Estimated shrinkage wc (mm)	0.226	0.46
Measured wc (mm)	0.12	0.1
Healing index	46%	78%

Figure 19. Steel fiber distribution for cast-in-place 6 cm wall (**a**), precast 3 cm slab (**b**), cast-in-place 20 × 20 cm column (**c**), shrinkage cracks distributed along the top portion of second basin (**d**), and the estimated versus measured shrinkage cracks for ORC (on left) and UHDC (on right) walls (**e**).

This uneven distribution of the fibers resulted also into higher proneness to shrinkage cracking: as a matter of fact, equally spaced shrinkage cracks were observed in the top part of the 60 mm-thick wall. These cracks, spaced about 250 mm, extend about 500 mm downward from the top edge of the wall, which corresponds to the depth of the zone where a lack of steel fibers was detected through a non-destructive survey (Figure 19a). In fact, these cracks did not appear in the third basin where the wall is made with precast elements and the steel fibers were well distributed. Since these cracks are located at the top, they are believed to be due to the shrinkage deformation restrained by the constraints provided by the lateral walls located at the ends as well as by the bottom layers, richer in fibers; less effective crack control was found in this region poorer in fibers.

Figure 19e shows the simple model adopted by Destrée et al. [42] to estimate the crack width induced by shrinkage deformation in the UHDC wall, whereas the methodology illustrated in EN1992-1-1 [32] to estimate the restrained shrinkage crack width in basin one ORC wall for the UHDC section the steel fiber contribution was eliminated in the top part from the crack width calculation based on the steel fibers content survey. The calculated crack widths were compared to the measured ones after the basin have been exposed for more than a year, and a significant reduction in crack width is noticed, which could be attributed to the potential self-healing promoters or autogenous healing in UHDC and ORC walls, respectively. Evidence of crack sealing in the ORC and UHDC wall, as observed during a pilot testing on 14 May 2021, are shown in Figure 20.

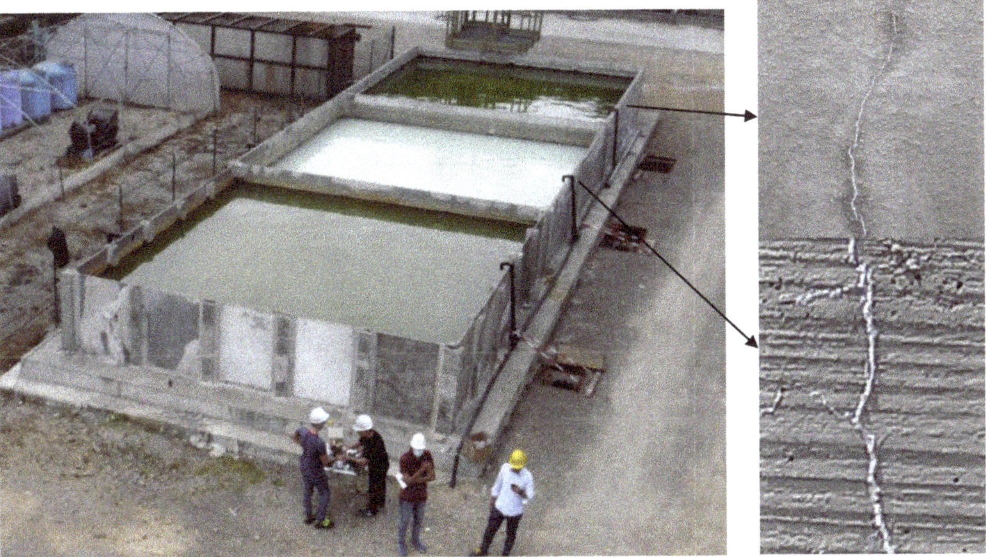

Figure 20. Evidence of crack sealing as monitored during a water filling test of the pilot on 14 May 2021.

5.2. Water Height–Displacement Relationship

In order to validate the assumption of the analysis for each section of the basin (30 mm UHPC/UHDC precast panels supported by a UHPC/UHDC cast-in-place column, 60 mm UHPC/UHDC wall and 100 mm reinforced concrete wall), the load or water height–displacement relationship was calculated for each section and compared to on-site measurements, where each cell of the basin was filled gradually and the displacement at the top and midpoint of the section was recorded, as shown in Figure 21. It can be also observed, where the numerical estimated water level height–displacement relationship is compared to the experimental obtained one, that the two relationships are in a good agreement for the third basin, which validates the assumption of an inverted T-beam hypothesis in the third basin (precast UHDC slabs supported by UHDC columns). However, for the first, the numerical analysis exhibited a stiffer behavior, which could be attributed to the restrained shrinkage cracks that developed at the base of the wall, and reduces the stiffness of the section that is particularly responsible for the overall behavior of the wall. For the second basin, the 60 mm-thick UHDC wall, the experimental curve exhibits a stiffer response than the numerical one, which could be attributed to the additional stiffening contribution from the transverse action brought by the relatively thin section element.

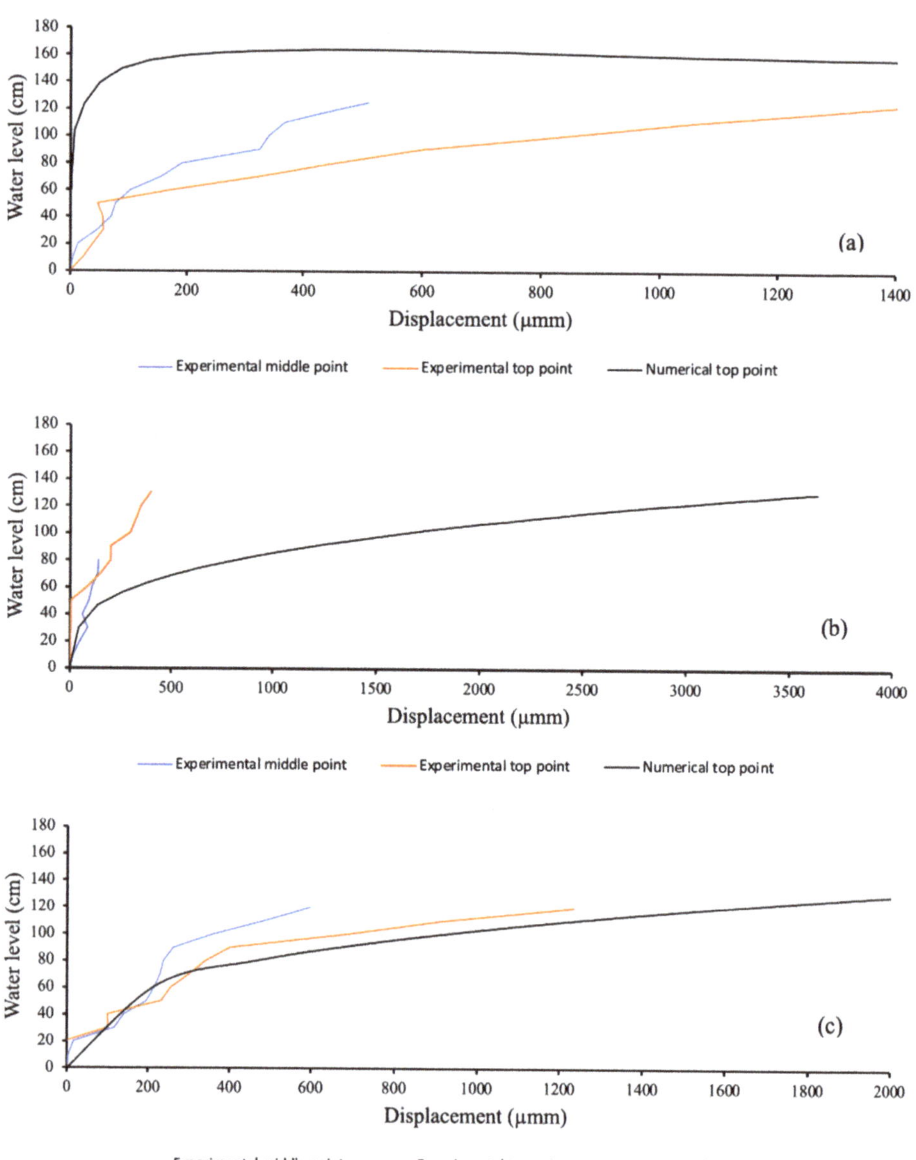

Figure 21. Water height versus displacement for the first (**a**), second (**b**) and third basin (**c**).

5.3. Crack Width and Water Tightness

As stated in the EN-1992—Part 3 [43], where liquid retaining and containment structures are addressed and reported in Table 6 below, the minimum thickness, minimum crack width and minimum depth of the compression zone were checked here for the basin. However, since the UHDC materials has the ability to control the crack width, accommodate residual tensile strain and healing/sealing of the up to 0.3 mm crack width under water immersion, this requirement is considered to be tolerable [44].

For instance, the requirements suggest a minimum thickness of the water retaining structure of 120 mm. This limited thickness is required to control the water tightness,

However, for the UHDC walls Basin 2, which are 60 mm thick, and Basin 3, which are only 30 mm thick, the sectional nonlinear analysis indicates that these sections are not cracked under the service load level, despite some healable shrinkage cracks at the top of the second basin. According to Table 5, and depending on the leakage requirements, the tightness class is classified from 0 to 3. Class 1 tightness indicates that some amount of leakage is permitted, such as surface staining and damp patches. In this case, the requirement is limited the strain under service condition to 0.00015 mm/mm, which is exactly the same strain level calculated under service actions for Basin 2 walls and much higher than the tensile strain for the case of Basin 3 under the service load level. In fact, during the validation visits, some surface staining and damp patches occurred at the first time of filling but, interestingly, some of these cracks were completely sealed and some were partially healed, as observed and detailed in Figure 20 above.

The other aspect is the thickness of the compression zone depth for the sections of the walls under the service load level, as reported in Figure 18; the compression zone depths of all sections satisfies the water tightness requirements (20% of the section thickness).

Table 6. Prescriptions regarding liquid retaining and containment structures for the different tightness classes as per EN 1992—Part 3.

Class	Requirements for Leakage	Specific Requirements
0	Some degree of leakage acceptable, or leakage of liquids irrelevant.	Silos holding dry materials may generally be designed with this class.
1	Leakage to be limited to a small amount. Some surface staining or damp patches acceptable.	Any cracks expected to pass through the full thickness should be limited to $w_{k1,}$ Healing may be assumed if the expected range of strain under service condition is less than 150×10^{-6}.
2	Leakage to be minimal. Appearance not to be impaired by cracks.	Cracks should not pass through the full width of a section, the design value of the depth of the compression zone should be at least $x_{min.}$ x_{min} = min (50 mm or 0.2 h—h is the element thickness).
3	No leakage permitted.	

Further validation of these findings could pave the way for a suitable extension of the tightness class requirements for UHPC structures, considering material-specific mechanical and durability properties as well as typically employed construction technologies (e.g., pre-casting).

6. Final Discussion and Conclusions

The pilot concrete structure analyzed in this paper represents a tank to contain geothermal water in a geothermal power plant; the pilot has been constructed with different structural/sectional concepts according to the different employed materials (ordinary reinforced concrete and cast-in-situ/precast Ultra-High-Performance/Ultra-High-Durability Concrete). Giving the applied loads represented by the hydrostatic water pressure acting on the cantilever walls, and the mechanical properties of the materials as identified through dedicated laboratory experimental campaigns, the structural systems were analyzed at both the ultimate and serviceability limit states in order to validate the design concepts. Upon entering the structure its service states (filled with 1.3 m of geothermal water), a series of full-scale field tests were also performed to validate the design's theoretical assumptions and the boundary conditions hypothesized, as also realized through the employed construction technologies. The most remarkable achievement is the reduction of the thickness from 100 mm in ordinary reinforced concrete to 60 mm in the case of UHDC cantilever walls and even less in the third basin (only 30 mm), where a solution based on precast slabs was implemented. The analysis at the ultimate limit state results in safety factors higher than 3 and 4 for the 60 mm-thick UHDC wall and 30 mm UHDC slabs supported by columns, respectively, as compared to 3.9 in the case of the 100 mm reinforced concrete section.

On the other hand, the serviceability analysis of the basin highlighted a very good performance in terms of crack-width limits and tightness, thanks to the mechanical performance of the UHDC materials. The serviceability analysis was also validated against the on-site observation and monitoring, where the filling and emptying of the basins were regularly carried out and no signs of leakages were observed.

Based on the evaluations presented in the paper, as corroborated by the on-site measurements as above, the following conclusions can be drawn:

- Among the employed UHDC structural solutions, the 30 mm-thick UHDC slabs supported by 200 × 200 mm UHDC columns performed better than the 60 mm UHDC cast-in-place wall in terms of steel fiber distribution, material consumption and structural performance under service and ultimate limit states.
- The nonlinear analysis carried out on Basin 2 and Basin 3 shows that the maximum expected crack width is very low, thanks to the signature tensile behavior of the employed UHPC/UHDC materials, to the benefit of the durability and overall long-term performance of the structure. The importance of the serviceability limit state conditions in governing the design of the structure is highlighted, as long as the superior mechanical and durability performance of the material (and the release of minimum cover/thickness constraints due to the complete elimination of conventional reinforcement replaced by dispersed fibers) do allow for a significant reduction in structural thickness.
- Load vs. displacements relationship as obtained by means of computational nonlinear analysis reasonably fit the corresponding curves obtained during the validation tests, confirming the reliability of the adopted construction processes and technologies and analysis methods.
- The site observation and monitoring proved the ability of the UHDC materials to heal and seal the small cracks after being filled with geothermal water, with a healing index close to 80%.
- The sections designed by using the UHDC materials exhibited very good serviceability performance although a significant reduction in the section thickness was used. The nonlinear analysis also allowed to confirm that the crack widths and compression zone thickness may fulfil the tightness class requirements, once suitable adaption to UHPC/UHDC materials can be made, exactly upon confirmation of the experimental and modelling findings herein highlighted.
- The numerical structural analysis results along with ultimate yield line mechanisms indicate that the second basin 60 mm UHDC section and the third section 30 mm UHDC section stiffened by 200 × 200 mm^2 UHDC columns can resist a hydrostatic pressure 1.5 times higher than the service load level before collapse.

Author Contributions: Preparation of the manuscript F.A., I.M., M.L. and S.S.; manage and supervision, F.L.M.; laboratory tests, P.B.; analysis of the yield line mechanisms, L.F.; defined the research concept, S.A.O. All authors have read and agreed to the published version of the manuscript.

Funding: The activity described in this paper has been performed in the framework of the project "Rethinking coastal defense and Green-energy Service infrastructures through enHancEd-durAbiLity high-performance cement-based materials-ReSHEALience", funded by the European Union Horizon 2020 research and innovation program under GA No 760824. The information and views set out in this publication are those of the authors and do not necessarily reflect the official opinion of the European Commission.

Institutional Review Board Statement: Not applicable.

Informed Consent Statement: Not applicable.

Data Availability Statement: This Data and further updates can be found here: [https://uhdc.eu/].

Conflicts of Interest: The authors declare no conflict of interest.

References

1. Younis, A.; Ebead, U.; Suraneni, P.; Nanni, A. Cost effectiveness of reinforcement alternatives for a concrete water chlorination tank. *J. Build. Eng.* **2020**, *27*, 100992. [CrossRef]
2. D3.2—Definition of Key Durability Parameters for Each Scenario—ReSHEALience. Available online: https://uhdc.eu/documentation/d3-2-definition-of-key-durability-parameters-for-each-scenario/ (accessed on 21 September 2020).
3. Thomas, M.; Bremner, T. Performance of lightweight aggregate concrete containing slag after 25 years in a harsh marine environment. *Cem. Concr. Res.* **2012**, *42*, 358–364. [CrossRef]
4. Simultaneous Structural and Environmental Loading of an Ultra-High Performance Concrete Component | FHWA. Available online: https://highways.dot.gov/research/projects/simultaneous-structural-environmental-loading-ultra-high-performance-concrete-component (accessed on 13 June 2021).
5. Krelani, V.; Krelani, V.; Moretti, F. Autogenous healing on the recovery of mechanical performance of High Performance Fibre Reinforced Cementitious Composites (HPFRCCs): Part 2—Correlation between healing of mechanical performance and crack sealing. *Cem. Concr. Compos.* **2016**, *73*, 299–315. [CrossRef]
6. Ferrara, L.; Krelani, V.; Moretti, F.; Roig Flores, M.; Serna Ros, P. Effects of autogenous healing on the recovery of mechanical performance of High Performance Fibre Reinforced Cementitious Composites (HPFRCCs): Part 1. *Cem. Concr. Compos.* **2017**, *83*, 76–100. [CrossRef]
7. Cuenca, E.; Ferrara, L. Self-healing capacity of fiber reinforced cementitious composites. State of the art and perspectives. *KSCE J. Civ. Eng.* **2017**, *21*, 2777–2789. [CrossRef]
8. Yildirim, G.; Alyousif, A.; Şahmaran, M.; Lachemi, M. Assessing the self-healing capability of cementitious composites under increasing sustained loading. *Adv. Cem. Res.* **2015**, *27*, 581–592. [CrossRef]
9. Woyciechowski, P.P.; Kalinowski, M. The influence of dosing method and material characteristics of superabsorbent polymers (SAP) on the effectiveness of the concrete internal curing. *Materials* **2018**, *11*, 1600. [CrossRef]
10. Wang, Y.S.; Alrefaei, Y.; Dai, J.G. Silico-aluminophosphate and alkali-aluminosilicate geopolymers: A comparative review. *Front. Mater.* **2019**, *6*, 106. [CrossRef]
11. Ferrara, L.; Ozyurt, N.; Di Prisco, M. High mechanical performance of fibre reinforced cementitious composites: The role of "casting-flow induced" fibre orientation. *Mater. Struct.* **2011**, *44*, 109–128. [CrossRef]
12. Li, V.C.; Stang, H.; Krenchel, H. Micromechanics of crack bridging in fibre-reinforced concrete. *Mater. Struct.* **1993**, *26*, 486–494. [CrossRef]
13. Al-Obaidi, S.; Bamonte, P.; Ferrara, L.; Luchini, M.; Mazzantini, I. Durability-based design of structures made with ultra-high-performance/ultra-high-durability concrete in extremely aggressive scenarios: Application to a geothermal water basin case study. *Infrastructures* **2020**, *5*, 102. [CrossRef]
14. Borg, R.P.; Cuenca, E.; Garofalo, R.; Schillani, F.; Nasner, M.L.; Ferrara, L. Performance Assessment of Ultra-High Durability Concrete Produced From Recycled Ultra-High Durability Concrete. *Front. Built Environ.* **2021**, *7*, 78. [CrossRef]
15. An Overview on H2020 Project "ReSHEALience" | Semantic Scholar. Available online: https://www.semanticscholar.org/paper/An-Overview-on-H2020-Project-\T1\textquotedblleftReSHEALience\T1\textquotedblright-Ferrara-Bamonte/9fc028915a762840ca6d11592b6acca013f4893b (accessed on 10 June 2021).
16. Alexander, M.G.; Ballim, Y.; Stanish, K. A framework for use of durability indexes in performance-based design and specifications for reinforced concrete structures. *Mater. Struct. Constr.* **2008**, *41*, 921–936. [CrossRef]
17. Alexander, M.G. Service life design and modelling of concrete structures—background, developments, and implementation. *Rev. ALCONPAT* **2018**, *8*, 224–245. [CrossRef]
18. Simons, B. Concrete Performance Specifications: New Mexico Experience. *Concr. Int.* **2004**, *26*, 68–71.
19. Perspective on Prescriptions. Available online: https://trid.trb.org/view/758581 (accessed on 13 June 2021).
20. Bickley, J.A.; Hooton, R.D.; Hover, K.C. Performance Specifications for Durable Concrete. *Concr. Int.* **2006**, *28*, 51–57.
21. Preparation of a Performance-Based Specification for Cast-in-Place Concrete. Available online: https://www.yumpu.com/en/document/read/25802659/preparation-of-a-performance-based-specification-for-cast-in-place- (accessed on 13 June 2021).
22. Lo Monte, F.; Ferrara, L. Tensile behaviour identification in Ultra-High Performance Fibre Reinforced Cementitious Composites: Indirect tension tests and back analysis of flexural test results. *Mater. Struct. Constr.* **2020**, *53*, 1–12. [CrossRef]
23. Cuenca, E.; D'Ambrosio, L.; Lizunov, D.; Tretjakov, A.; Volobujeva, O.; Ferrara, L. Mechanical properties and self-healing capacity of Ultra High Performance Fibre Reinforced Concrete with alumina nano-fibres: Tailoring Ultra High Durability Concrete for aggressive exposure scenarios. *Cem. Concr. Compos.* **2021**, *118*, 103956. [CrossRef]
24. Cuenca, E.; Mezzena, A.; Ferrara, L. Synergy between crystalline admixtures and nano-constituents in enhancing autogenous healing capacity of cementitious composites under cracking and healing cycles in aggressive waters. *Constr. Build. Mater.* **2021**, *266*, 121447. [CrossRef]
25. Materazzi, A.L.; Ubertini, F.; D'Alessandro, A. Carbon nanotube cement-based transducers for dynamic sensing of strain. *Cem. Concr. Compos.* **2013**, *37*, 2–11. [CrossRef]
26. Cuenca, E.; Rigamonti, S.; Gastaldo Brac, E.M.; Ferrara, L. Crystalline admixture as healing promoter in concrete exposed to chloride-rich environments: An experimental study. *ASCE J. Mater. Civ. Eng.* **2020**, in press. [CrossRef]
27. Li, V.C.; Leung, C.K.Y. Steady-State and Multiple Cracking of Short Random Fiber Composites. *J. Eng. Mech.* **1992**, *118*, 2246–2264. [CrossRef]

28. Naaman, A.E.; Reinhardt, H.W. Proposed classification of HPFRC composites based on their tensile response. *Mater. Struct. Constr.* **2006**, *39*, 547–555. [CrossRef]
29. Shi, C.; Chen, B. (Eds.) Proceedings of the 2nd International Conference on UHPC Materials and Structures (UHPC2018-China), Fuzhou, China, 7–10 November 2018; RILEM Pubs: Fuzhou, China, 2018; p. 832.
30. Li, V.C.; Mishra, D.K.; Wu, H.C. Matrix design for pseudo-strain-hardening fibre reinforced cementitious composites. *Mater. Struct.* **1995**, *28*, 586–595. [CrossRef]
31. Upgrading the Concept of Uhpfrc for High Durability in the Cracked State: The Concept of Ultra High Durability Concrete (UHDC) in the Approach of the h2020 Project Reshealience—Core. Available online: https://core.ac.uk/display/195747484 (accessed on 13 June 2021).
32. *EN 1992-1-1: Eurocode 2: Design of concrete structures—Part 1-1: General rules and rules for buildings*; CEN: Brussels, Belgium, 2004.
33. *NTC 2018: Norme tecniche per le costruzioni (Italian Guidelines for Constructions)*; Gazzeta Ufficiale: Rome, Italy, 2018; Volume 20.
34. Lo Monte, F.; Ferrara, L. Characterization of the Tensile Behaviour of Ultra-High Performance Fibre-Reinforced Concrete. In Proceedings of the NBSC2019 Workshop, IMReady. Milan, Italy, 19–20 September 2019; pp. 45–54.
35. Di Prisco, M.; Ferrara, L.; Lamperti, M.G.L. Double edge wedge splitting (DEWS): An indirect tension test to identify post-cracking behaviour of fibre reinforced cementitious composites. *Mater. Struct. Constr.* **2013**, *46*, 1893–1918. [CrossRef]
36. Timoshenko, S.P.; Woinowsky-Krieger, S. *Theory of Plates and Shells*; McGraw-Hill: New York, NY, USA, 1959.
37. Mander, J.B.; Priestley, M.J.N.; Park, R. Theoretical Stress-Strain Model for Confined Concrete. *J. Struct. Eng.* **1988**, *114*, 1804–1826. [CrossRef]
38. Abdalla, H.M.; Karihaloo, B.L. A method for constructing the bilinear tension softening diagram of concrete corresponding to its true fracture energy. *Mag. Concr. Res.* **2004**, *56*, 597–604. [CrossRef]
39. Ferrara, L.; Cremonesi, M.; Faifer, M.; Toscani, S.; Sorelli, L.; Baril, M.A.; Réthoré, J.; Baby, F.; Toutlemonde, F.; Bernardi, S. Structural elements made with highly flowable UHPFRC: Correlating computational fluid dynamics (CFD) predictions and non-destructive survey of fiber dispersion with failure modes. *Eng. Struct.* **2017**, *133*, 151–171. [CrossRef]
40. Ferrara, L.; Faifer, M.; Toscani, S. A magnetic method for non destructive monitoring of fiber dispersion and orientation in steel fiber reinforced cementitious composites-part 1: Method calibration. *Mater. Struct. Constr.* **2012**, *45*, 575–589. [CrossRef]
41. Ferrara, L.; Faifer, M.; Muhaxheri, M.; Toscani, S. A magnetic method for non destructive monitoring of fiber dispersion and orientation in steel fiber reinforced cementitious composites. Part 2: Correlation to tensile fracture toughness. *Mater. Struct. Constr.* **2012**, *45*, 591–598. [CrossRef]
42. Destrée, X.; Yao, Y.; Mobasher, B. Sequential Cracking and Their Openings in Steel-Fiber-Reinforced Joint-Free Concrete Slabs. *J. Mater. Civ. Eng.* **2016**, *28*, 04015158. [CrossRef]
43. EN 1992-1-1. *Eurocode 2-Design of Concrete Structures-Part 3: Liquid Retaining and Containment Structures*; CEN: Brussels, Belgium, 1992; Volume 4.
44. Cuenca, E.; Tejedor, A.; Ferrara, L. A methodology to assess crack-sealing effectiveness of crystalline admixtures under repeated cracking-healing cycles. *Constr. Build. Mater.* **2018**, *179*, 619–632. [CrossRef]

MDPI
St. Alban-Anlage 66
4052 Basel
Switzerland
Tel. +41 61 683 77 34
Fax +41 61 302 89 18
www.mdpi.com

Sustainability Editorial Office
E-mail: sustainability@mdpi.com
www.mdpi.com/journal/sustainability

www.ingramcontent.com/pod-product-compliance
Lightning Source LLC
LaVergne TN
LVHW071009170726
843515LV00006B/1115

* 9 7 8 3 0 3 6 5 4 4 6 1 8 *